A Publication of the
National Research Council of Canada
Monograph Publishing Program

The Systematics of New World *Clinocera* Meigen (Diptera: Empididae: Clinocerinae)

Bradley J. Sinclair

Entomology – Ontario Plant Laboratories,
Canadian Food Inspection Agency,
Ottawa, Ontario, Canada

NRC·CNRC

NRC Research Press
Ottawa 2008

Printed in Canada on acid-free paper. ♾

ISBN 978-0-660-19800-2
NRC No. 46860
Date of publication October 10, 2008

Library and Archives Canada cataloguing in publication data

Sinclair, Bradley James

The Systematics of New World *Clinocera* Meigen (Diptera: Empididae: Clinocerinae) / Bradley J. Sinclair.

Issued by: National Research Council Canada.

Includes bibliographical references.
ISBN 978-0-660-19800-2

1. *Clinocera*--Classification. 2. *Clinocera*--Identification. 3. *Clinocera*--North America.
4. *Clinocera*--South America. 5. Empididae.
I. National Research Council Canada II. Title.

QL537.E4S56 2008 595.77'3 C2008-980187-3

NRC Monograph Publishing Program

Inquiries: Monograph Publishing Program, NRC Research Press, National Research Council of Canada, Ottawa, Ontario K1A 0R6, Canada. Web site: http://pubs.nrc-cnrc.gc.ca

Correct citation for this publication: Sinclair, B.J. 2008. *The Systematics of New World* Clinocera *Meigen (Diptera: Empididae: Clinocerinae).* NRC Research Press, Ottawa, Ontario, Canada. viii+245 pp.

Table of Contents

Abstract

The New World species of *Clinocera* Meigen (Diptera: Empididae: Clinocerinae) are revised and include 67 species, of which 44 and 23 occur in the Nearctic Region and Neotropical Region, respectively. The following 49 new species of *Clinocera* are described from the New World: *C. amarilla* **sp. nov.**, *C. appalachia* **sp. nov.**, *C. arizonica* **sp. nov.**, *C. arnaudi* **sp. nov.**, *C. attenuata* **sp. nov.**, *C. borkenti* **sp. nov.**, *C. brevicauda* **sp. nov.**, *C. brunnea* **sp. nov.**, *C. bulbosa* **sp. nov.**, *C. caerulea* **sp. nov.**, *C. chilensis* **sp. nov.**, *C. collini* **sp. nov.**, *C. colombiana* **sp. nov.**, *C. coloradensis* **sp. nov.**, *C. cordillerana* **sp. nov.**, *C. cummingi* **sp. nov.**, *C. daniellae* **sp. nov.**, *C. disjuncta* **sp. nov.**, *C. femorata* **sp. nov.**, *C. guttipennis* **sp. nov.**, *C. insularis* **sp. nov.**, *C. isabellina* **sp. nov.**, *C. lapazensis* **sp. nov.**, *C. longicauda* **sp. nov.**, *C. macdonaldi* **sp. nov.**, *C. madera* **sp. nov.**, *C. marshalli* **sp. nov.**, *C. melanderi* **sp. nov.**, *C. meridana* **sp. nov.**, *C. mexicana* **sp. nov.**, *C. miradorana* **sp. nov.**, *C. naomina* **sp. nov.**, *C. nigriscutata* **sp. nov.**, *C. notialis* **sp. nov.**, *C. obunca* **sp. nov.**, *C. pacifica* **sp. nov.**, *C. parana* **sp. nov.**, *C. penai* **sp. nov.**, *C. rainiericola* **sp. nov.**, *C. robinsoni* **sp. nov.**, *C. sabroskyi* **sp. nov.**, *C. schlinkerti* **sp. nov.**, *C. setitibia* **sp. nov.**, *C. stigma* **sp. nov.**, *C. sublineata* **sp. nov.**, *C. substagnalis* **sp. nov.**, *C. subtrunca* **sp. nov.**, *C. wallowa* **sp. nov.**, and *C. woodi* **sp. nov.**

The following new synonyms are proposed: *C. taos* Melander and *C. undulata* Melander = *C. conjuncta* Loew, *C. brevitibia* Melander = *C. fuscipennis* Loew, and *C. genualis* Coquillett = *C. stagnalis* (Haliday). *Aclinocera* Yang & Yang **syn. nov.** is newly synonymized with *Clinocera* and its type species is also transferred as *C. sinica* (Yang & Yang) **comb. nov.** Lectotypes are designated for *C. brevitibia, C. lecta, C. lineata*, and *C. oriunda*. Identification keys to genera of Clinocerinae of North and South America are included and keys are provided to species-groups and all known New World species of *Clinocera*. The distribution of each species is illustrated and a species-group phylogeny given. Distributions are used to support the hypothesis that southward dispersals into South America through Mesoamerica occurred once during the Late Cretaceous and again during the Pliocene.

Résumé

Les espèces du genre *Clinocera* Meigen (Diptera: Empididae: Clinocerinae) du Nouveau Monde sont révisées. *Clinocera* renferme maintenant 67 espèces, dont 44 et 23 se trouvent respectivement dans les régions néarctique et néotropicale. Parmi les espèces répertoriées, les 49 espèces suivantes sont nouvelles et issues du Nouveau Monde : *C. amarilla* **sp. nov.**, *C. appalachia* **sp. nov.**, *C. arizonica* **sp. nov.**, *C. arnaudi* **sp. nov.**, *C. attenuata* **sp. nov.**, *C. borkenti* **sp. nov.**, *C. brevicauda* **sp. nov.**, *C. brunnea* **sp. nov.**, *C. bulbosa* **sp. nov.**, *C. caerulea* **sp. nov.**, *C. chilensis* **sp. nov.**, *C. collini* **sp. nov.**, *C. colombiana* **sp. nov.**, *C. coloradensis* **sp. nov.**, *C. cordillerana* **sp. nov.**, *C. cummingi* **sp. nov.**, *C. daniellae* **sp. nov.**, *C. disjuncta* **sp. nov.**, *C. femorata* **sp. nov.**, *C. guttipennis* **sp. nov.**, *C. insularis* **sp. nov.**, *C. isabellina* **sp. nov.**, *C. lapazensis* **sp. nov.**, *C. longicauda* **sp. nov.**, *C. macdonaldi* **sp. nov.**, *C. madera* **sp. nov.**, *C. marshalli* **sp. nov.**, *C. melanderi* **sp. nov.**, *C. meridana* **sp. nov.**, *C. mexicana* **sp. nov.**, *C. miradorana* **sp. nov.**, *C. naomina* **sp. nov.**, *C. nigriscutata* **sp. nov.**, *C. notialis* **sp. nov.**, *C. obunca* **sp. nov.**, *C. pacifica* **sp. nov.**, *C. parana* **sp. nov.**, *C. penai* **sp. nov.**, *C. rainiericola* **sp. nov.**, *C. robinsoni* **sp. nov.**, *C. sabroskyi* **sp. nov.**, *C. schlinkerti* **sp. nov.**, *C. setitibia* **sp. nov.**, *C. stigma* **sp. nov.**, *C. sublineata* **sp. nov.**, *C. substagnalis* **sp. nov.**, *C. subtrunca* **sp. nov.**, *C. wallowa* **sp. nov.** et *C. woodi* **sp. nov.**

Voici une liste des nouveaux synonymes proposés : *C. taos* Melander et *C. undulata* Melander = *C. conjuncta* Loew, *C. brevitibia* Melander = *C. fuscipennis* Loew et *C. genualis* Coquillett = *C. stagnalis* (Haliday). *Aclinocera* Yang & Yang **syn. nov.** est nouvellement mise en synonymie avec *Clinocera* et son espèce type devient *C. sinica* (Yang & Yang) **comb. nov.** Des lectotypes sont offerts pour *C. brevitibia, C. lecta, C. lineata* et *C. oriunda*. L'article renferme des clés d'identification pour les genres de la sous-famille Clinocerinae de l'Amérique du Nord et de l'Amérique du Sud, de même que des clés pour divers groupes d'espèces et toutes les espèces connues du genre *Clinocera* du Nouveau Monde. On y présente également des illustrations de la répartition de chaque espèce et une description phylogénique du groupe. Les répartitions sont utilisées pour corroborer l'hypothèse voulant que la dispersion vers le sud pour atteindre l'Amérique du Sud par la Méso-Amérique se soit produite à deux reprises — une fois au Crétacé tardif et l'autre au cours du Pliocène.

Acknowledgments

I am indebted to the following curators and their respective institutions for the loan of specimens: D. Grimaldi (AMNH); D. Azuma (ANSP); M. Barták (BARC); T. Saigusa (BLKU); J. Chainey (BMNH); N.L. Evenhuis (BPBM); R.W. Baumann (BYU); P.H. Arnaud (CAS); C.W. Young (CMNH); J.M. Cumming (CNC); S. Fitzgerald (CSU); E.R. Hoebeke (CUIC); S.A. Marshall (DEBU); G.J. Steck (FSCA); A.C. Pont (HMOU); R.E. Roughley (JBWM); K. Ueda (KMNH); C.L. Hogue and B.V. Brown (LACM); D. Furth (MCZ); C. Daugeron (MNHN); F.W. Stehr (MSU); R.L. Hurley (MSUB); U. Kallweit (MTD); C.J.E. Lamas (MZUSP); J.K. Barnes (NYSM); N.H. Anderson (OSUC); J.F. MacDonald (PERC); E. Masteller (PSBC; now housed in ZFMK); E.J. Rogers (ROGC); G.E. Wiggins and D.C. Currie (ROM); J.S. Ashe and E.F. Smith (SEMC); S. Cannings (UBC); J.A. Chemsak (UCB); S. Heyden (UCD); S.I. Frommer (UCR); F.W. Merickel (UIM); L. Cloutier (UMQ); R.J. Clausen (UMSP); J.F. Burger (UNH); N.E. Woodley (USNM); W.J. Hanson (USUL); S.R. Shaw (UWL); R. Danielsson (UZIL); W.J. Turner and R.S. Zack (WSU); P. Vilkamaa (ZMH); T. Pape (ZMUC); M. Kotrba (ZSM). Their patience and assistance were greatly appreciated.

Initial stages of this project were financed through the assistance of an Entomological Society of Canada Graduate Research/Travel Grant and a Natural Sciences and Engineering Research Council of Canada Postgraduate Scholarship. During this period (1988–1993) laboratory facilities were generously provided by the Diptera Section of the Canadian National Insect Collection (Ottawa). I am sincerely indebted to Jeff Cumming, Monty Wood, Dick Vockeroth, Jim O'Hara, Bruce Cooper, Harold Walther, and the late Guy Shewell and Herb Teskey. Ed Masteller provided SEM photographs from his collection and donated his large holdings of emergence-trap material. Rüdiger Wagner provided advice concerning European species. Norm Woodley kindly assisted in locating type specimens and an Alexander Koenig Stiftung assisted fieldwork in 2004. All mites were identified by Ian M. Smith (CNC). Jim O'Hara and Steve Marshall provided photographs of habitats and flies and Dirk Rohwedder (ZFMK) and Scott Brooks (CNC) assisted in producing habitus photographs of pinned specimens. Jeff Cumming greatly encouraged and supported all stages of this project. Steve Marshall and John MacDonald kindly critically reviewed the manuscript and their efforts are sincerely appreciated.

My family, especially my wife, helped immeasurably and were of great support throughout the various stages of this project.

Introduction

General biology of Clinocerinae

The aquatic dance fly subfamily Clinocerinae (Empididae) is one of the major groups of predacious Diptera in running water or lotic habitats. Adults are frequently encountered sitting on emergent rocks in springs, seepages, streams, and rivers (Figs. 1*a*, *d*). The subfamily is most common in the Nearctic and Palearctic regions and becomes significantly less common at lower latitudes and south of the equator. In Australia and New Zealand the dominance of this micro-habitat is taken over by the subfamily Ceratomerinae (Sinclair 2003*b*).

Among adult flies, those of the genus *Clinocera* Meigen are among the most

Fig. 1. Photographs of living individuals of species of Clinocerinae: *a*, *Trichoclinocera* sp. gathered on a rock in a creek; *b*, *Clinocera fuscipennis*; *c*, *Clinocera* sp. from Chile; *d*, *C. wallowa* gathered on rocks in a river; *e*, *C. binotata*, mating pair. (Photographs *a–c* and *e* by S.A. Marshall, reproduced with permission.)

commonly encountered in seepage habitats and are also encountered resting on rocks and moss in headwater streams. In the southern Appalachians and the western Cordillera, *Oreothalia* Melander is also found in seepages and small first-order streams. *Clinocera* is normally rare in large, open-canopy streams, creeks, and rivers, where *Trichoclinocera* Collin and *Wiedemannia* Zetterstedt predominate. The genus *Dolichocephala* Macquart occurs in wet seepage soil and moss and wet depressions near streams and very rarely in running-water habitats.

Overall, species of Clinocerinae are confined to relatively clean, cool running water with reduced silt content. Consequently the potential for the use of clinocerines as indicator species for water-quality studies is high, but this has not been well investigated. Only a few widespread clinocerine species (e.g., *C. fuscipennis* Loew, *C. stagnalis* (Haliday), *C. nigra* Meigen) inhabit organically enriched streams or streams where the percentage of silt is significant.

In North America the Clinocerinae range from beyond treeline in northern Canada to the southern humid canyons of New Mexico and Arizona. Few records are known from the central region (prairies) of North America, but this region has not been surveyed for clinocerines and present records are for widespread species, e.g., *C. lineata* Loew. This was the case for the isolated population of *C. lineata* from the Cypress Hills in southeastern Alberta, an isolated forested region surrounded by arid prairie.

Both larval and adult clinocerines are predacious, apparently feeding primarily on other flies. Both flying insects and larval prey have been reported (Table 1). Some are known to also feed on insect carrion trapped in algal mats on water surfaces. Larvae and adults occupy the same habitats.

During the collection of material for a revision of the Clinocerinae, red water mites were observed attached to adult clinocerines and other flying insects associated with creeks, streams, springs, and seepages. Twelve species of *Clinocera* were found to be hosts of larval water mites in this survey. The water mite genus *Hydrovolzia* Thor was observed parasitizing clinocerine empidids almost exclusively, and often 2 or 3 mites were found attached to their hosts, primarily to the legs.

Hydrovolzia is widespread in the Holarctic and Oriental regions and there are five described Nearctic species and a number of undescribed species (Smith 1991). These numerous records of *Hydrovolzia* parasitizing almost exclusively empidids of the subfamily Clinocerinae are the first confirmed records of this mite genus parasitizing Empididae, which may be the primary host of *Hydrovolzia*. Water mites are usually opportunistic generalists and only a few genera are limited primarily to a specific host. It would appear that most species of *Hydrovolzia* prefer clinocerines. The parasitic larval stage is the primary mode of dispersal for water mites (Smith and Oliver 1986), but the impact of parasitism on their hosts is unknown.

Adult clinocerines were also heavily parasitized by several genera of Hydryphantidae. This mite group appears to be very opportunistic and is also

Table 1. Partial summary of feeding records of adult and larval Clinocerinae.

Predator	Prey	Source
Clinocera fuscipennis Loew (A)	Chironomidae (A)	This study
C. olivacea Melander (A)	Chironomidae (L)	This study
C. stagnalis (Haliday) (A)	Flies trapped on surface	Lindroth 1931, cited in Downes and Smith 1969
Clinocera sp. (A)	Thaumaleidae (L)	Sinclair and Marshall 1987
Clinocera sp. (A)	Chironomidae (L and A)	Sinclair 1995
Clinocera sp. (A)	Simuliidae (L)	Peterson 1960
Roederiodes junctus Coquillett (A)	Simuliidae (dead and emerging P and A)	Peterson and Davies 1960
R. wirthi Chillcott (L and A)	Simuliidae (L; 1st instars)	Sinclair and Harkrider 2004
R. wirthi (L)	*Rheotanytarsus* sp. (Chironomidae) (L)	Sinclair and Harkrider 2004
Trichoclinocera hamifera (Melander) (A)	Chironomidae (A)	B.J. Sinclair, unpubl. data
T. ozarkensis Sinclair (A)	Chironomidae and Ceratopogonidae (A)	B.J. Sinclair, unpubl. data
T. rupestris Sinclair (A)	Chironomidae (A)	B.J. Sinclair, unpubl. data
Wiedemannia bistigma (Curtis) and *W. lota* Walker (A)	Dead insects (flies, caddis-flies) on water film	Laurence 1953
Wiedemannia ouedorum Vaillant (L and A)	Simuliidae (L and P)	Vaillant 1952*a*
Wiedemannia sp. (A)	Simuliidae (L)	Wirth and Stone 1956
Wiedemannia (*s. s.*) sp. (A)	Simuliidae (L)	D. Werner, pers. comm., 2006

Note: A, adults; P, pupae; L, larvae.

known to parasitize Chironomidae, Tipulidae, Simuliidae, and other aquatic orders of insects (Smith and Oliver 1986).

Introduction to Clinocera

The genus *Clinocera* is generally confined to cool headwater streams, springs, and seepages and is occasionally found in larger, sunny, exposed streams and rivers (Figs. 2, 3). Adults are very numerous in such habitats, standing on the seepage or madicolous (= hygropetric) zone of springs and remaining very close to rock faces. The genus is most diverse in the Cordillera of western North America (see Biogeography section). There are some 60 species worldwide in addition to 49 new species described in this study. Species diversity of *Clinocera* is much greater in North America (44 species) than in Europe (13 species). In addition, two species-groups are restricted to North America.

Fig. 2. Photographs of *Clinocera* habitats: *a*, falls, Columbia River Gorge, Oregon; *b*, mountain stream crossing the Illecillewaet trail, Glacier National Park, British Columbia; *c*, Evan Thomas Creek, Kananaskis Country, Alberta; *d*, above Paradise River in early July, Mount Rainier National Park, Oregon; *e*, Halfway River, British Columbia; *f*, creek at Anna Ruby Falls, Georgia.

Adults range in wing length from about 3 mm to slightly more than 5 mm. They are often dark to yellowish brown with distinct patterns of bluish pruinescence (Figs. 1*b, c*). The pruinescence, together with narrow wings, long legs, pubescent eyes, and pulvilliform empodia, are important adaptations to living in the spray zone and on wet rocks associated with streams and waterfalls. The adults sit on emergent rocks in streams and seepages, where they have been observed feeding on larvae of seepage midges (Thaumaleidae) and larval and adult chironomid midges (Sinclair and Marshall 1987; Sinclair 1995). Vaillant (1956) listed seven species of *Clinocera* that were predacious (both adult and larval stages) on various larvae living in madicolous (or hygropetric) habitats. Adults of *C. stagnalis* have been observed feeding on flies trapped in the surface film (Lindroth 1931, cited in Downes and Smith 1969). Adults of *Clinocera* have also been observed feeding on black fly larvae and adults (see Table 1; Peterson 1960; Werner and Pont 2003).

Mating also occurs directly on the wet rock faces and is similar to that described for *Trichoclinocera* (Sinclair 1994) and *Wiedemannia* (Rogers 1981). No premating behaviour was noted; the male simply jumps on top of the female. In the "false male above position" (Huber et al. 2007), the male's abdomen and

Fig. 3. Photographs of *Clinocera* habitats: *a*, roadside spring and seepage, Inglis Falls Conservation Area, Ontario (photograph by S.A. Marshall, reproduced with permission); *b*, Mae Cotira River, Brazil, type locality for *C. parana* and *C. cummingi* (photograph by J.E. O'Hara, reproduced with permission); *c*, roadcut seepage, Jolly Cut, Ontario.

genitalia pass over the left side of the female's abdomen (Fig. 1*e*). The forelegs of male *C. binotata* close about the base of the female's wing, clamping down over the costal margin. The detailed position and coupling of the genitalia in *Clinocera* have not been observed, but are presumably somewhat similar to the description for *Empis borealis* L. (Ulrich 1972).

A number of surveys of the empidid fauna of streams are available and *Clinocera* is normally reported (Caspers and Wagner 1982; Harper 1980; Joost 1982; Landry and Harper 1985; Wagner 1980, 1982). In a 11-year survey of a stream in Germany, Wagner and Gathmann (1996) discussed phenological data, abundance and distribution patterns, annual variations, and effects of environmental conditions on the dominant species.

The adults are present throughout the spring and summer and often occur in much colder months also. For example, in rocky springs of southern Ontario (Canada), adults may be found throughout the winter months on the underside of emergent rocks (Sinclair and Marshall 1987). In Europe, *C. wesmaeli* also has been observed to overwinter as an adult (Chvála 1983; Niesiołowski 1992). In the present study, many species of the *C. fuscipennis* group were observed most frequently during autumn and winter.

Larvae and pupae of *Clinocera* occur in the same habitat, often being found in saturated mats of filamentous algae or submerged clumps of moss. Larvae prey on other aquatic larvae, especially chironomids. References and additional information concerning immature stages of *Clinocera* were summarized in Sinclair (1995, p. 692) and Wagner (1997).

Species of *Clinocera* are readily recognized by external colour patterns, but unequivocal determination usually requires examination of the male terminalia. Often the terminalia are exposed and visible above the abdomen, but it is usually best to dissect and clear the genitalia for species determination. The male terminalia of all described species have been illustrated in the literature and few taxonomic problems remain concerning these taxa.

Many species in eastern Asia remain undescribed, and the phylogenetic relationships among the species or species-groups of *Clinocera* have never been analyzed. This, combined with the poor knowledge of New World species, makes a revision of this group a priority among aquatic empidids.

Taxonomic history of Clinocera

The first Nearctic species of *Clinocera* were described by Loew (1860, 1862, 1876). Species were added by Melander (1902) with the first key to species of *Clinocera* (11 species). This key includes many species no longer assigned to *Clinocera*. An additional six species were later described by Melander (1928) in his monumental monograph on Nearctic Empididae, which remains the primary source for identification of many groups. He also provided a key to 27 species of

Clinocera s. lat., which included species of *Wiedemannia* and the then unrecognized genus *Trichoclinocera*. No additional taxonomic work on Nearctic *Clinocera* has appeared since, except for an unpublished thesis (Farr 1958) that included one additional new manuscript name. Consequently there are no reliable and up-to-date keys to North American species of *Clinocera*. The species definitions are not reliable, especially for western species, which handicaps identification in the west with its tremendous diversity of species.

The descriptions of the five known Neotropical species are scattered in the literature and no keys to their identification are available. The first description of a Neotropical species of *Clinocera* was by Bezzi (1909). Collin (1933) and Smith (1962) described the other four known species.

Prior to this study there were 17 Nearctic (Melander 1965) and 5 Neotropical (Smith 1967; *C. oriunda* was omitted from the list) described species. This is the first study to attempt to compile data for all available specimens and continues a series of revisions of the New World fauna of the Clinocerinae, which includes reviews of *Trichoclinocera* and *Wiedemannia* (Sinclair 1994, 1998).

Materials and methods

Source of material

This study is based on the examination of approximately 13 500 adult specimens of *Clinocera* and other clinocerines borrowed from or deposited in 50 institutions or individual collections. These collections are listed below, with the names of the curators and (or) assistants in parentheses. Collection codens in the text indicate the location of specimens and are based mostly on the list provided in the Website "The Insect and Spider Collections of the World" (http://hbs.bishopmuseum.org/codens/).

AMNH American Museum of Natural History, New York, USA (D. Grimaldi)
ANSP Department of Entomology, The Academy of Natural Sciences, Philadelphia, Pennsylvania, USA (D. Azuma)
BARC M. Barták collection [private], Prague, Czech Republic (M. Barták)
BLKU Biosystematics Laboratory, Kyushu University, Fukuoka, Japan (T. Saigusa)
BMNH The Natural History Museum, London, England (J. Chainey and D. Notton).
BPBM Bernice P. Bishop Museum, Honolulu, Hawai'i, USA (N.L. Evenhuis)
BYU Monte L. Bean Life Sciences Museum, Brigham Young University, Provo, Utah, USA (R.W. Baumann)
CAS California Academy of Sciences, San Francisco, California, USA (P.H. Arnaud)
CMNH Section of Invertebrate Zoology, The Carnegie Museum of Natural History, Pittsburgh, Pennsylvania, USA (C.W. Young)

CNC Canadian National Collection of Insects, Ottawa, Ontario, Canada (J.M. Cumming)

CSU Colorado State University, Fort Collins, Colorado, USA (S. Fitzgerald and B. Kondratieff)

CUIC Department of Entomology, Cornell University, Ithaca, New York, USA (E.R. Hoebeke and J.K. Liebherr)

DEBU University of Guelph, Guelph, Ontario, Canada (S.A. Marshall)

FSCA Florida State Collection of Arthropods, Gainesville, Florida, USA (G.J. Steck)

HMOU Hope Entomological Collections, Oxford University, Oxford, England (A.C. Pont)

JBWM J.B. Wallis Museum, University of Manitoba, Winnipeg, Manitoba, Canada (R.E. Roughley)

KMNH Kitakyushu Museum of Natural History and Human History, Kitakyushu, Japan (K. Ueda)

LACM Natural History Museum of Los Angeles County, Los Angeles, California, USA (C.L. Hogue and B.V. Brown)

MCZ Museum of Comparative Zoology, Harvard University, Cambridge, Massachusetts, USA (D. Furth and P. Perkins)

MNHN Muséum national d'Histoire naturelle, Paris, France (C. Daugeron)

MNHNS Museo Nacional de Historia Natural, Santiago, Chile (M. Elgueta)

MSU Entomology Museum, Michigan State University, East Lansing, Michigan, USA (F.W. Stehr)

MSUB Department of Entomology, Montana State University, Bozeman, Montana, USA (R.L. Hurley)

MTD Museum für Tierkunde, Dresden, Germany (U. Kallweit)

MZUSP Museu de Zoologia da Universidade de São Paulo, São Paulo, Brazil (C.J.E. Lamas)

NYSM Biological Survey, New York State Museum, Albany, New York, USA (J.K. Barnes)

OSUC Department of Entomology, Oregon State University, Corvallis, Oregon, USA (N.H. Anderson)

PERC Purdue Entomological Research Collection, Purdue University, West Lafayette, Indiana, USA (J.F. MacDonald)

PSBC Behrend College, Pennsylvania State University at Erie, Erie, Pennsylvania, USA (now housed in ZFMK) (E. Masteller)

ROGC E.J. Rogers collection [private], Villa Grande, California, USA (E. Rogers)

ROM Royal Ontario Museum, Toronto, Ontario, Canada (G.E. Wiggins and D.C. Currie)

SEMC Snow Entomological Collections, Natural History Museum and Biodiversity Research Center, University of Kansas, Lawrence, Kansas, USA (J.S. Ashe)

UBC Spencer Entomological Museum, University of British Columbia, Vancouver, B.C., Canada (S. Cannings)
UCB Essig Museum of Entomology, University of California, Berkeley, California, USA (J.A. Chemsak)
UCD R.M. Bohart Museum of Entomology, University of California, Davis, California, USA (S. Heyden)
UCR Department of Entomology, University of California, Riverside, California, USA (S.I. Frommer)
UIM Department of Entomology, University of Idaho, Moscow, Idaho, USA (F.W. Merickel)
UMQ Département de sciences biologiques, Université de Montréal, Montréal, Québec, Canada (L. Cloutier)
UMSP Department of Entomology, University of Minnesota, St. Paul, Minnesota, USA (P.J. Clausen)
UNH Department of Entomology, University of New Hampshire, Durban, New Hampshire, USA (J.F. Burger)
USNM United States National Museum of Natural History, Washington, D.C., USA (N.E. Woodley)
USUL Department of Biology, Utah State University, Logan, Utah, USA (W.J. Hanson)
UWL University of Wyoming, Laramie, Wyoming, USA (S.R. Shaw)
UZIL Museum of Zoology, Lund University, Lund, Sweden (R. Danielsson)
WSU Department of Entomology, Washington State University, Pullman, Washington, USA (W.J. Turner and R.S. Zack)
WOOC N.E. Woodley collection [private], Washington, D.C., USA (N.E. Woodley)
ZFMK Zoologisches Forschungsmuseum Alexander Koenig, Bonn, Germany
ZMH Zoological Museum, University of Helsinki, Helsinki, Finland (P. Vilkamaa)
ZMUC Zoological Museum, University of Copenhagen, Denmark (T. Pape)
ZSM Zoologische Staatssammlung München, Munich, Germany (M. Kotrba)

Abbreviations

The following abbreviations are used in the descriptions and material examined sections:

alc. – alcohol
Br. – Branch
Ck/ck – creek
Co. – county
Cons. – Conservation
Cpgd/cpgd – campground
CR – county road
Cyn/cyn – canyon
dc – dorsocentral bristle
EF – Experimental Forest
emerg. tp – emergence trap
Exp. – Experiment
Fk/fk – Fork
For./for. – forest
Gov't – Government
GSMNP – Great Smoky Mountains National Park
Hwy – highway

Is. – Island/Islands
jct. – junction
Lk./lk. – Lake/lake
Mt/Mtn(s) – Mount/Mountain(s)
MT – Malaise trap
N – north
NF – National Forest
NM – National Monument
NP – National Park
npl – notopleural bristle
nr – near
pal – postalar bristle
Pk – Park
Pkwy – Parkway
PP – Provincial Park
pprn – postpronotal bristle
presut spal – presutural supra-alar bristle
Prov. – Provincial
psut spal – postsutural supra-alar bristle
Pt – Point
PT – pan trap
R. – River/Rivière
Rd/rd – road
Rec.– Recreation
Res. – Reserve
Rt. – Route
sctl – scutellar bristle
SF – State Forest
SP – State Park
Spr./spr. – spring
Stn – station
Str./str. – stream
Tr. – trail
trib. – tributary
Twp – Township
ypans – yellow pan trap

The following abbreviations for the most frequent collectors are used in the "Material examined" sections: AB – A. Borkent; AF – A. Foley; ALM – A.L. Melander; BH – B. Heming; BJS – B.J. Sinclair; BVB – B.V. Brown; BVP – B.V. Peterson; CBDG – C.B.D. Garrett; CHC – C.H. Curran; CHM – C.H. Mann; CPA – C.P. Alexander; CRN – C.W. Nelson; CWJ – C.W. Johnson; CS – C.W. Sabrosky; DDW – D.D. Wilder, DEH – D.E. Hardy; DMW – D.M. Wood; DSC – D.S. Chandler; ECM – E.C. Masteller; EJR – E.J. Rogers; EMF – E.M. Fisher; EST – E.S. Tucker; ETC – E.T. Cresson; FCH – F.C. Harmston; FME – F. and M. Edwards; GCS – G.C. Steyskal; GEB – G.E. Bohart; GES – G.E. Shewell; GFK – G.F. Knowlton; GJS – G.J. Spencer; HBL – H.B. Leech; HER – H.E. Robinson; HJT – H.J. Teskey; HRF – H.R. Foxlee; JES – J.E. Swann; JFM – J.F. MacDonald; JFMC – J.F. McAlpine; JGC – J.G. Chillcott; JMA – J.M. Aldrich; JMC – J.M. Cumming; JRV – J.R. Vockeroth; LEP – L.E. Pena; LM – L. Masner; MCVD – M.C. Van Duzee; MTJ – M.T. James; NEW – N.E. Woodley; OB – O. Bryant; OSF – O.S. Flint; OY – O. Yata; PHA – P.H. Arnaud, Jr.; PPH – P.P. Harper; RAC – R.A. Cannings; REL – R.E. Leach; REO – R.E. Orth; RER – R.E. Roughley; RES – R. and E. Shannon; RLH – R.L. Hurley; RP – R. Pilfrey; RSZ – R.S. Zack; RWB – R.W. Baumann; SAM – S.A. Marshall; SF – S. Fitzgerald; SGC – S.G. Cannings; TS – T. Saigusa; TWF – T.W. Fisher; WJB – W.J. Brown; WJH – W.J. Hanson; WJT – W.J. Turner; WMW – W.M. Wheeler; WRMM – W.R.M. Mason; WRR – W.R. Richards; WWM – W.W. Moss; WWW – W.W. Wirth.

Collecting methods and specimen preservation and preparation

Adults were most readily collected by aspirating specimens directly from emergent rocks in riffle zones and rocky seepages or by skimming a sweep net slightly above the water surface. Generally the adults remain very close to the substrate and consequently aspirating them directly is much more efficient than sweep-netting. Malaise traps set next to or across streams have been found to be effective in collecting some aquatic empidids, including clinocerines (J.F. MacDonald, pers. comm., 2007). Yellow pans rarely attract clinocerines and consequently were found to be not as effective as hand-collecting.

Freshly collected specimens were killed in tubes containing potassium cyanide or directly added into 70%–75% ethanol. Specimens from the killing tubes were either mounted the same day and stored at below 0 °C for a period of 2 months or laid out flat and dried in paper triangles. Specimens collected into alcohol were later critical-point-dried or dried in ethyl acetate following the methods of Vockeroth (1966). Abdomens and larval head capsules were macerated in crucibles of hot 85% lactic acid to examine terminalia and mouthparts. A strong base used as a clearing agent (e.g., KOH) was found to destroy the delicate membranous tip of the phallus.

Wings were mounted on microscope slides for photography. A wing was detached from the body and run through 2-propanol and then equal parts 2-propanol and cedarwood oil before being mounted directly from the last mixture into Canada balsam (or euparal). Light pressure with fine forceps on the cover slip eliminated bubbles and helped orientate the wing.

Drawings and wing measurements

Details of macerated specimens were viewed under differential interference microscopy (Nomarski illumination). To make illustrations, specimens were mounted temporarily in glycerin on depression slides, using a small amount of shellac to anchor the specimen. Figures were drawn using a drawing device attached to a compound microscope. Before being inked on Mylar, line drawings were checked for accuracy under a dissecting microscope. Specimens were later removed from the shellac and stored in plastic microvials pinned under the adult specimen.

Measurements were made through a dissecting microscope at a low magnification. Body size is extremely variable in *Clinocera*, and among individuals it can vary depending on the age of the specimen when collected and whether it was air-dried, critical-point-dried, or dried directly from ethyl acetate. Consequently, to estimate the size of a species, the length of a wing was measured from the extreme base in a straight line along the costal margin to a line perpendicular to the tip of the wing.

Descriptive format

References to the original publication and junior synonyms are listed in order of date of description. Label data for primary types are cited in full, following the style used in previous revisions (e.g., Sinclair 1994). Labels are listed from the top downward, with data from each label in quotation marks. Labels are cited in full, with original spelling, punctuation, and date, and lines are delimited by a slash (/). Additional information is included in [square] brackets. The repository of each type is given in parentheses. Secondary type data are abridged and listed alphabetically.

Additional material examined for previously described species is also listed, but the number of individual specimens per locality is omitted. All known previously described species are included in a list under each species-group. Types of many non-New World species examined are indicated. All valid species are preceded by one or more of the following letters indicating their biogeographic region of distribution:

AF Afrotropical, *sensu* Crosskey and White (1977)
NE Nearctic, *sensu* Griffiths (1980)
OR Oriental, *sensu* Delfinado and Hardy (1973)
PA Palearctic, *sensu* Soós and Papp (1984)
NT Neotropical, *sensu* Griffiths (1980)
AU Australasian, *sensu* Evenhuis (1989)

Cladistic methods

A preliminary cladistic analysis of the species-group relationships of *Clinocera* was performed using the parsimony program PAUP* version 4.0b10 (Swofford 2002) with the following settings: branch-and-bound search, all characters unordered and of equal weight, Multrees option in effect. Character evolution, character-state distributions, and alternative tree topologies were examined using the program MacClade version 4 (Maddison and Maddison 2003). The data matrix (Table 3) consists of 14 morphological characters (Table 2) compiled for 12 species of *Clinocera*. Ten binary and 4 multistate characters are treated. Exemplars from the genera *Trichoclinocera* and *Oreothalia* were used as outgroup taxa, as they are assumed to be closely related to *Clinocera* (see Sinclair 1995).

Morphology of *Clinocera*

Terms used for adult structures primarily follow those of McAlpine (1981). The term bristle is used for differentiated large setae on the head, mesonotum, and legs bearing a particular name or for one of a series with a particular name, e.g., notopleural bristle, dorsocentral bristle, anteroventral bristle.

Head and mouthparts

The head is attached slightly above the middle of the eye as seen in lateral view. The eye is slightly oval, strongly convex, and completely clothed with dense hairs or ommatrichia (Fig. 8*b*). The eye facets are subequal in size, not

Table 2. Characters used in the analysis of the species-groups of *Clinocera.*

1. *Face*: (0) flat; (1) with ventral tubercle; (2) convex
2. *Lower margin of face*: (0) cleft; (1) straight, not cleft
3. *Postsutural supra-alar bristle*: (0) one; (1) two; (2) absent
4. *Fifth tarsomere*: (0) truncate; (1) with dorsoapical extension
5. *Stigma*: (0) present; (1) absent
6. *Wing*: (0) infuscate only; (1) with distinct spots
7. *Phallus*: (0) lacking whip-like distiphallus; (1) with whip-like distiphallus
8. *Distiphallus*: (0) whip-like, flexible; (1) stiff and strongly recurved apically
9. *Distiphallus*: (0) continuous with phallic shaft; (1) sharply arched at apex of phallic shaft
10. *Apex of phallic shaft*: (0) narrow; (1) with wide membranous expansion, twice width of shaft
11. *Apex of surstylus*: (0) rounded; (1) beak-like; (2) truncate; (3) with shallow notch
12. *Surstylus size*: (0) short and narrow; (1) greatly expanded and lengthened
13. *Clasping cercus*: (0) truncate apically; (1) digitiform; (2) expanded apically with small anteroventral nipple-like projection
14. *Spermatheca*: (0) straight or wavy tube; (1) loosely coiled tube

enlarged anteriorly or dorsally. Both sexes are dichoptic, with the eyes separated by approximately 3 times the width of the ocellar triangle. The ocellar triangle bears a pair of normally long ocellar bristles inserted anterior to the lateral ocelli. Postocular setae are well developed and rather stout on the upper half of the eye. A pair of setae on the vertex posteromedial to the postocular row of setae could be viewed as vertical bristles as observed in Cyclorrhapha. Slender and pale genal setae arise ventral to the postocular row of setae. The gena is narrow, separated

Table 3. Data matrix for *Clinocera* species-groups, listing 14 terminal taxa and 14 characters used in the analysis.

	Character	
	1 1234567890	1111 1234
Trichoclinocera hamifera	0010100??0	0010
Oreothalia pelops	0110000??0	0100
Clinocera tripunctata	0120000??1	0111
Clinocera caerulea	0100001010	2111
Clinocera conjuncta	0100011000	3101
Clinocera olivacea	0100001000	3101
Clinocera fuscipennis	1110001000	0121
Clinocera gubernans	1110011000	0121
Clinocera lineata	0100101101	1111
Clinocera madera	0100101101	1111
Clinocera nigra	0100101000	1111
Clinocera aucta	0100101000	0111
Clinocera stagnalis	2101111000	0111
Clinocera oriunda	2101111000	0111

Note: For character descriptions see Table 2. Character states were scored 0–3; ? denotes inapplicable data.

from the face by the frontoclypeal suture, which extends from the base of the palpus to the margin of the eye.

The face is flat, or somewhat convex (in some species of the *C. stagnalis* group), or distinctly swollen above the lower facial margin (in many species of the *C. fuscipennis* group), and usually lacking setulae (except in some species of the *C. lineata* group). The face often bears species-specific pruinescent patterns that can be viewed laterally or anteriorly. The lower margin of the face is entire and not divided by a median membranous cleft or notch. The width of the face is the same in both sexes and the clypeus is not differentiated.

The terminology of the antenna is taken from Stuckenberg (1999). The antenna is very uniform in this genus, dark and lacking any distinctive coloration, inserted above the middle of the eye, and directed upward. The scape (segment 1) is short and cylindrical, bearing several setulae. The pedicel (segment 2) is oval, bearing a preapical fringe of setulae. The postpedicel (first flagellomere) is short and onion-shaped and produced apically, bearing a long bi-articled, arista-like stylus. The basal article is nearly square and the apical article is normally rather stout, only slightly tapered, and lacking an apical mechanoreceptor (Fig. 6*a*).

The proboscis is short and sucker-like, only the labellum and palpi projecting beyond the head (Figs. 6*d*, 8*b*). The palpus is short and straight, lying close to the mouthparts, not projecting free (Fig. 6*c*). The palpus bears many slender setae, but lacks stout or distinctive setation. The clypeal ridge is articulated with the labrum and cibarium at the labrofulcral articulation point (Fig. 6*e*). The labrum is broadly triangular with the apex sometimes rather blunt. A long slender apodeme extends from the inner dorsal margin of the labrum. The epipharyngeal blades are very pronounced and separated from the labrum, bearing a series of stout mechanoreceptors. The lacinia is not free but secondarily fused to the inner labial sclerite and labial paraphysis. The inner apex of the lacinia bears a rod-shaped stipes, which is greatly lengthened to more than twice the length of the lacinia. The hypopharynx is needle-like, with a broad base. The labellum is held in a horizontal position, is sucker-like, and lacking pseudotracheae. The prey is held mostly by the labrum and hypopharynx, in a similar fashion to that observed in Dolichopodidae.

Wing

The homology of wing veins and cells in Diptera (Figs. 29*a*, *b*) follows the interpretation of Wootton and Ennos (1989) and Saigusa (2006)[1]. The wings of *Clinocera* are infuscate, often with cloudy spots or darkened areas along crossveins. The stigma is elongate, faint, and rarely dark, but is absent in several species-groups. At the base of the wing is a long stout seta and in some species of the *C. conjuncta* group there are often several basal pairs, with additional long setae extending distally along the costal margin. All veins except the costa are bare,

[1]T. Saigusa. 2006. Homology of wing venation of Diptera. Unpublished handout distributed at the 6th International Congress of Dipterology, Fukuoka, Japan.

lacking macrosetae. The costa bears stout erect setae from the apex of vein R_1 to the apex of R_{4+5} (see Sinclair 2000*b*, figs. 5, 6). The subcosta (Sc) is complete and R_1 joins the costa near the midlength of the wing. Vein R_{2+3} is normally straight, occasionally gently arched beneath the stigma. An auxiliary crossvein is sometimes present between R_{2+3} and R_4 (Fig. 29*g*). R_4 is strongly or weakly divergent from R_5, straight or bell-shaped. Cell dm is variable in shape, either truncate or produced apically; the bases of M_1 and M_2 are broadly separated and three veins are emitted from cell dm (M_1, M_2, M_4). Cell cua (anal cell or cell cu*p*) is slightly narrower than cell bm and rounded apically. The faint anal vein arising subapically from this cell (CuP+CuA or A_1+CuA_2) is usually reduced in thickness, mostly appearing simply as a streak, and does not extend to the wing margin.

Thorax

The thorax is compact and densely clothed in pruinescence. The prosternum is fused laterally to the proepisternum, forming a complete precoxal bridge. The scutum is often patterned dorsally, with a pair of median vittae or a single vitta that vary in colour (Fig. 5*b*). The notopleuron and prescutellar depression may also be differentiated with bluish pruinescence. The chaetotaxy is reduced to rows of distinct bristles (Fig. 5*b*): several acrostichal setulae anterior to the first dorsocentral bristles; normally 5 dorsocentral bristles (occasionally as many as 10), usually stout and well developed, increasing in length posteriorly; 1 postpronotal bristle; 1 presutural supra-alar bristle; 2 or 3 notopleural bristles; 1 or 2 postsutural supra-alar bristles; 1 postalar bristle; 2 slightly divergent scutellar bristles.

The laterotergite bears a patch of pale setae and generally there are few additional pleural setae. There are often some pale setulae along the posterior margin of the mesepisternum and on the proepisternum.

Legs

The legs are long and slender with mostly only evenly scattered setae. The legs vary from dark-coloured to yellowish brown, with the ventral face of the femora often distinctly paler in various views. In the *C. stagnalis* group the apex or "knee" of the femora is often pale, in contrast to the remaining portion.

The anterior face of the forecoxa bears long soft setae, while the mid and hind coxae possess long lateral setae. The male forefemur usually possesses a double row of spine-like setae beneath, and a preapical anterior comb (Fig. 8*d*) is also usually present in both sexes. In males the foretibia bears an anterior apical comb and a biserial row of setae beneath. The hind tibia bears a posterior apical comb and dorsal and ventral erect setae of various lengths and strengths on the apical half. The second tarsomere of the foreleg is much longer than the third and fourth tarsomeres and the dorsal apex of the fifth tarsomere bears a prolongation

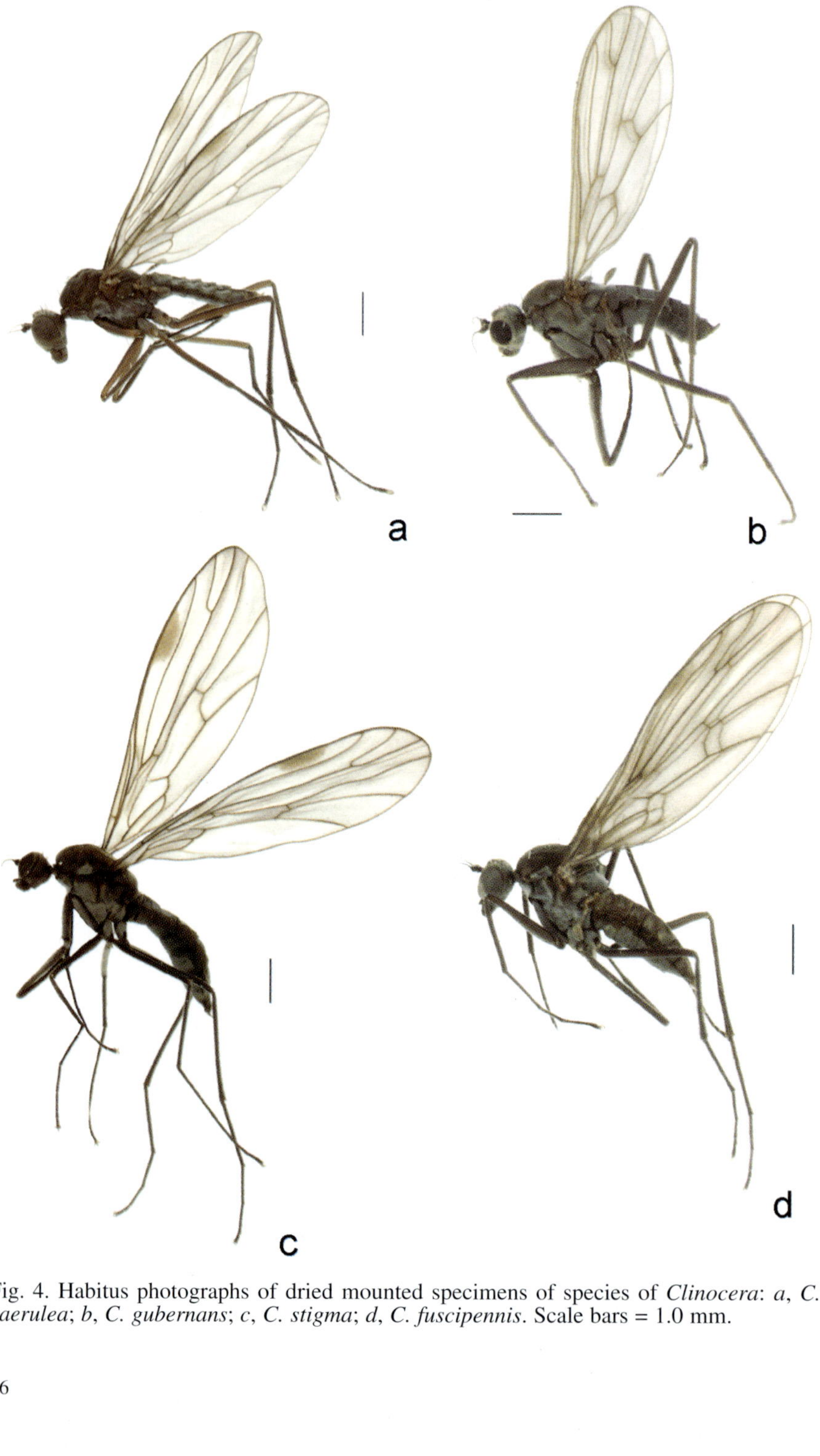

Fig. 4. Habitus photographs of dried mounted specimens of species of *Clinocera*: *a*, *C. caerulea*; *b*, *C. gubernans*; *c*, *C. stigma*; *d*, *C. fuscipennis*. Scale bars = 1.0 mm.

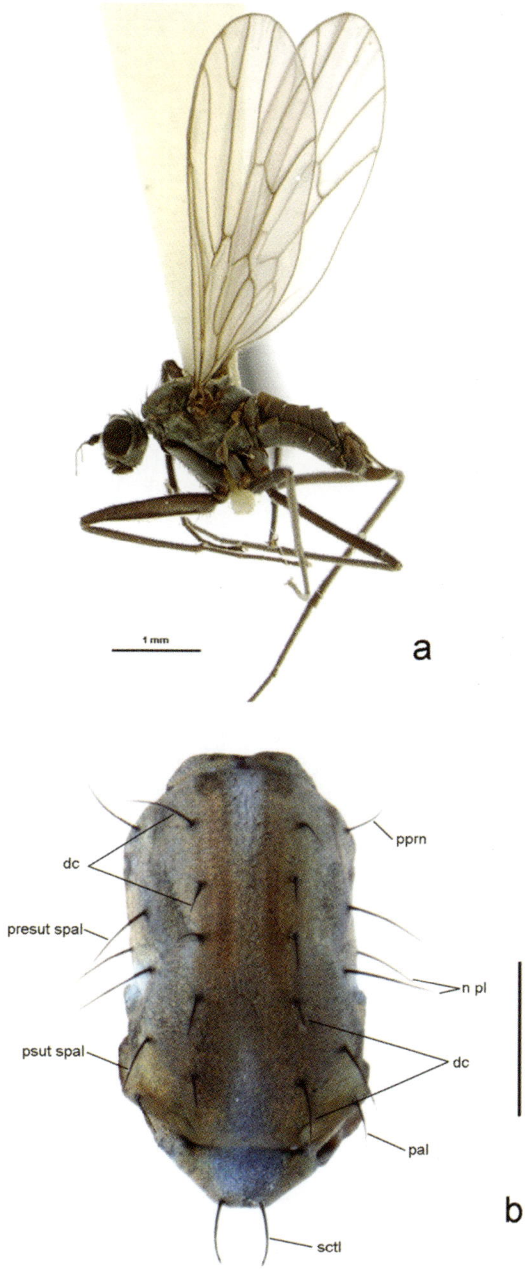

Fig. 5. Photographs of *Clinocera*: *a*, habitus of a dried mounted specimen of *C. schlinkerti*; *b*, *C. conjuncta*, thorax, dorsal view. Scale bar = 0.3 mm. (dc, dorsocentral bristle; n pl, notopleural bristle; pal, postalar bristle; pprn, postpronotal bristle; presut spal, presutural supra-alar bristle; psut spal, postsutural supra-alar bristle; sctl, scutellar bristle.)

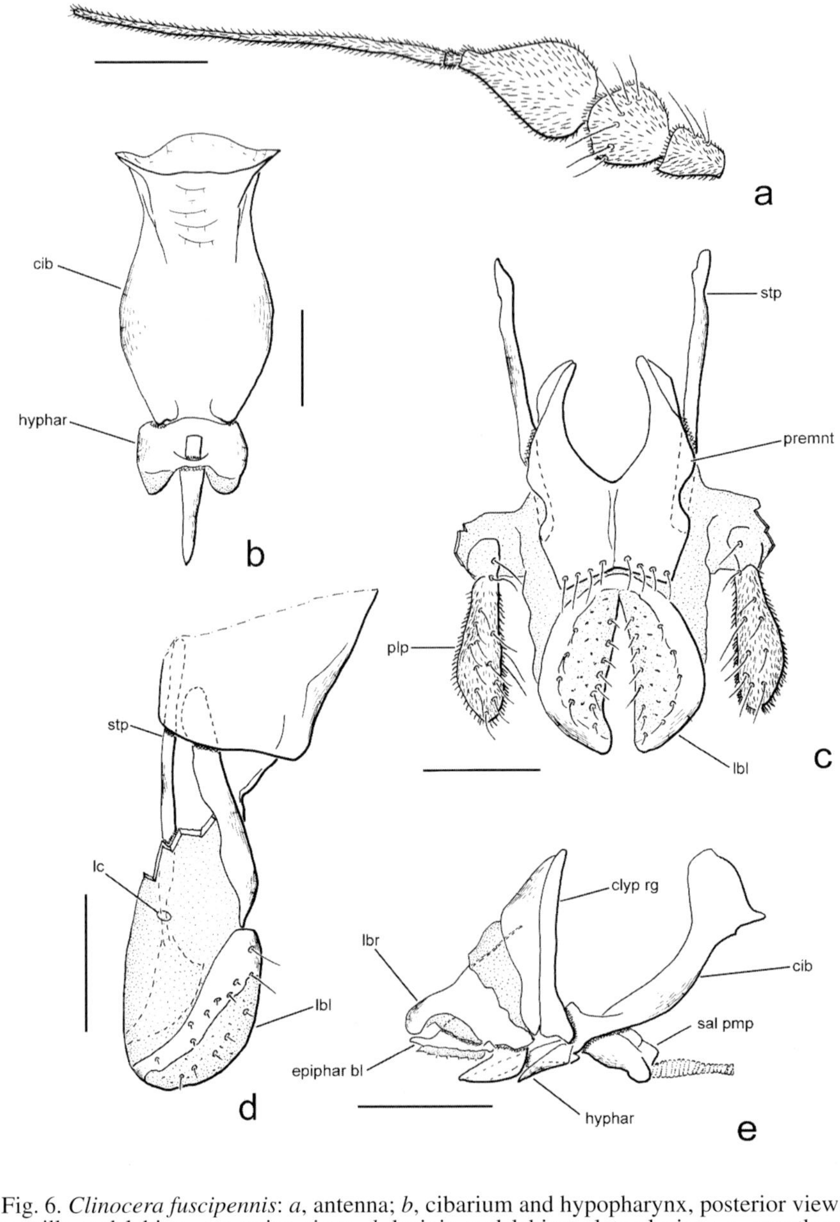

Fig. 6. *Clinocera fuscipennis*: *a*, antenna; *b*, cibarium and hypopharynx, posterior view; *c*, maxilla and labium, posterior view; *d*, lacinia and labium, lateral view; *e*, mouthparts, maxillae and labium removed, lateral view (cib, cibarium; clyp rg, clypeal ridge; epiphar bl, epipharyngeal blade; hyphar, hypopharynx; lbl, labellum; lbr, labrum; lc, lacinia; plp, palpus; premnt, prementum; sal pmp, salivary pump; stp, stipes). Scale bars = 0.1 mm.

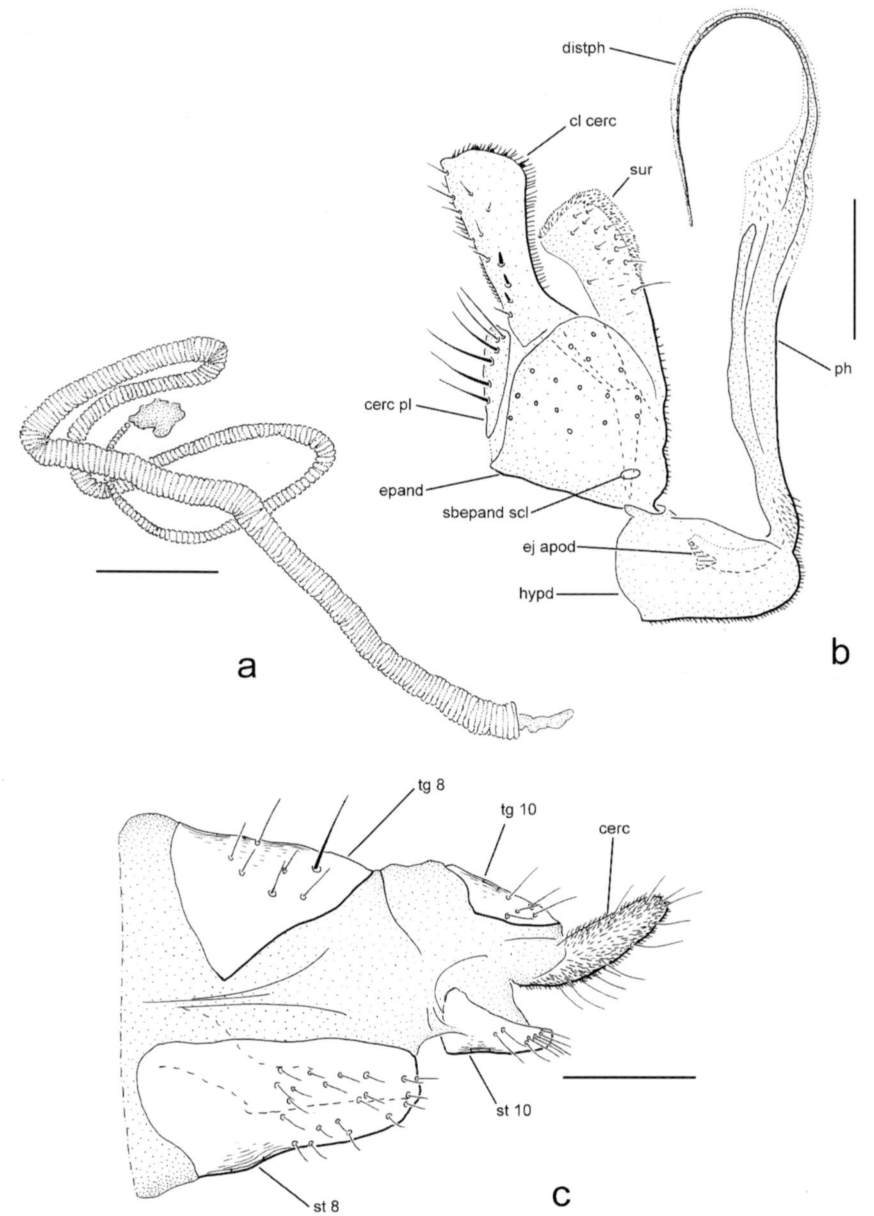

Fig. 7. *Clinocera fuscipennis*: *a*, spermathecal duct; *b*, male terminalia, lateral view; *c*, female terminalia, lateral view (cerc, cercus; cerc pl, cercal plate; cl cerc, clasping cercus; distph, distiphallus; ej apod, ejaculatory apodeme; epand, epandrium; hypd, hypandrium; pgt, postgonite; ph, phallus; sbepand scl, subepandrial sclerite; st, sternite; sur, surstylus; tg, tergite). Scale bars = 0.1 mm.

or extension in the *C. stagnalis* and *C. gubernans* groups (Fig. 9*b*). The pulvilli and empodium are both well developed and pulvilliform, while the tarsal claws are strongly curved and vary in length (Fig. 8*c*).

Abdomen and male terminalia

Homology of the male terminalia follows that of Sinclair (1994, 2000*a*) and Cumming et al. (1995), except that the apical phallic filament is referred to as the

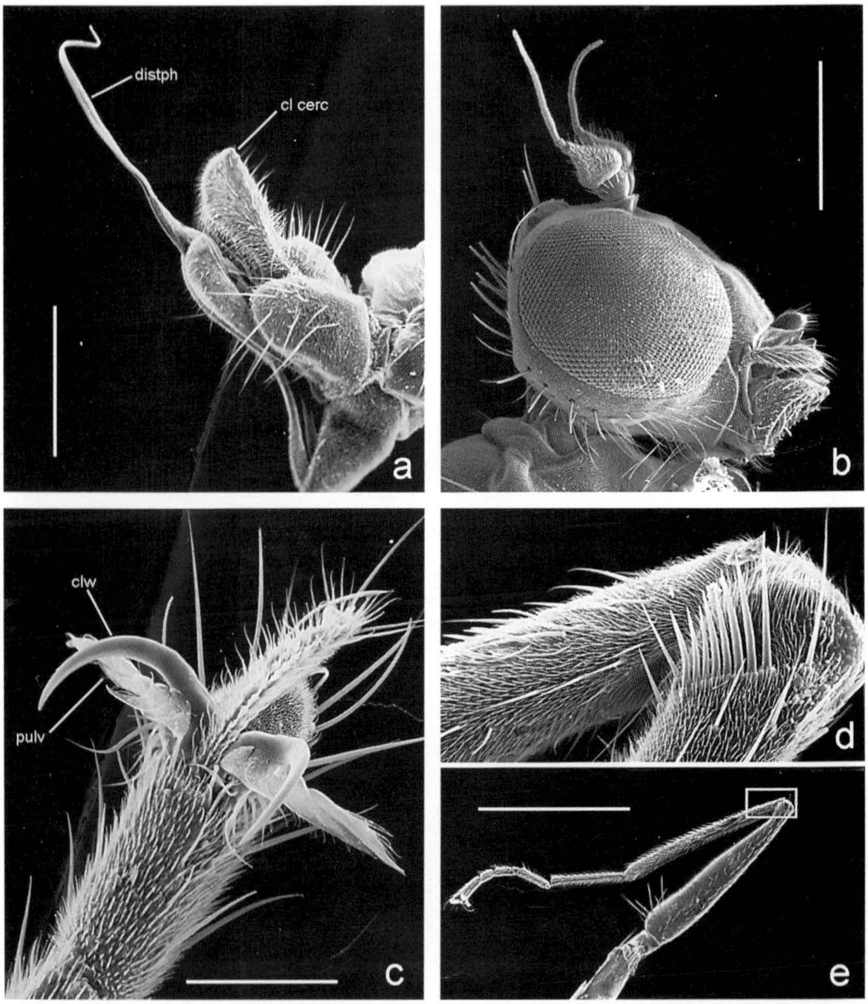

Fig. 8. Scanning electron photographs of *Clinocera*: *a*, *C. fuscipennis*, male terminalia, lateral view. Scale bar = 176 µm; *b*, *C. maculata*, head, lateral view. Scale bar = 0.30 mm; *c*, *C. sublineata*, fifth tarsomere, ventral view. Scale bar = 43 µm; *d*, *C. maculata*, enlargement of apex of femur (from *e*) and preapical comb, lateral view; *e*, *C. maculata*, foreleg. Scale bar = 0.75 mm (cl cerc, clasping cercus; clw, claw; distph, distiphallus; pulv, pulvillus.)

distiphallus. Abdominal tergites 1–7 lack modified setae or tubercles and tergite 8 is narrow, less than half as long as the preceding segment. This modification allows the terminalia to arch anteriorly and lie or rest on the terminal segments of the abdomen. The hypandrium is very small, cone-shaped, and usually lacking macrosetae. The phallus arises from the apex of the hypandrium and bears a whip-like apical filament or distiphallus that arches anteriorly (Fig. 7*b*). The ejaculatory apodeme is small, oval, and inconspicuous, removed from the base of

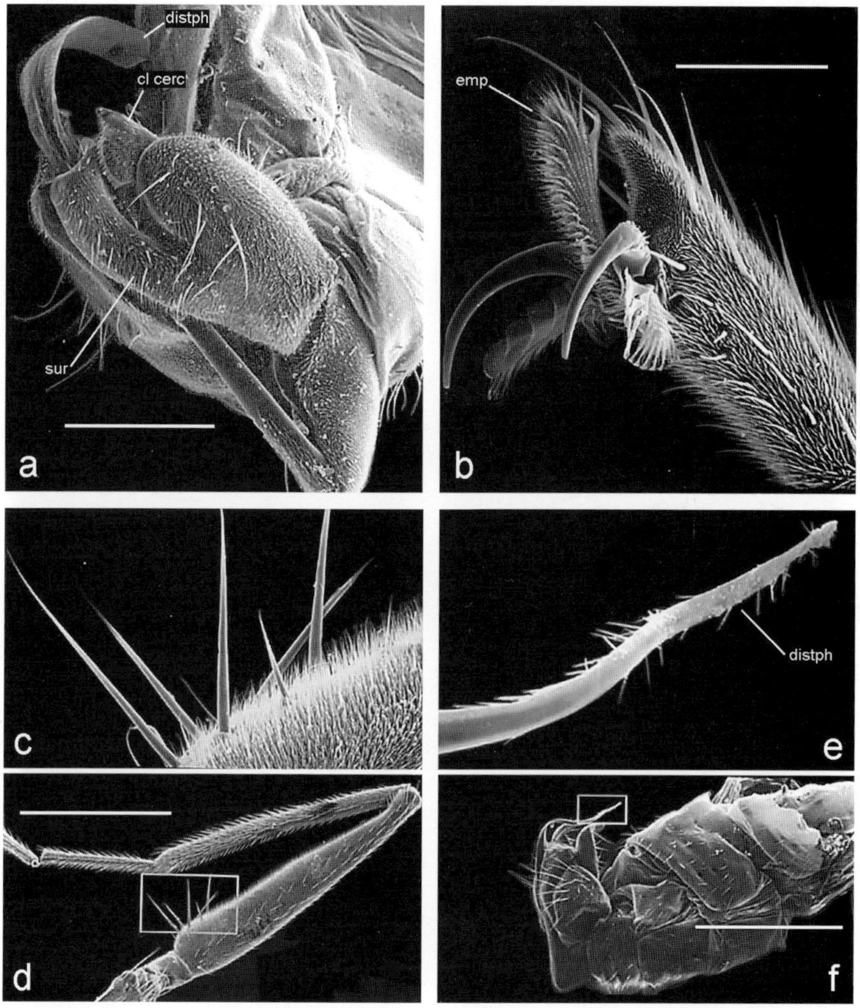

Fig. 9. Scanning electron photographs of *Clinocera*: *a*, *C. binotata*, male terminalia, oblique posterior view. Scale bar = 176 μm; *b*, *C. binotata*, fifth tarsomere, oblique ventral view. Scale bar = 50 μm; *c*, *C. maculata*, enlargement of base of forefemur (from *d*), lateral view; *d*, *C. maculata*, foreleg. Scale bar = 0.50 mm; *e*, *C. maculata*, enlargement of distiphallus (from *f*), lateral view; *f*, *C. maculata*, male terminalia. Scale bar = 0.43 mm. (cl cerc, clasping cercus; distph, distiphallus; emp, empodium; sur, surstylus.)

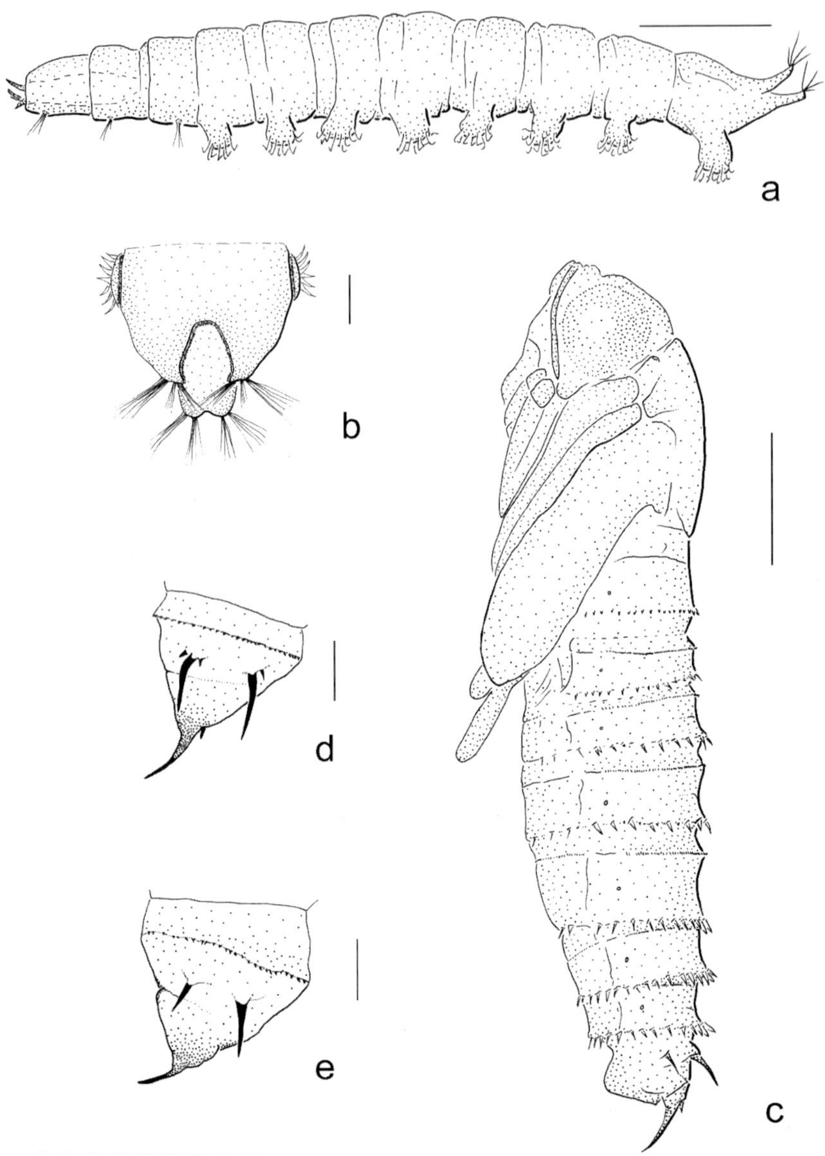

Fig. 10. Larva and pupa of *Clinocera*: *a*, larva, lateral view; *b*, larval caudal segment, dorsal view; *c*, pupa, lateral view; *d*, *C. lineata*, caudal end of pupa, lateral view; *e*, *C. fuscipennis*, caudal end of pupa, lateral view. Scale bars = 0.5 mm.

phallus, lying within the hypandrium, and connected to the phallus by a narrow membranous endophallic tube. The endophallus or sperm sac is also readily visible above the apodeme. The shape of the phallus and the form of the apex of the phallus are very important in species sorting and identification.

The epandrium is divided into a pair of expanded lamellae connected by a narrow bridge, and articulates ventrolaterally with the base of the hypandrium. The outer face of the epandrial lamella bears long setae, which usually fall off during preparation of the genitalia and do not appear to be useful for species differentiation. The cercus is subdivided into a flattened cercal plate that surrounds the anus and possesses long, evenly distributed setae and an upper component, known as the clasping cercus, that lies above the anus. The clasping cercus is much larger than the cercal plate. It is modified for clasping the female abdomen during mating, with the inner face of the clasping cercus bearing large patches of short pointed or peg-like setae. In addition to the phallus, the shape of the clasping cercus is important in species identification.

The subepandrial sclerite is fused to the base of clasping cercus and extends medially between the epandrial lamellae. The large, wide surstylus usually extends beyond the epandrium and functions mostly as a phallic guide. The phallus is often found positioned between the surstyli in dried, unprepared specimens (Fig. 9*a*).

Female terminalia

The female abdomen is telescopic and the terminalia lack clear species-specific characters. The length of female segment 8 is equal to or less than that of the preceding segment. Tergite 10 and sternite 10 are split medially and tergite 10 is unmodified, lacking acanthophorite spine-like setae. The cercus is narrow and straight (Fig. 7*c*). The spermatheca is loosely to tightly coiled and trachea-like, and the apex often bears a sphere of thinly sclerotized glandular tissue (Fig. 7*a*). A small triangular genital fork is often present below the posterior apex of the spermatheca (Sinclair 1995, fig. 22).

Immature stages

The larvae and pupae of *Clinocera* in North America can be identified using the keys in Sinclair and Harkrider (2004). The larva and pupa of several European species of *Clinocera* have been described or illustrated (e.g., Brindle 1964, 1973; Niesiołowski 1992; Vaillant 1952*a*, 1952*b*; Wagner 1997). Larvae and pupae of Nearctic *Clinocera* are identical in habitus with those of these European species (Steyskal and Knutson 1981) (Figs. 10*a*, *c*). As in all Clinocerinae, the larvae are apneustic and characterized by eight pairs of well-developed, crochet-bearing abdominal prolegs (Fig. 10*a*). The terminal abdominal segment ends in one pair of dorsolateral processes, equal or less in length to their width, and a single median process with the apex indistinctly to distinctly divided into a pair of short lobes bearing setae (Fig. 10*b*).

The pupa of *Clinocera* is characterized by a pair of curved caudal hooks, two or fewer pairs of mesonotal setae, two transverse rings of abdominal chitinous projections, and a pair of short antennal sheaths (Fig. 10*c*). Two eastern Nearctic species of *Clinocera* were reared from pupae collected from the madicolous zone of roadcut seeps and cascading streams. The pupa of *C. lineata* is characterized by biserial abdominal tergal spicules, a caudal segment with two long lateral spicules subequal in length, and a pair of short spicules between the caudal hooks (Fig. 10*d*). The pupa of *C. fuscipennis* is characterized by uniserial tergal spicules, a caudal segment with the two long lateral spicules of unequal lengths, and no spicules at the base of the caudal hooks (Fig. 10*e*).

Taxonomy

Key to Nearctic genera of Clinocerinae

1 R_{4+5} unbranched 2
— R_{4+5} branched 3

2 Wing cell dm lacking (crossvein dm-cu absent); R_{2+3} branched . . . . **Asymphyloptera** Collin
— Wing cell dm present; R_{2+3} unbranched . . . . **Oreothalia** Melander

3 Sc evanescent apically . . . . **Proclinopyga** Collin
— Sc ending in costa 4

4 Proboscis as long as height of head; labrum long and slender; labellum not sucker-like . . . . **Roederiodes** Coquillett
— Proboscis shorter than height of head; labrum subtriangular; labellum sucker-like. 5

5 Neck arising high on occiput, head extending obliquely forward; wing normally with white irrorations or circular spots . . . . **Dolichocephala** Macquart
— Neck arising near centre of occiput, head hypognathous; wing lacking white irrorations or circular spots 6

6 R_1 with macrotrichia extending along dorsal surface; forefemur of both sexes with stout setae beneath; female abdomen truncate apically . . . . **Trichoclinocera** Collin
— R_1 lacking macrotrichia; forefemur lacking modified setae or with only simple biserial row of short setae, often only on males; female abdomen attenuated apically 7

7 Lower margin of face (clypeal region) lacking notch (Sinclair 1995, fig. 57); crossvein dm-cu normally long, widely separating M_1 and M_2; phallus normally with non-articulated apical distiphallus ..... ***Clinocera*** Meigen

— Lower margin of face (clypeal region) with notch (Sinclair 1995, fig. 62); crossvein dm-cu often short, veins M_1 and M_2 arising closely approximated; phallus normally with articulated apical distiphallus............ .... ***Wiedemannia*** Zetterstedt

Key to Neotropical genera of Clinocerinae

1 R_{4+5} unbranched; cell dm lacking (crossvein dm-cu absent); R_{2+3} branched .... ***Asymphyloptera*** Collin

— R_{4+5} branched; cell dm present; R_{2+3} unbranched 2

2 Sc evanescent apically .... ***Proagomyia*** Collin

— Sc ending in costa 3

3 Proboscis as long as height of head; labrum long and slender; labellum not sucker-like................................ ***Roederiodes*** Coquillett

— Proboscis shorter than height of head; labrum subtriangular; labellum sucker-like.. 4

4 Neck arising high on occiput, head extending obliquely forward; wing normally with white irrorations or circular spots .... ***Dolichocephala*** Macquart

— Neck arising near centre of occiput, head hypognathous; wing lacking white irrorations or circular spots .... ***Clinocera*** Meigen

Clinocera Meigen

Clinocera Meigen, 1803: 271. Type species, *C. nigra* Meigen, 1804 (subsequent monotypy by Meigen, 1804: 292).

Atalanta Meigen, 1800: 31. Suppressed by the International Code of Zoological Nomenclature, 1963: 339.

Paramesia Macquart, 1835: 656. Type species, *P. wesmaeli* Macquart (des. Coquillett, 1903: 255).

Hydrodromia Macquart, 1835: 658. Type species, *Heleodromia stagnalis* Haliday (des. Coquillett, 1903: 251).

Aclinocera Yang & Yang, 1995: 236. **syn. nov.** Type species, *A. sinica* Yang & Yang (original designation).

Notes on synonymy

Yang and Yang (1995) described a new genus, *Aclinocera*, from China (Zhejiang Province) on the basis of a male specimen with "unusual" terminalia and two ventral bristles on the hind tibia. I have seen very similar species from Taiwan and Nepal, all of which are easily assigned to the *C. lineata* group, which is closely related to *C. stuckelbergi* (Vaillant). This small subgroup is characterized by the form of the clasping cercus and the distinctive facial setulae. The latter feature was overlooked by Yang and Yang (1995). Although facial setulae are normally diagnostic for *Kowarzia* Mik, a number of eastern Asian species of the *C. lineata* group possess such very fine setulae (Sinclair 1995). The variation of the clasping cercus is within the variation observed in the *C. lineata* group, especially in the Oriental Region, and in the genus as a whole. Consequently *Aclinocera* is viewed as a junior synonym of *Clinocera*.

Recognition

In North America, adult *Clinocera* are most similar to adult *Oreothalia*, but are distinguished by the branched R_{4+5}. *Proclinopyga* possesses broader wings, Sc is incomplete, and the face is protruded (see Sinclair 1995, fig. 56). In *Trichoclinocera* and *Wiedemannia* the face has a distinct notch and the gena is wide. In the former genus R_1 bears macrotrichia, the apex of the female abdomen is truncate, and there is usually a series of stout ventral setae on the forefemur. In *Wiedemannia* the apex of cell dm is often narrowed, such that the M veins arise closely approximated and acrostichal setulae are usually more developed than in *Clinocera*.

Clinocera can be identified using the key to genera of the Clinocerinae in Sinclair (1995) or the genus-level key to North American Empididae in Steyskal and Knutson (1981), in addition to the new, updated keys to genera in North and South America provided above.

Diagnosis

Face usually lacking setulae; lower margin entire, lacking median membranous cleft. Gena narrow, separated from face by frontoclypeal suture extended from base of palpus to margin of eye. Antenna inserted above middle of eye; arista-like style bi-articulated. Labrum broadly triangular. Lacinia fused to labial sclerite and labial paraphysis. All wing veins lacking macrotrichea or setulae; Sc complete; vein R_{4+5} branched. Forefemur often with stout setae beneath; forefemur usually with preapical anterior comb; fifth tarsomere with dorsoapical extension often present. Phallus usually with whip-like distiphallus. Cercal plate with long setae evenly distributed; clasping cercus relatively large, with modified setae on inner face. Subepandrial sclerite fused to base of clasping cercus. Surstylus large, wide, usually extended beyond epandrium. Female tergite 10 split medially, lacking modified setae; cercus narrow, straight; spermatheca loosely to tightly coiled, trachea-like, often with thinly sclerotized oval glandular tissue at apex.

Classification

Sinclair (1995) confirmed the monophyly of this genus and elevated three of its former subgenera (*Kowarzia* Mik, *Bergenstammia* Mik, and *Phaeobalia* Mik) to genus level because of their closer phylogenetic relationship with other genera than with *Clinocera*. All remaining subgeneric divisions in *Clinocera* (e.g., *Hydrodromia* Macquart) were synonymized because of their weak definition when species beyond Europe were considered. Subsequently, several species-groups were introduced as replacements for these subgeneric divisions. These new species-groups will facilitate freedom of movement of species among these groups until the world fauna is better understood.

The definitions of these species-groups were simply introduced and not discussed by Sinclair (1995), consequently each species-group is herein clearly defined and its worldwide distribution and assigned species are discussed in the subsequent section. The groups were erected primarily on the basis of synapomorphies of the male terminalia, e.g., the shape of the clasping cercus and surstylus or the form of the apex of the phallus. Additional defining characters include wing venation (e.g., loss of stigma), legs (e.g., dorsoapical extension of the fifth tarsomere), and robustness of the head and shape of the face. There are presently seven species-groups recognized worldwide, including the South African group (Sinclair 1999*b*, 2003*a*), which was not previously recognized. Six species-groups are present in the Nearctic Region, of which the *C. conjuncta* and *C. caerulea* groups are endemic.

Distribution

Clinocera is the most widely distributed genus of Clinocerinae, with records from the northern coast of Alaska and Baffin Island to the southern tip of South America. It is known from Campbell Island and eastern Australia in the Australasian Region, but is unknown from the Oceanian Region. In Africa this genus is restricted to South Africa. It is widespread in the Palearctic Region, from Greenland and Iceland in the west to Japan, and extending south into the Oriental Region as far south as Java (Sinclair and Saigusa 2005).

Key to species-groups of *Clinocera*

1 Stigma absent (Figs. 31*g*, 32*e*); face flat or convex, never with tubercle . . . 2

— Stigma in middle of cell r_1, elongate and usually distinct (Figs. 30*i*, *h*); if absent, face with distinct tubercle 4

2 Face broad, width at midpoint distinctly wider than width of antennal sockets or ocellar triangle; surstylus normally rounded apically (Figs. 21*c*, 22*c*) 3

— Face narrow, middle only slightly wider than width of antennal sockets or ocellar triangle; surstylus usually with beak-like ventroapical projection and expanded and broadened dorsoapically (Fig. 19*a*)... ***C. lineata* group**

3 Fifth tarsomere often with dorsoapical extension (Fig. 9*b*); cell dm usually truncate distally (Fig. 32*d*); face often brightly coloured and shiny; wings often with spots present on major crossveins (Figs. 32*d–j*); phallus usually split at base of distiphallus (Figs. 22, 23) .... ***C. stagnalis* group**

— Fifth tarsomere lacking extension; cell dm produced distally (Fig. 32*c*); face often with small pale pruinescent area at each lower corner of face; wing spots present or absent; phallus slender, lacking split at base of distiphallus (Figs. 21*b*, *c*) .... ***C. nigra* group**

4 Four dorsocentral bristles present; postsutural supra-alar bristles absent; wings with small round white spots (South Africa) ... ***C. tripunctata* group**

— Five or more dorsocentral bristles present, one or more postsutural supra-alar bristles present; wings variable in coloration, but lacking small round white spots 5

5 Two or more (rarely one, or sometimes very slender and inconspicuous) postsutural supra-alar bristles; facial tubercle on lower face sometimes present; distiphallus arising from narrow shaft; apex of clasping cercus tapered to ventroapical nipple-like tip (Fig. 16) .... ***C. fuscipennis* group**

— One postsutural supra-alar bristle (Fig. 5*b*); facial tubercle lacking; distiphallus arising from expanded apex of shaft; apex of clasping cercus truncate (Fig. 13) or tapered to slender round tip (Fig. 11) 6

6 Scutum lacking pair of black stripes; clasping cercus tapered to round tip, inner stout setae distributed on apical half; surstylus truncate apically, lacking notch (Fig. 11); specimens usually reddish brown .... ***C. caerulea* group**

— Scutum with pair of black stripes (bivittate); apical margin of clasping cercus truncate, bearing stout setae on apical third of inner face; surstylus often with subapical notch (Figs. 13*a*, *d*); specimens generally greyish, often with extensive blue pruinescence .... ***C. conjuncta* group**

New World species of *Clinocera*

Clinocera caerulea group

Diagnosis

This species-group is characterized by a small truncate head; a short truncate surstylus; and a slender clasping cercus with short stout setae on the inner face. The distiphallus is generally heavily sclerotized, reducing flexibility, and the hypandrium is small in comparison with the epandrium.

Remarks

The heavily sclerotized and stiff distiphallus characterizes three of the four species of this group. The fourth species (*C. longicauda*) is unique among *Clinocera* in having a long and coiled distiphallus. The monophyly of this group is supported on the basis of the following apomorphies: distiphallus sharply arched at apex of phallus and apex of surstylus truncate.

This species-group is perhaps most closely related to the *C. lineata* group, with which it shares a truncate surstylus and small hypandrium.

Distribution

This species-group is confined to North America. Three of the four species assigned to the *C. caerulea* group are restricted to the west, and the fourth species (*C. brunnea*) has widely disjunct populations in eastern and western North America.

Included species

This species-group includes the following four new species:

NE *Clinocera brunnea* **sp. nov.**
NE *Clinocera caerulea* **sp. nov.**
NE *Clinocera daniellae* **sp. nov.**
NE *Clinocera longicauda* **sp. nov.**

Key to species of the *Clinocera caerulea* group

1 Face brown, lacking blue pruinescence 2
— Face with blue pruinescence. 3

2 Head dark, in contrast to thorax, especially pleura; hind first tarsomere with 2 or 3 stout erect setae beneath; distiphallus long and coiled (Fig. 11*d*) (western) . . . . ***C. longicauda*** sp. nov
— Head concolorous with thorax; hind first tarsomere lacking erect setae beneath; distiphallus shorter than phallic shaft (Fig. 11*a*) (widespread) . . . . ***C. brunnea*** sp. nov.

3 Face with central blue pruinescent spot; lower half of pleura with bluish pruinescence; distiphallus arising from C-shaped arch; apex of clasping cercus straight to slightly curved (Fig. 11*b*); R_{2+3} mostly straight beneath stigma (western). . . . . ***C. caerulea*** sp. nov
— Face mostly with blue pruinescence, except immediately below antenna; pleura with faint pruinescence; distiphallus arising straight from phallic shaft; apex of clasping cercus distinctly curved (Fig. 11*c*); R_{2+3} arched beneath stigma (western) . . . . ***C. daniellae*** sp. nov.

Clinocera brunnea sp. nov.

Figs. 11*a*, 29*a*; Map 1

Clinocera sp. Landry and Harper, 1985: 1381.

Type material. HOLOTYPE, male labelled "ONT: Owen Sound/ Inglis Falls - spring/ 29.XI.1985/ B. Sinclair"; "HOLOTYPE/ Clinocera/ brunnea/ Sinclair [red label]" (CNC).

Paratypes: **CANADA. Ontario:** Dundas, Rock Chapel, 10.viii.1986, BJS (1 ♀, DEBU-reared); Dundas, Webster's Falls, 12.vii.1986, BJS (2 ♂♂, DEBU); Eugenia Falls Cons. Area, spr., 2.x.1985, BJS (1 ♂, 3 ♀♀, DEBU); Grey Co., Owen Sound, Inglis Falls Cons. Area, spr., 24.i.1986, 19.ii.1986, 4, 6.v.1986, 12.v.1985, 4.vi.1985, 29.vii.1990, 26.x.1985, 25.xi.1984, 29.xi.1985, 17.xii.1985, 12.vii.2002, BJS, SAM (25 ♂♂, 27 ♀♀, CNC, DEBU, ZFMK); Tobermory, Dunk's Bay, 30.x.1996, SAM (1 ♂, DEBU).

Additional material examined. CANADA. Alberta: Banff, 28.ix.1932, OB (1 ♂, CAS); Richardson Hwy, mi. 206, beaver pond, SAM (1 ♀, DEBU). **British Columbia:** Terrace, Eliza Ck, 22.vi.1960, GES (2 ♂♂, CNC); Glacier NP, 1250 m, str., 30.vi.1989, BJS (1 ♂, CNC); Manning PP at Hwy 3, Goodfellow Ck, 8.x.1979, DDW (1 ♂, 3 ♀♀, CAS); 12 km N Revelstoke, Hwy 23, St. Cyr Ck, 2.vii.1989, BJS (1 ♀, CNC). **Quebec:** Mt Orford, 9.ix.1937, GES (1 ♂, 1 ♀, CNC); Mt St. Hilaire, 500–700′, 4.vi.1963, JRV (1 ♀, CNC): R. du Castor, trib. Lac Nathalie, 25, 28.ix.1974, J.L. Frechette (4 ♂♂, 5 ♀♀, UMQ, in alc.); R. du Castor, James Bay, emerg. tp, 28.viii., 25, 28.ix.1974, J.L. Frechette (4 ♂♂, 6 ♀♀, UMQ). **USA. California:** Plumas Co., Lassen NF, N Fk Feather R., 28.v.1977, DDW (1 ♂, CAS). **Colorado:** Monarch Pass, 8000′, 21.vi.1940, ALM (1 ♂, USNM). **Idaho:** Latah Co., 15 mi ENE Potlatch, 2900′, Strychnine Ck, 8.vi.1982, WJT (1 ♀, WSU). **New Hampshire:** White Mtns, Morrison (1 ♀, USNM). **Michigan:** Alger Co., Munising, Alger Falls, 9.ix.1982, R. Hurley (1 ♂, MSUB). **Oregon:** Klamath Co., 13 mi. NE Bly, Deming Ck, 16.ix.1960, J. Schuh, (1 ♀, WSU); Lane Co., Siuslaw NF, 10 mi. S Yachats, Rock Ck, 15.vii.1989, BJS (1 ♂, CNC); Lane Co., Willamette NF, McKenzie Bridge Cpgd, MacKenzie R., 21.vi.1978, DDW (3 ♂♂, 3 ♀♀, CAS); Lane Co., 11 mi. NE Blue R., H.J. Andrews EF, 800 m, Mack Ck, 29.vi.1982, GWC (1 ♂, OSU); Marion Co., Silver Falls SP, str. Cyn Tr., 12.vii.1989, BJS (1 ♀, CNC); Multnomah Co., Lower Wahkeenah Falls, 25.xi.1984, D.C. McClain (1 ♀, NCSU); Wallowa Co., Wallowa Lk. SP, 6 mi. S Enterprise, 3500–4000′, 28–30.viii.1973, WJT (1 ♀, WSU). **Utah:** Cache Co., W Hodges Cyn, 14–27.ix.1976, GFK, WJH (3 ♂♂, 2 ♀♀, USUL); Summit Co., Provo R., Hwy 150, Provo R. Falls, 1.v.1981, RWB and S.M. Clark (1 ♂, BYU); Utah Co., Sundance Ski Area, 10.v.1979, G.P. Poulsen (1 ♂, BYU); Utah Co., S Fk American Fk R., 1 mi. above N Fk, 11.ix.1987, CRN and Franklin (1 ♂, BYU); Wellsville, 23.v.1948, R.S. Bailey (1 ♀, USUL). **Washington:** Clallam Co., Bogachiel, Bogachiel R., 29.vi.1974, PHA (1 ♂, CAS); Greys Harbor Co., Olympic NF, Willaby Cpgd, 21.vii.1978,

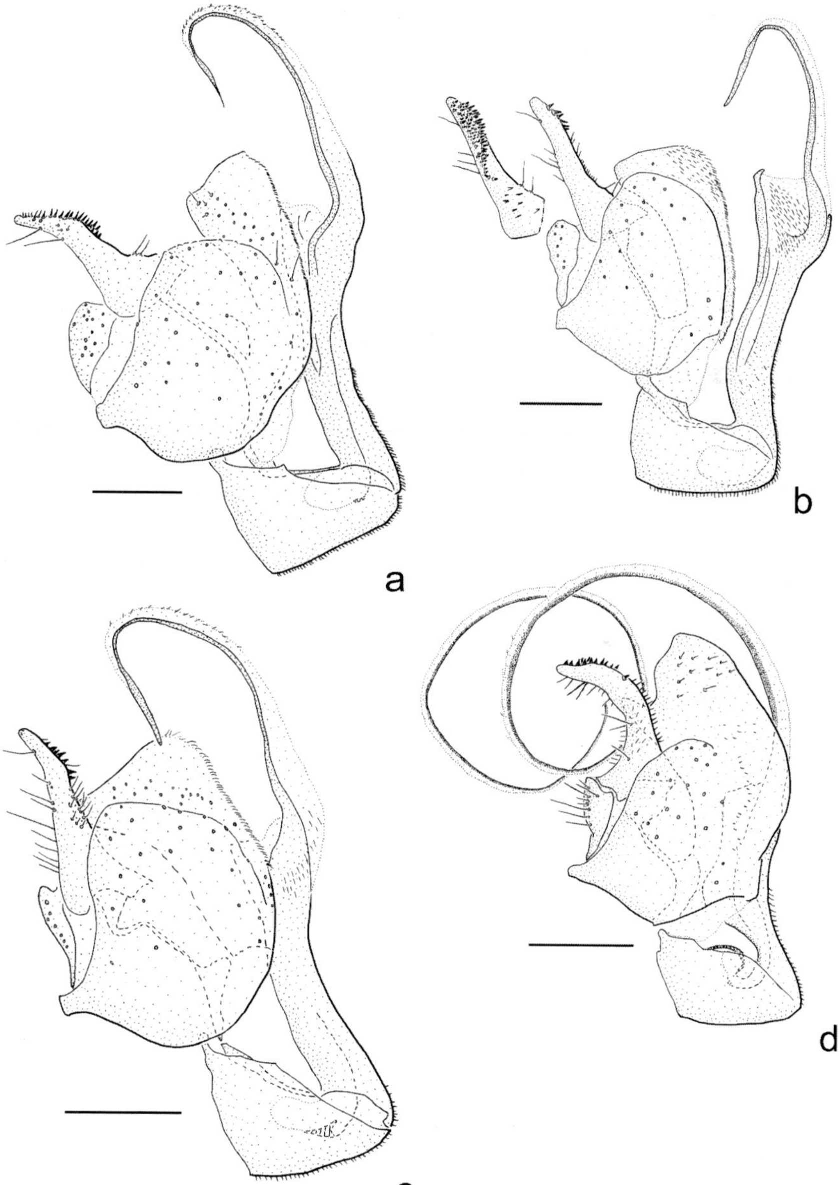

Fig. 11. Male terminalia of *Clinocera*: *a*, *C. brunnea*; *b*, *C. caerulea* with inner view of clasping cercus; *c*, *C. daniellae*; *d*, *C. longicauda*. Scale bars = 0.1 mm.

DDW (10 ♂♂, 2 ♀♀, CAS); Lewis Co., 23 mi. W Chehalis, Rainbow Falls SP, 300′, 13–14.vii.1977, WJT (1 ♂, WSU).

Possible additional material. Arizona: Cochise Co., Miller Cyn, 6000′, Huachuca Mtns, 5.viii.1952, H.B. Leech and J.W. Green (1 ♀, CAS).

Recognition. This is a dark–brown species lacking black vittae on the scutum, distinguished from *C. daniellae* by the dark–brown face and strongly arched clasping cercus.

Description (wing length 3.9–4.4 mm).

Male. Face entirely dull brown, lacking blue spot; remaining head dark brown; face flat. Arista-like stylus short and slender. Postocular setae dark, upper 3 or 4 stout.

Scutum with two pale, dull, brown median stripes. Pleura dark brown, lower half with faint pruinescence when viewed from above. Several short acrostichals anterior to first dc; 5 dc; 1 pprn; 1 presut spal; 2 npl, lower shorter and slender; 1 psut spal; 1 pal; 2 sctl. Laterotergite with patch of short pale setae. Posterior margin of anepisternum lacking setulae.

Wing (Fig. 29*a*) darkly infuscate, with clouding or spots absent; single basal costal bristle; stigma beyond R_1 faint, elongate; costal spines present. Veins straight, R_{2+3} not dipped below stigma; auxiliary crossvein between R_4 and R_{2+3} absent; stump vein absent; R_4 bell-shaped; anal vein reduced to short streak; cell dm produced apically. Halter brown.

Coxae and femora reddish brown (paler ventrally and apically), remaining segments dark. Forecoxa with long dark setae. Forefemur with preapical anterior comb; ventral surface with short dense white pile; antero- and postero-ventral row of short fine setae. Foretibia with biserial row of short fine setae beneath. Hind tibia with erect dorsal and ventral bristles on apical half. Fifth tarsomere lacking dorsoapical extension; empodium light brown laterally.

Abdomen dark, brownish. Hypandrium broadly triangular. Phallus heavily sclerotized, straight; distiphallus of phallus heavily sclerotized and strongly recurved apically. Apical third of clasping cercus strongly arched, inner face bearing short spines on apical half. Surstylus slightly extended beyond epandrium; apex truncate (Fig. 11*a*).

Female. Similar to male except ventral setae of foretibia and forefemur much less pronounced.

Geographic distribution and seasonal occurrence. This species occurs in both eastern and western North America. In the east this species is recorded from southern Ontario, Quebec, Michigan, and the White Mountains of New Hampshire. In the west it is much more widespread, with localities ranging from the Rockies to the coastal region (Map 1).

Remarks. This species corresponds to a series of unidentified specimens of *Clinocera* recorded from the Rivière du Castor drainage of Quebec (Landry and Harper 1985) (see also *C. fuscipennis*). This species was frequently collected in the springs at Inglis Falls, Ontario (type locality; Fig. 3*a*), and also corresponds to the undetermined species of *Clinocera* in Sinclair and Marshall (1987).

Etymology. The specific name is from the Latin *brunneus* ("dusky", "dark"), referring to the overall dark-brown colour of this species.

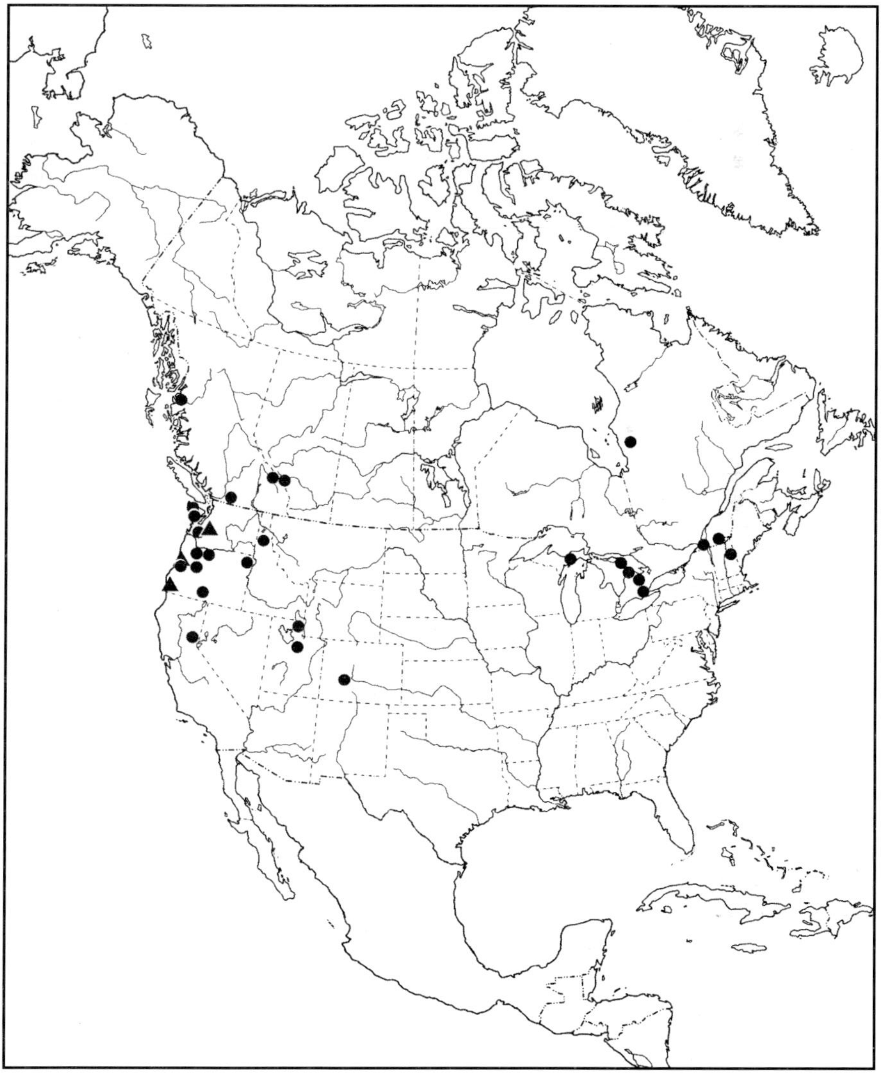

Map 1. Known distribution of *Clinocera brunnea* (●); *C. daniellae* (▲).

Clinocera caerulea sp. nov.

Figs. 4*a*, 11*b*, 29*b*; Map 2

Type material. HOLOTYPE, male labelled "USA OREGON: Siuslaw/ Nat.For, Mary's Peak/ Benton Co. 14.VII.1989/ Parkers Ck. & Falls/ BJ Sinclair"; "HOLOTYPE/ Clinocera/ caerulea/ Sinclair [red label]" (CNC).

Paratypes: **CANADA. Alberta:** Jasper NP, Jasper, Cabin Ck, 27.vi.1989, BJS (1 ♂, CNC). **British Columbia:** 7.3 km N Armstrong, 27.vi–4.vii.1997, AB (1 ♂, CNC); 49 mi. W Terrace, Slickenslide Ck, 9.vii.1960, JGC (2 ♂♂, 1 ♀, CNC); 4 km W Glacier NP, roadcut stream, 1.vii.1989, BJS (2 ♂♂, 1 ♀, CNC); Ketchcum Lk., 3600′, 58°22′N, 131°45′W, 23.viii.1960, WWM (1 ♂, CNC). **USA. Oregon:** Benton Co., Mary′s Peak, 2000′, Connection Ck, 21.vi.1982, N.H. Anderson (1 ♂, 2 ♀♀, OSU); Benton Co., Siuslaw NF, Mary's Peak, Parkers Ck and Falls / Alder Falls, 14.vii.1989, BJS (4 ♂♂, 3 ♀♀, CNC); Curry Co., Siskiyou NF, Pt Orford, 24 km E Elk R. Rd, roadcut seep, 17.vii.1989, BJS (4 ♂♂, 4 ♀♀, CNC); Curry Co., Siskiyou NF, 3 mi. S Yachats, roadcut seep, 15.vii.1989, BJS (6 ♂♂, 9 ♀♀, CNC); Hood R. Co., Mt Hood NF, nr Sahalie Falls, Mitchell Ck, 27.vi.1978, DDW (1 ♂, CAS); Hood R. Co., Starvation Ck SP, below falls, 10.vii.1989, BJS (1 ♀, CNC); Lane Co., 2 mi. N Blue R., 490 m, Quartz Ck, 26.vi., 6.x.1982, GWC (5 ♂♂, 1 ♀, OSU); Lane Co., 11 mi. NE Blue R., 800 m, Mack Ck (emerg. tp), GWC (1 ♂, 1 ♀, OSU); Lane Co., Siuslaw NF, 3 mi. S Yachats, Gwynn Ck Tr., 15.vii.1989, BJS (1 ♂, CNC); Wasco Co., Warm Springs Indian Res., Beaver Ck, 23.vi.1978, DDW (1 ♂, 2 ♀♀, CAS). **Utah:** Cache Co., Blacksmith Fk Cyn, 1.viii.1963, 1.vii.1977, WJH (1 ♂, 1 ♀, USUL); Utah Co., Mapleton, MT, 16–23.vii.1985 (1 ♀, USUL); Utah Co., 12 mi. E Thistle, Hwy 6, MT, 15.ix–20.x.1994, CRN (1 ♀, BYU); Wasatch Co., Gerber Spring nr Midway, 23.xi.1984, CRN, White, Wells (1 ♀, BYU).

Additional material examined. MEXICO. Baja California: Norte Arroyo Valladeres, 24 km E Meling Ranch, 18.ix.1980, DDW (18 ♂♂, 20 ♀♀, CAS). **USA. Alaska:** Seward Hwy, mi. 100, nr Indian, 9.vi.1960 (as larvae), emerged 2–4.vii.1960, K.M. Sommerman (2 ♂♂, 2 ♀♀, USNM, in alc.). **Arizona:** Cochise Co., Chiricahua Mtns, ck nr Ash Spring, 5800–6100′, 31°52.3′N, 109°14.0′W, 17.viii.2007, JMC, S.E. Brooks (10 ♂♂, 10 ♀♀, CNC); Cochise Co., Chiricahua Mtns, Cave Creek Cyn, Cave Ck at Herb Martyr Dam, 3.vi.1991, BJS (1 ♂, 1 ♀, CNC); Cochise Co., Chiricahua Mtns, East Turkey Ck, 6500′, 12.vi.1991, BJS (3 ♂♂, 5 ♀♀, CNC); Cochise Co., Coronado NF, Pinery Cyn Cpgd, 7000′, 2.vi.1991, BJS (6 ♂♂, 2 ♀♀, CNC); Cochise Co., Coronado NF, Huachuca Mtns, Miller Cyn, Miller Ck, 3.vi.1991, BJS (1 ♂, 1 ♀, CNC); Grand Cyn NP, Bert's Cyn, CR 2, 10–21.v.1998, JFM (1 ♀, PERC); Grand Cyn NP, CR 6, Han's Ck, 10–21.v.1998, JFM (3 ♂♂, PERC); Graham Co., Hospital Flat, Gabriel Mtns, 9050′, 19.viii.1952, HBL, J.W. Green (3 ♂♂, 2 ♀♀, CAS); Ramsey Cyn, 15 mi. S Sierra Vista, Huachuca Mtns, 28.ix.1967, Stemitzky (1 ♀, CNC); Santa Cruz Co., Coronado NF, Santa Rita Mtns, Madera Cyn, Old Balby

Tr., 4.vi.1991, BJS (1 ♀, CNC). **California:** Alamedo Co., Sunol Regional Wilderness, Indian Joe Ck, 29.iii.1979, DDW (18 ♂♂, 22 ♀♀, CAS); Big Sur, 5.x.1946, ALM (1 ♀, USNM); Cathedral City, 22.xi.1944, ALM (16 ♂♂, 34 ♀♀, USNM); Contra Costa Co., Briones Regional Pk, Toyon Cyn, 26.v.1979, DDW (3 ♂♂, 3 ♀♀, CAS); Contra Costa Co., 2 mi. SE Lafayette, MT, 10.iv.1968, WJT (1 ♀, WSU); Contra Costa Co., Mt Diablo SP, 520 m, 24.viii.1974, PHA (19 ♂♂, 33 ♀♀, CAS); Contra Costa Co., Mt Diablo SP, Donner Cyn, 29.vi.1979, DDW (8 ♂♂, 12 ♀♀, CAS); Contra Costa Co., Mt Diablo SP, Mitchell Cyn, 13.vi. 1979, DDW (5 ♂♂, 6 ♀♀, CAS); Crestline, 4.vii.1942, ALM (1 ♀, USNM); Donner Pass, 6.xii.1991, SAM (2 ♀♀, DEBU); Fresno Co., Dinkey Ck, 22.vii.1984, JFM (1 ♀, PERC), Rancheria Cyn, Huntington Lk., 10, 20.viii.1984, JFM (2 ♀♀, PERC), Sierra NF, nr Kaiser Pass, 2700 m, 1–2.viii.1979, DDW (8 ♂♂, 7 ♀♀, CAS), Sierra NF, nr Forebay Lk., 31.vii.1979, DDW (1 ♀, CAS); Kent Co., Mt Pinos, 8000′, 3.vi.1992, J. Skevington, A. Goering (1 ♀, DEBU); Los Angeles Co., Angeles NF, Windy Spring, 27.vi.1974, DDW (1 ♀, CAS), Hwy 39, 3 mi. S Coldspring Stn, San Gabriel R., 2.v.1978, DDW (56 ♂♂, 78 ♀♀, CAS), San Dimias Cyn, 16.i.1999, J.R. Harkrider (2 ♀♀, ZFMK), San Gabriel Mtns, nr Hidden Springs, 2.iii.1955, C.L. Hogue (1 ♂, LACM); Madera Co., SE slope Green Mtn, gate rd to Mark Mine, 7600′, 20.viii.1971, HBL (2 ♂♂, CAS); Marin Co., below Alpine Dam, Laquanitas Ck, 28.v.1979, DDW (42 ♂♂, 55 ♀♀, CAS), Mill Valley, 22.v.1952, HBL (2 ♀♀, CAS), Mill Valley, nr Cascade Reservoir, Mill Ck, 21.v.1978, DDW (4 ♂♂, 6 ♀♀, CAS), Mt Tamalpais, Little Carson Tr., nr Kent Lk., 15.vii.1978, DDW (15 ♂♂, 20 ♀♀, CAS); Mono Co., Tepaz Lk., 1.xi.1948, WWW (1 ♀, USNM); Monterey Co., Los Padres NF, below Escondido, Antonio R., 22.v.1977, DDW (3 ♂♂, 3 ♀♀, CAS); Mt Diablo, 2000′, 23.viii.1951, F.X. Williams (2 ♂♂, 1 ♀, CAS); Muir Woods, 7.viii.1915, ALM (1 ♂, USNM); Napa Co., Oakville, 25.iv.1971, PHA (1 ♂, USNM); Plumas Co., Lassen NF, Domingo Spr., 31.v.1976, DDW (1 ♂, CAS), Lassen NF, N Fk Feather R., 28.v.1977, DDW (1 ♀, CAS); Plumas Co., Grizzly Ck, above Hwy 70, 14.ii.1985, RWB and CRN (♂,♀, BYU, in alc.); Riverside Co., Palm Springs, Tahqitz Cyn, 7.i.1953. PHA (3 ♂♂, 2 ♀♀, USNM); San Benita Co., S Hollister, Limekiln Cyn, 3.iv.1979, DDW (24 ♂♂, 18 ♀♀, CAS); San Diego Co., Palomar Observatory Cpgd, 1524 m, 26.vi.1968, PHA (10 ♂♂, 2 ♀♀, USNM); Santa Clara, Alum Rock Pk, 5.v.1939, T.G.H. Aitken (2 ♂♂, 2 ♀♀, CAS); Santa Clara Co., Uvas Cyn Co. Pk, 26.xi.1975, D.G. Denning (1 ♂, UCD); Santa Cruz Co., Ben Lomond, 12.viii.1951, A.A. Huber (1 ♀, SEMC); Santa Cruz Co., Big Basin SP, Opal Ck, 14.vi.1978, DDW (11 ♂♂, 14 ♀♀, DDW); Santa Cruz Co., Sanborn Skyline Co. Pk, 13.xii.1986, BJS (1 ♀, DEBU); Trinity Co., Shasta NF, Hwy 3, Scott Mtn Ck, 20.vii.1980, DDW (1 ♀, CAS); Trinity Co., Shasta NF, Trinity R., 2.vii.1978, DDW (1 ♀, CAS); Tulare Co., Quaking Aspen, 7.viii.1984, D.J. Burdick (1 ♀, PERC). **Colorado:** Larimer Co., Sheep Ck nr jct Buckhorn Ck, 15.x.1995, SF, AF (15 ♂♂, 23 ♀♀, CSU); Larimer, Greyrock Tr., str., 16.iv.1996, SF, AF (3 ♂♂, 1 ♀, CSU). **Idaho:** Nez Perce Co, Juliaetta Falls, 11.iv.1975, D.F. Veirs (1 ♂, WSU); Lk. Waha, 9.vi.1938, ALM (1 ♀, USNM). **Nevada:** Clark Co.,

Spring Mtns, Deer Ck, 21.i.1985, RWB and CRN (1 ♂, BYU, in alc.); White Pine Co., N Fk Cleve Ck above Cleve Ck Cpgd, 25.v.1976, RWB and Logan (2 ♂♂, BYU in alc.). **New Mexico:** Catron Co., Gila Cliff Dwellings NM, Cliff Dweller Spr., 29.v.1991, BJS (6 ♂♂, 3 ♀♀, CNC); Grant Co., Gila NF, 14 mi. NE Silver City, Cherry Ck, 7400′, 5.vi.1991, BJS (1 ♂, 3 ♀♀, CNC); Grant Co., Iron Ck nr Wright's Cabin Cpgd, 7600′, 32°54.8′N, 107°46.8′W, 14.viii.2007, JMC (9 ♂♂, 5 ♀♀, CNC); Lincoln Co., 9.7 km W Alto Lincoln NF, 21–24.vi.1995, S.D. Gaimari (1 ♂, CNC). **Washington:** Asatin Co., Fields Spring SP, 30.vi.1975, WJT (4 ♂♂, 2 ♀♀, WSU); Asatin Co., 4 mi. S Anatone, 3500–4000′, 24.v.1975, NEW (1 ♂, 2 ♀♀, WSU); Mt Baker NF nr Glacier Ck Rd, 1100 m, Grouse Ck, 4.x.1979, DDW (1 ♂, 3 ♀♀, CAS); Mt Rainier NP, Paradise Pk, Edith Ck, 1650

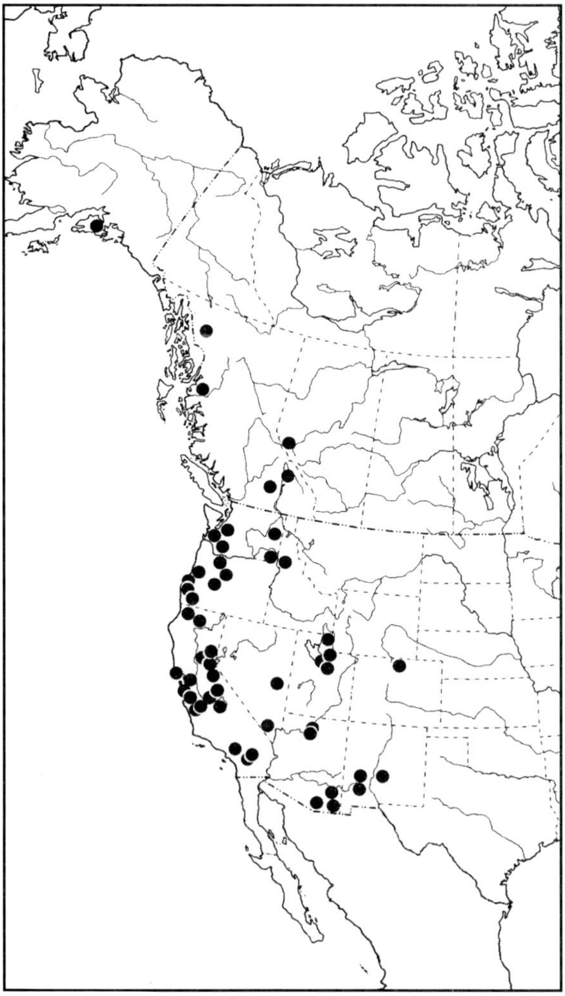

Map 2. Known distribution of *Clinocera caerulea*.

m, 2.viii.1977, DDW (1 ♂, 1 ♀, CAS); Mt Rainier NP, Comet Falls, 8.vii.1989, BJS (1 ♂, 2 ♀♀, CNC); Skamania Co., Gifford Pinchot NF, Big Spring Ck, 20 mi. N Trout Lk., 9.vii.1989, BJS (1 ♂, 1 ♀, CNC); Spokane Co., Mt Spokane SP nr Bald Knob Cpgd, 4800–5200′, 25.vi.1979, WJT (1 ♀, WSU); Toledo, 27.vi.1935, ALM (1 ♂, USNM).

Recognition. This species is distinguished by the large bluish central spot on the face, the lower half of the pleura with blue pruinescence, and the large distinct stigma on the wing.

Description (wing length 4.3–4.8 mm).

Male. Face flat, with large central blue patch of pruinescence, remaining head dark brown. Arista-like stylus short and slender. Postocular setae mostly pale, except upper 3–5 setae.

Scutum dark brown with faint black opaque median stripe. Lower half of pleura with faint blue pruinescence (Fig. 4*a*). Several inconspicuous acrostichals anterior to first dc; 5 dc; 1 pprn; 1 presut spal; 2 npl, lower more slender than upper; 1 psut spal; 1 pal; 2 sctl. Laterotergite with patch of pale setae. Posterior margin of anepisternum lacking setulae.

Wing (Fig. 29*b*) infuscate, with clouding or spots absent; stigma beyond R_1 distinct, elongate and elliptical; 1 long basal costal seta; costal spines present. R_{2+3} dipped only slightly below stigma; remaining veins straight; auxiliary crossvein between R_4 and R_{2+3} absent; stump vein absent; R_4 straight, diverging from R_5; anal vein a long streak; cell dm produced apically. Halter dark.

Apex of coxa, trochanter, immediate base, and "knee" of forefemur pale; remaining leg segments brown to reddish brown (sometimes legs often quite brightly pale). Forecoxa with dark setae. Forefemur with anterior preapical comb; ventral surface with short dense white pile; antero- and postero-ventral row of short setae extending length of forefemur, becoming stouter and shorter apically. Foretibia with biserial row of short stout setae beneath. Hind tibia with dorsal and ventral erect bristles on apical half. Fifth tarsomere lacking dorsoapical extension; empodium dark laterally.

Abdomen dark, brownish. Hypandrium broadly triangular. Phallus stout, slightly curved; distiphallus arising from C-shaped arch at apex of shaft of phallus; distiphallus heavily sclerotized and strongly recurved. Clasping cercus straight and slender; occasionally apex slightly inflated and arched; inner face with short stout spines on apical half. Surstylus short and truncate, not projecting free from epandrium; dorsal margin with matted microsetae (Fig. 11*b*).

Female. Similar to male except ventral setae of forefemur and foretibia reduced and inconspicuous.

Geographic distribution and seasonal occurrence. This species is widespread in the western Cordillera, ranging from southern Alaska and northern British Columbia to southeastern Arizona (Map 2). It is apparently much less common in its northern range (i.e., Alberta and British Columbia) than in the southwestern USA. Adults have been reared from larvae collected in algae in a waterfall in Alaska (material in USNM).

Remarks. Variation in leg coloration (light to dark) is found in some southwestern populations; for example, New Mexico (Gila Cliff) and Grand Canyon populations possess very brightly coloured legs. In addition, the clasping cercus is variable, from straight to slightly arched apically.

Etymology. The specific name is from the Latin *caeruleus* ("sky blue"), referring to the large blue facial spot.

Clinocera daniellae sp. nov.

Figs. 11*c*, 29*c*; Map 1

Type material. HOLOTYPE, male labelled "USA: OREGON: Siskiyou/ Nat.For. 24 kmE/ Elk R.Rd. Pt Orford/ Curry Co.17.VII.1989/ roadcut seep BJ Sinclair"; "Wing/ Slide/ No. 9927"; "HOLOTYPE/ Clinocera/ daniellae/ Sinclair [red label]" (CNC).

Paratypes: **USA. Oregon:** Benton Co., Parker Ck Falls, 12.iii.1963 (1 ♂, OSU); Benton Co., Siuslaw NF, Marys Peak, roadcut seep, 14.vii.1989, BJS (1 ♂, CNC); ditto, Alder Falls (1 ♀, CNC); Curry Co., same data as holotype (1 ♂, 2 ♀♀, CNC); same locality as previous except 17.6 km E (1 ♀, CNC).

Possible additional material. USA. Washington: Mt Rainier NP, str. nr Comet Falls, 8.vii.1989, BJS (1 ♀, CNC).

Recognition. This species is distinguished by the mostly blue pruinescent face, head broad and dark, clasping cercus weakly bent, and stigma distinct with R_{2+3} dipped below.

Description (wing length 3.9–4.6 mm).

Male. Face flat with blue pruinescence, except below antennae, contrasting with dark-brown head. Arista-like stylus short and slender. Postocular setae pale, upper 3 or 4 stout.

Scutum with pair of pale-brown median strips. Pleurae with faint blue pruinescent tinge. Several short acrostichals anterior to first dc; 5 dc; 1 pprn; 1 presut spal; 2 npl, lower short and fine; 1 psut spal; 1 pal; 2 sctl. Laterotergite with patch of pale setae. Posterior margin of anepisternum lacking setulae.

Wing (Fig. 29*c*) infuscate, with clouding or spots absent; stigma beyond R_1 distinct, elongate, and elliptical; 1 long basal costal seta; costal spines present. R_{2+3} dipped below stigma; remaining veins straight; auxiliary crossvein between R_4 and R_{2+3} absent; stump vein absent; R_4 straight, strongly divergent from R_5; anal vein short streak; cell dm produced apically. Halter dark.

Inner and ventral margin of mid and hind coxae and femora pale yellowish brown, remaining segments dark. Anterior face of forecoxa yellowish brown with dark, stiff setae; outer surface with long, slender setae. Forefemur with anterior preapical comb; ventral surface with short dense white pile; antero- and postero-ventral row of dark fine setae, progressively shorter and stouter distally. Foretibia with single row of short setae beneath. Hind tibia with erect dorsal and ventral bristles on apical half. Fifth tarsomere lacking dorsoapical extension; empodium dark laterally.

Abdomen dark, brownish. Hypandrium broadly triangular. Phallus heavily sclerotized, strongly indented at midlength; distiphallus heavily sclerotized and strongly recurved. Apical third of clasping cercus slightly arched, inner face bearing short spines on apical half. Surstylus wide, apex truncate; only partially extending beyond epandrium; dorsal margin with matted microsetae (Fig. 11*c*).

Female. Similar to male except ventral setae of foretibia and forefemur reduced; blue pruinescence of coxae and abdomen somewhat more distinct.

Geographic distribution and seasonal occurrence. This species is known only from Oregon and tentatively from Washington (male specimens are required for verification) (Map 1).

Etymology. The name *daniellae* is given in recognition of the author's wife's extensive support throughout this study.

Clinocera longicauda sp. nov.

Figs. 11*d*, 29*d*; Map 3

Type material. HOLOTYPE, male labelled "USA: OREGON: Siskiyou/ Nat. For. 17.6 kmE/ Elk Rd. Pt. Orford/ Curry Co. 17.VII.1989/ stream BJ Sinclair"; "HOLOTYPE/ Clinocera/ longicauda/ Sinclair [red label]" (CNC).

Paratypes: **USA. Oregon:** Benton Co., 5 mi. NW Corvallis, MacDonald For., intermittent str., 18.vi., 15.vii.1988, M. Dieterich (1 ♂, 2 ♀♀, OSU); Curry Co., Siskiyou NF, 7 km W Agness, 18.vii.1989, BJS (3 ♀♀, CNC); same data as holotype (3 ♂♂, 1 ♀, CNC); Hood R. Co., Starvation Ck SP, 10.vii.1989, BJS (1 ♂, CNC); Hood R. Co., Viento SP, 9.vii.1989, BJS (1 ♀, CNC); Lane Co., 11 mi. NE Blue R., H.J. Andrews EF, Mack Ck, 800 m, 20.vii, 11.viii., 15.ix., 15.x.1982, GWC (1 ♂, 5 ♀♀, OSU); Lane Co., 2 mi. N Blue R., Quartz Ck, 490 m, 25.iv., 26.vi, 6, 21, 27.x.1982, GWC (3 ♂♂, 3 ♀♀, OSU); Marion Co., Silver Falls SP, 12.vii.1989, BJS (1 ♂, CNC).

Additional material examined. CANADA. British Columbia: Cowichan Lk., 18.ix., 1.x.1964, J.A. Chapman (2 ♀♀, CNC); Mt Revelstoke NP, 620 m, Inspiration Woods Tr., 12.ix.2003, BJS (1 ♀, ZFMK); Vancouver Is., 2.3 km from Lk. Cowichan, cold spr., MT, 19–28.vii.1986, I.M. Smith (1 ♀, CNC). **USA. California:** El Dorado Co., Kyburz, 4080′, 16.vi.1978, E. Rogers (1 ♀, ROGC); Humboldt Co., Humboldt, Redwoods SP, Bull Ck Flats Rd, Cabin Ck, 17.v.1978,

DDW (1 ♀, CAS); Marion Co., Lily Pond, Alpine Lk., 1500′, MT, iv–v.1970, 7–28.viii.1970, 18.iii.–1.iv.1971, 17–25.v.1971, D.D. Munroe (3 ♂♂, 3 ♀♀, CNC); Marion Co., below Alpine Dam, Lagunitas Ck, 28.v.1979, DDW (1 ♀, CAS); Marion Co., Bolinas Ridge, Bolinas/Fairfax Rd, 10.iv.1978, DDW (1 ♀, CAS); Marion Co., Liberty Gulch / Lily Lk., Mt Tamalpais, 31.v.1978, DDW (2 ♀♀, CAS); Marin Co., Mill Valley, Blithedale Ridge, 110 m, 15–20.i.1968, PHA (1 ♂, 3 ♀♀, USNM); Nevada Co., Hwy 20 at Drum Canal, cold seep/str., 7.xii.1986, BJS (1 ♀, DEBU); Santa Clara Co., Uvas Cyn Co. Pk, 26.xii.1975, D.G. Denning (1 ♀, UCD); Santa Cruz Co., Big Basin SP, Opal Ck, 14.v.1978, DDW (1 ♂, CAS); Sierra Co., Cold Ck, Hwy 89, Cold Ck Cpgd, 8.xii.1986 (1 ♀, BYU). **Idaho:** Kootenai Co., Hwy 95, 3.6 mi. S jct. Hwy 90 nr Lk. Coeur d'Alene, 2600′, 25.iv.1992, W.J. Turner (1 ♀, WSU). **Washington:** Lk. Cushman, 27.vi.1917, H.G. Dyne (1 ♀, USNM); Quilcene, 24.vii.1917, ALM (1 ♂, USNM); Skagit Co., 13 km NE Marblemount at Damnation Ck, 2.vii.1988, W.E. Steiner (1 ♂, USNM).

Recognition. This is a small species, recognized by the deflection of R_{2+3} below the stigma, orange–brown body and face, hind first tarsomere with 2 or 3 spine-like setae, and the uniquely long, coiled distiphallus.

Description (wing length 3–3.5 mm).

Male. Generally small orange–brown species. Face flat, dark brown. Arista-like stylus short and slender. Postocular setae pale, upper 3 or 4 stout.

Scutum dark brown, postpronotal lobe pale. Pleura yellowish brown on elevated regions. Several acrostichals anterior to first dc; 5 dc; 1 pprn; 1 presut spal; 2 npl, upper strong, lower narrow; 1 psut spal; 1 thin pal; 2 sctl. Laterotergite with patch of dark setae. Posterior margin of anepisternum with setulae.

Wing (Fig. 29*d*) infuscate; single long basal costal bristle; stigma beyond R_1, distinct, broad. R_{2+3} strongly deflected below stigma, remaining veins straight; costal spines present. Crossvein at apex of cell dm, separating M_1 and M_2, clear; auxiliary crossvein between R_4 and R_{2+3} absent; stump vein absent; R_4 curved, divergent from R_5; anal vein short streak; cell dm produced apically. Halter knob dark, base of stalk pale.

Forecoxa yellowish; inner and ventral faces of coxae and femora light brown, remaining segments brown. Forecoxa with long erect setae. Forefemur with antero- and postero-ventral row of short setae, basal setae significantly longer than others; anterior preapical comb small; ventral surface with short dense white pile. Foretibia with two rows of short setae beneath. Hind tibia with erect dorsal and ventral bristles on apical third. Ventral surface of hind first tarsomere with long, spine-like basal, mid, and apical setae. Fifth tarsomere lacking dorsoapical extension; empodium yellowish laterally.

Abdomen dark, brownish. Hypandrium small, broadly triangular. Phallus with very long, coiled distiphallus. Epandrium much larger than hypandrium;

surstylus very broad, with narrow ventral apical lobe. Clasping cercus strongly arched, bearing short stout spines on inner face (Fig. 11*d*).

Female. Similar to male except ventral setae of forefemur weak, not in distinct rows.

Geographic distribution and seasonal occurrence. This species is mostly found in the far-western mountain ranges (Map 3). The outlying localities known from Mount Revelstoke (British Columbia) and Idaho are based on female specimens, and collection of males from these sites is required for verification. Apparently the range of *C. longicauda* has not penetrated very far north into previously ice-covered glaciated regions.

Remarks. *Clinocera longicauda* is most similar to species from Costa Rica on the basis of the pale-brown coloration.

Etymology. The specific name is from the Latin *longus* ("long") and *caudus* ("tail"), referring to the long, coiled phallus.

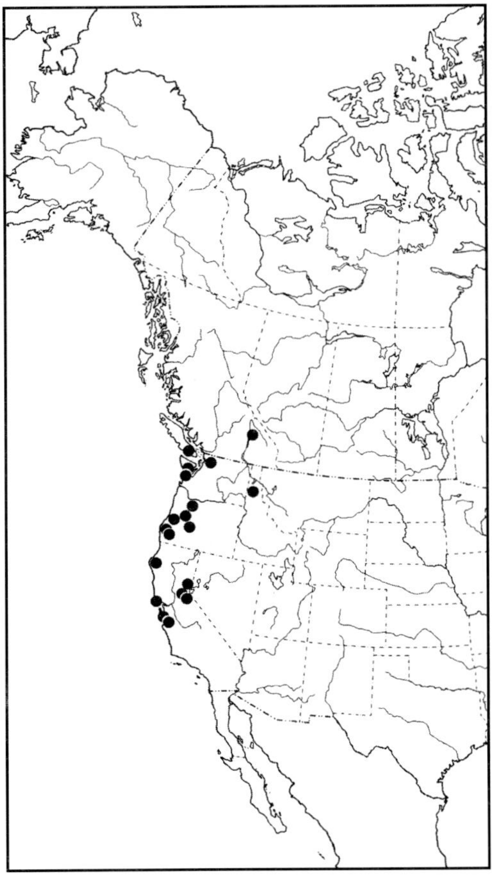

Map 3. Known distribution of *Clinocera longicauda*.

Clinocera conjuncta group

Diagnosis

This species-group is recognized by the broad face and truncate mouth; facial tubercle lacking; one postsutural supra-alar bristle; scutum bivittate; apical margin of clasping cercus truncate, bearing stout spines on apical third of inner face; surstylus with subapical notch; phallus often with slender sclerotized margin on subapical membranous region at base of distiphallus.

Remarks

The monophyly of this group is supported on the basis of the following apomorphies: clasping cercus truncate apically and surstylus with a shallow apical notch.

Distribution

This species-group is currently known only from North America, being primarily restricted to the western Nearctic Region, with only the transcontinental species *C. conjuncta* Loew occurring in the east.

Included species

This species-group was recognized by Sinclair (1995) and includes the following species:

NE *Clinocera amarilla* **sp. nov.**
NE *Clinocera arnaudi* **sp. nov.**
NE *Clinocera borkenti* **sp. nov.**
NE *Clinocera coloradensis* **sp. nov.**
NE *Clinocera conjuncta* Loew, 1860: 80.
NE *Clinocera cordillerana* **sp. nov.**
NE *Clinocera melanderi* **sp. nov.**
NE *Clinocera olivacea* Melander, 1928: 233.
NE *Clinocera pacifica* **sp. nov.**
NE *Clinocera prasinata* Melander, 1928: 232.
NE *Clinocera rainiericola* **sp. nov.**
NE *Clinocera sabroskyi* **sp. nov.**
NE *Clinocera setitibia* **sp. nov.**
NE *Clinocera subtrunca* **sp. nov.**
NE *Clinocera trunca* Melander, 1928: 230.

Key to species of the *Clinocera conjuncta* group

1 Legs and halter bright to pale yellowish brown 2
— Legs and halter dark reddish brown to black 3

2 Legs and forecoxa bright orange; mid and hind coxae clothed in pale-blue pruinescence; phallus extending slightly beyond clasping cercus, lacking apical distiphallus (Fig. 12*a*) (western). . . . . ***C. amarilla*** sp. nov.
— Legs, including coxae, brownish orange; all coxae clothed in pale-blue pruinescence; phallus extending well beyond clasping cercus, with short, straight apical distiphallus present (Fig. 14*a*) (western). . . . . ***C. pacifica*** sp. nov.

3 Costa with two or more basal costal bristles, often with a series of long costal setae distal to basal costal bristles; males lacking distiphallus 4
— Costa with one basal costal bristle, lacking additional long costal setae; males rarely lacking distiphallus 10

4 Seven or more dorsocentral bristles (or numerous setae interspersed); apex of phallus with small membranous flap and pair of small subapical plates (Fig. 12*d*); female face brown (western) . . . . ***C. coloradensis*** sp. nov.
— Five dorsocentral bristles, rarely more; apex of phallus lacking small membranous flap and subapical plates (Figs. 12*c*, 13*b*); female face brown or blue 5

5 Male foretibia with long erect setae beneath, longer than half width of tibia; lower half of female face with blue pruinescence (western) . . . . ***C. setitibia*** sp. nov.
— Male foretibia with short erect setae beneath, shorter than half width of tibia; female face brown or entire face with blue pruinescence. 6

6 Male 7
— Female 9

7 Clasping cercus short and truncate; apex of phallus with pair of narrow converging plates (Fig. 12*c*) (western) . . . . ***C. borkenti*** sp. nov.
— Clasping cercus long and arched (Figs. 12*b*, 13*b*); apex of phallus with single plate. 8

8 Apex of phallus with outwardly arching, spear-shaped, pointed tip; clasping cercus arched medially to opposing cercus, apex rounded and bearing stout, pointed setae (Fig. 12*b*) (western) . . . . ***C. arnaudi*** sp. nov.
— Apex of phallus with inwardly bent, narrow tip; clasping cercus arched dorsally, lacking pointed apical setae (Fig. 13*b*) (western). . . . . ***C. cordillerana*** sp. nov.

9 Female face with blue pruinescence (western). . . . ***C. cordillerana*** sp. nov.
— Female face brown (western) . . . . ***C. borkenti*** sp. nov. and ***C. arnaudi*** sp. nov.

10 Auxiliary crossvein between R_{2+3} and R_4 present on at least one wing (Figs. 29*h*, 30*e*) 11

— Auxiliary crossvein between R_{2+3} and R_4 absent (Fig. 30*b*) 14

11 Lower half of face with blue pruinescence 12

— Lower half of face brown 13

12 Wing with numerous spots on crossveins (Figs. 29*g*, *h*); R_{4+5} usually with stump vein near midlength; distiphallus with basal notch; apex of phallic shaft lacking ventral process; surstylus with apical notch (Fig. 13*a*) (widespread) . . . . ***C. conjuncta*** Loew

— Wing lacking spots (Fig. 30*e*); R_{4+5} lacking stump vein; distiphallus lacking basal notch; apex of phallic shaft with triangular ventral process; surstylus complete, lacking notch (Fig. 15*c*) (western). . . . . ***C. trunca*** Melander

13 Body with golden iridescent sheen; distiphallus present; surstylus with subapical notch (Fig. 15*b*) (western) . . . . ***C. subtrunca*** sp. nov.

— Body brown, lower half of pleura with faint blue pruinescence; distiphallus absent, phallus with hooked apex; surstylus with pointed apex, lacking subapical notch (Fig. 14*c*) (western). . . . . ***C. rainiericola*** sp. nov.

14 Male face with blue pruinescence, except brown just below antenna; notopleuron with blue pruinescence; female face brown; female forefemur lacking rows of ventral setae. 15

— Male and female face brown or with blue pruinescence on lower half only; notopleuron brown; female forefemur with biserial row of short setae beneath 16

15 Wing length 4–4.5 mm; clasping cercus truncate; distiphallus bent back at right angles at base; phallic shaft long, slender, and gently curved toward apex; lacking expansion at base of male clasping cercus (Fig. 14*b*) (western) . . . . ***C. prasinata*** Melander

— Wing length 3–3.5 mm; clasping cercus rounded apically; distiphallus straight; basal expansion of male clasping cercus covered with stout setae (Fig. 13*c*) (western) . . . . ***C. melanderi*** sp. nov.

16 Face brown; distiphallus arising subapically from phallus; clasping cercus with notch on posterior margin (Fig. 13*d*) (western). . . . . ***C. olivacea*** Melander

[Note: some specimens of *C. subtrunca* with auxiliary crossvein lacking will key out here, but are distinguished by a golden iridescent sheen and small size (wing length 4–4.5 mm)]

— Lower half of face with blue pruinescence; distiphallus arising dorsally from phallus; clasping cercus with straight posterior margin (Fig. 14*d*) (known only from Colorado). . . . . ***C. sabroskyi*** sp. nov.

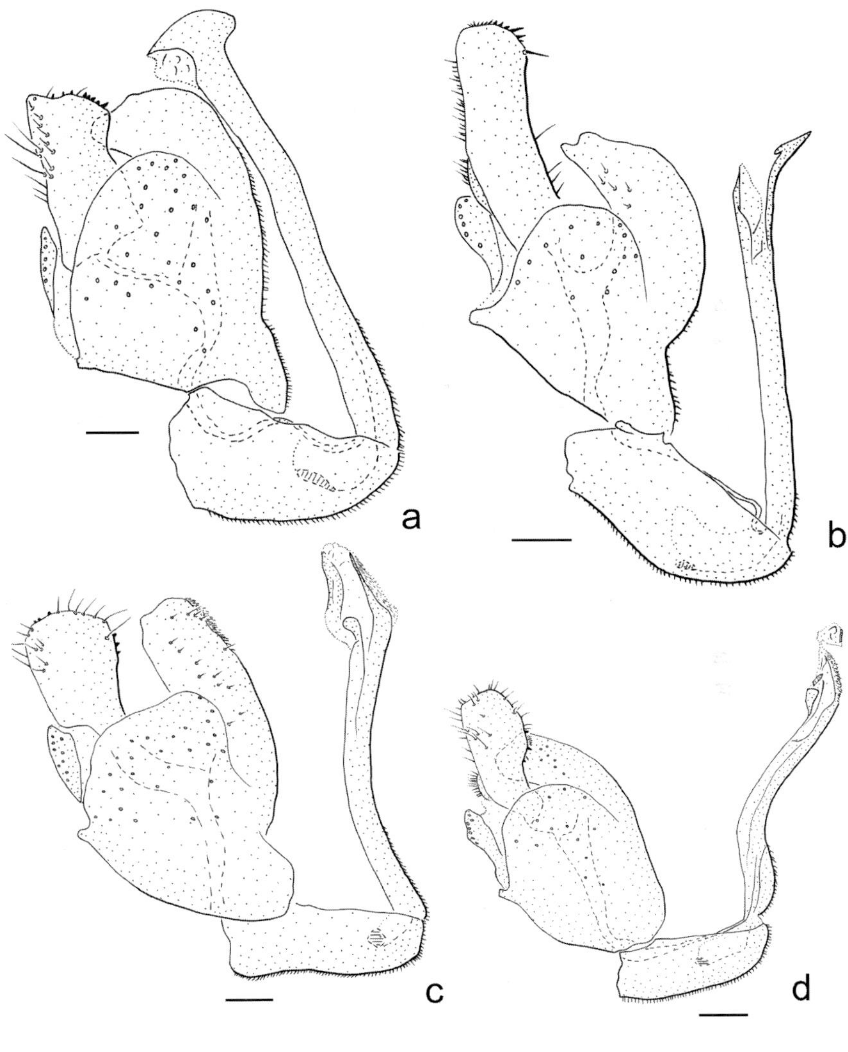

Fig. 12. Male terminalia of *Clinocera*: *a*, *C. amarilla*; *b*, *C. arnaudi*; *c*, *C. borkenti*; *d*, *C. coloradensis*. Scale bars = 0.1 mm.

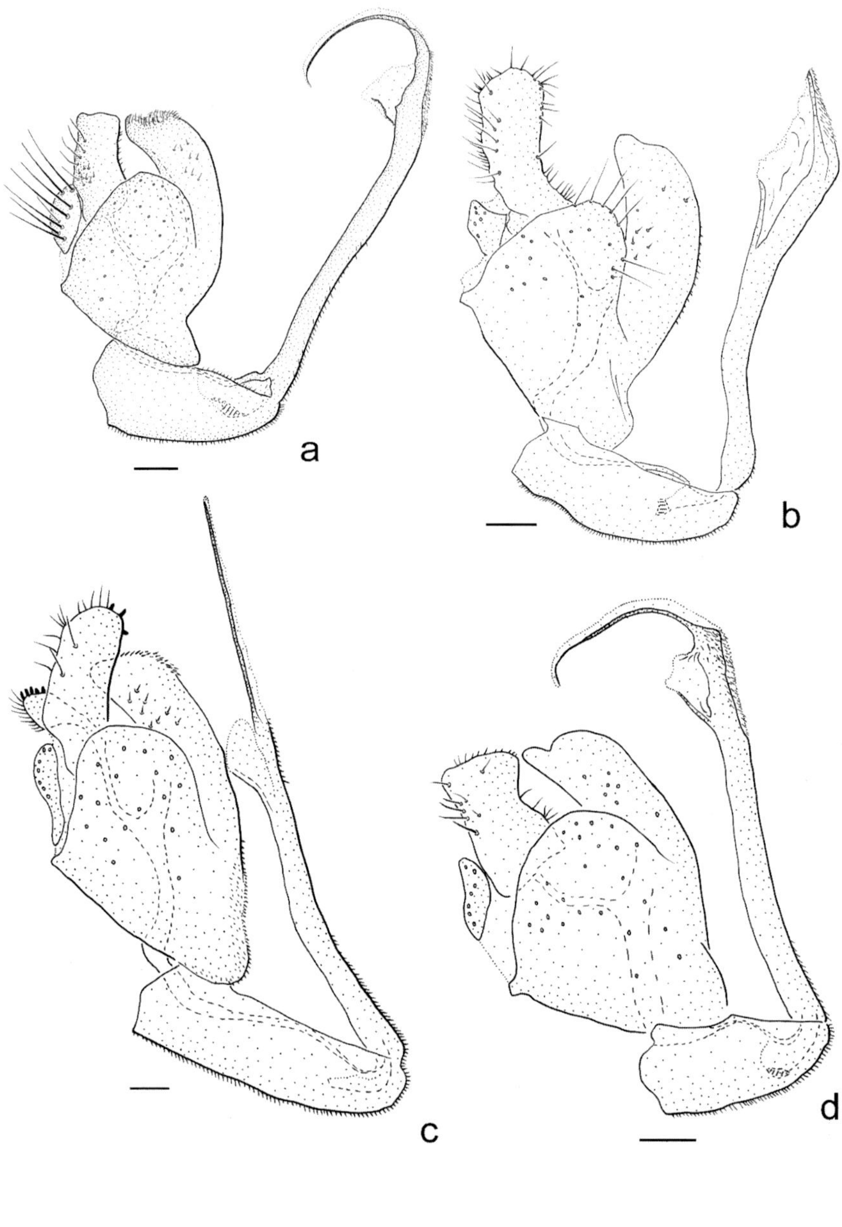

Fig. 13. Male terminalia of *Clinocera*: *a*, *C. conjuncta*; *b*, *C. cordillerana*; *c*, *C. melanderi*; *d*, *C. olivacea*. Scale bars = 0.1 mm.

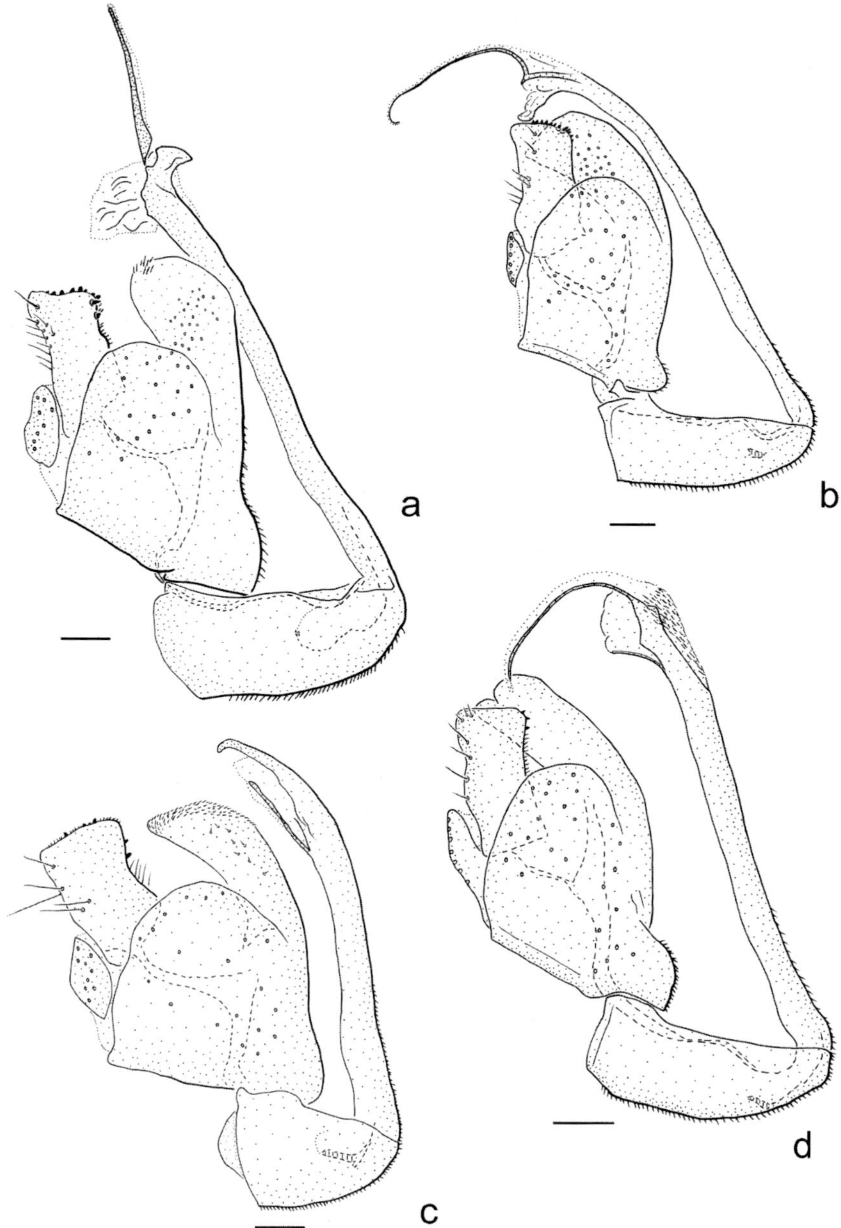

Fig. 14. Male terminalia of *Clinocera*: *a*, *C. pacifica*; *b*, *C. prasinata*; *c*, *C. rainiericola*; *d*, *C. sabroskyi*. Scale bars = 0.1 mm.

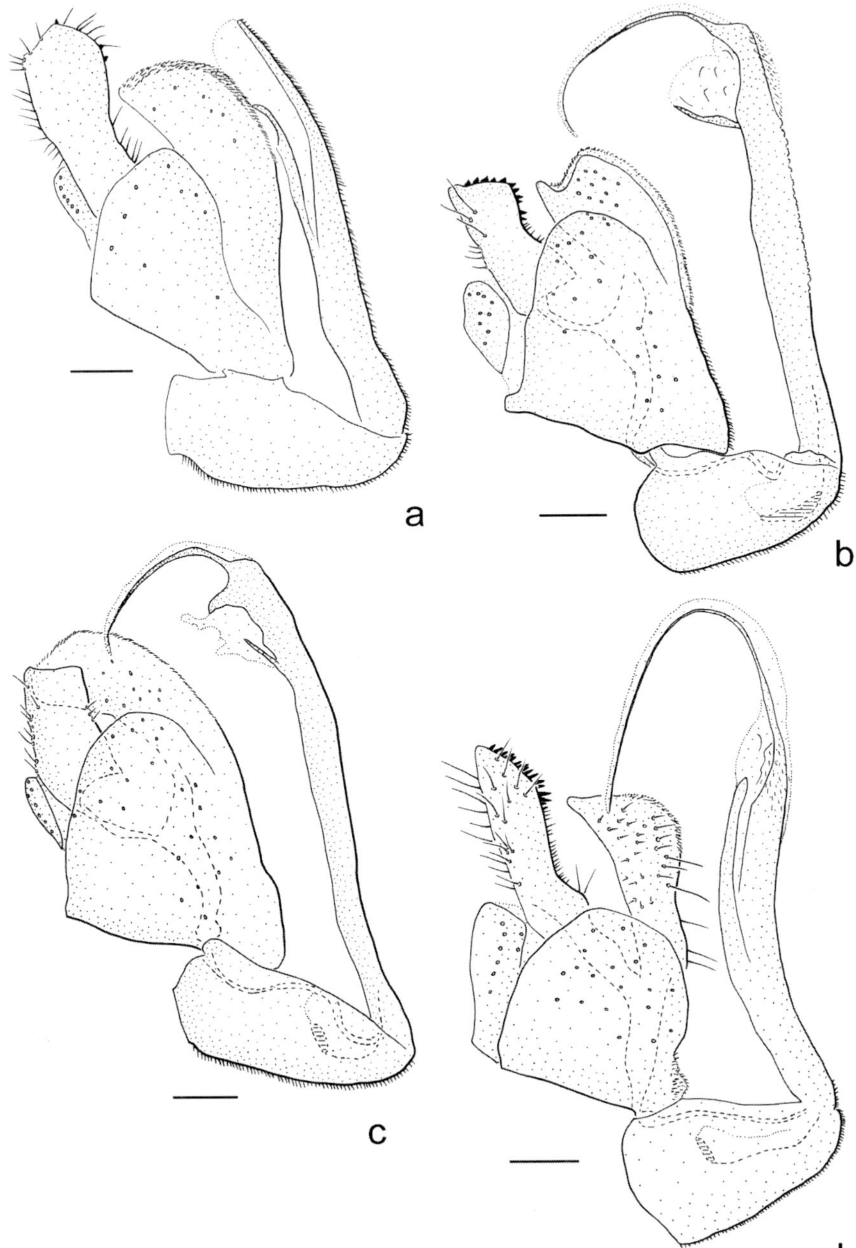

Fig. 15. Male terminalia of *Clinocera*: *a*, *C. setitibia*; *b*, *C. subtrunca*; *c*, *C. trunca*; *d*, *C. bulbosa*. Scale bars = 0.1 mm.

Clinocera amarilla sp. nov.

Figs. 12*a*, 29*e*; Map 4

Type material. HOLOTYPE, male labelled "CANADA:BRITISH COLUMBIA/ Manning Prov. Pk. 4.VII.1989/ Strawberry Flats, small/ stream, BJ Sinclair"; "HOLOTYPE/ Clinocera/ amarilla/ Sinclair [red label]" (CNC).

Paratypes: **CANADA. Alberta:** Waterton Lks. NP, 14–20.vii.1980, HJT (1 ♀, CNC). **British Columbia:** same data as holotype (5 ♂♂, 1 ♀, CNC); 14 km SW Bella Coola, 10–11.vii.1992, AB (1 ♀, CNC); Squamish, Diamond Head Tr., 4000′, 27.viii.1953, WRMM (1 ♀, CNC); 23 mi. N Terrace, Mt Allard, 11.viii.1960, CHM (1 ♀, CNC); 22 km E Vernon, Silver Star Prov. Rec. Area, str., 3.vii.1989, BJS (12 ♂♂, 12 ♀♀, CNC).

Additional material examined. USA. California: Tuolumne Co., Brightman Flat, 11.vii.1977, DDW (1 ♀, CAS). **Oregon:** Linn Co., Monument Peak, Guard Stn, 4000′, 28.vii.1969, E.M. Fisher (1 ♀, CAS); Mt Hood NF, 13 km E Gov't Camp + Hwy 35, BJS (1 ♂, 1 ♀, CNC). **Washington:** Lewis Co., Mt Rainier NP, Stevens Cyn Cpgd, 4000–4500′, Stevens Ck, 24.viii.1973, WJT (6

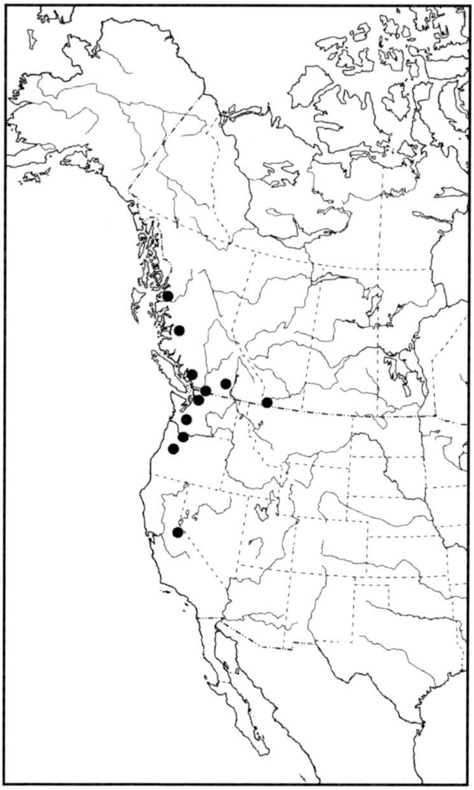

Map 4. Known distribution of *Clinocera amarilla*.

♂♂, 9 ♀♀, WSU); Mt Rainier NP, Comet Falls Tr. above van Trump Ck, 4500′, 13.viii.1977, WJT (9 ♂♂, 16 ♀♀, WSU), 10.ix.1977, E.J. Davis (4 ♀♀, WSU); Mt Rainier NP, Comet Falls (8.vii.1989), Carter Falls, seep (7.vii.1989), BJS (5 ♂♂, 13 ♀♀, CNC); Mt Rainier NP, Whitewater Cpgd, 31.viii.1981, PHA (1 ♂, CAS); Mt Rainier NP, West Side Rd, 910 m, Fish Ck, 1.viii.1977, DDW (5 ♂♂, 8 ♀♀, CAS); Mt Rainier NP, Christine Falls (16.viii.1917), Paradise (viii.1921), van Trump Ck (1.ix.1917), White R. (23.vii.1924), ALM (6 ♂♂, 13 ♀♀, USNM); Whatcom Co., Mt Baker Tr., 29.viii.1973, A.B. Gurney (4 ♂♂, USNM).

Recognition. The bright yellowish brown legs, including the coxae, contrasting with its dark body clearly distinguish this species from all other species of the *C. conjuncta* group.

Description (wing length 4.5–5 mm).

Male. Face flat, lower half with blue pruinescence. Gena and posterior region of head with tinge of blue pruinescence; vertex dark. Arista-like stylus short and slender. Postocular setae pale.

Scutum faintly bivittate. Pleura, notopleuron, and postpronotal lobe with greenish pruinescence. Several short pale acrostichals near collar; 5 dc; 1 pprn; 1 presut spal; 2 npl, lower short and slender; 1 psut spal; 1 pal; 2 sctl. Laterotergite with patch of long pale setae; posterior margin of anepisternum with pale setulae.

Wing (Fig. 29*e*) hyaline, infuscate along veins; 1 long basal costal seta; stigma beyond R_1 distinct, elongate elliptical, slightly deflecting R_{2+3}; costal spines present. Auxiliary crossvein between R_4 and R_{2+3} lacking; stump vein absent; R_4 straight, strongly divergent from R_5; anal vein dark streak; cell dm produced apically. Halter yellowish brown.

Legs (including coxae) yellowish brown, increasingly darker on tarsi; occasionally mid and hind coxae with blue pruinescence. Forecoxa with long pale setae. Forefemur with antero- and postero-ventral row of short setae along entire length; preapical anterior comb present; ventral surface with short dense white pile. Foretibia with anteroventral and anterodorsal rows of stout setae. Mid and hind femora lacking preapical setae. Hind tibia with erect dorsal setae on apical half. Hind first tarsomere with 3 long basal ventral bristles and 3 or 4 stout, short ventral bristles along entire length. Fifth tarsomere lacking dorsoapical extension; empodium pale brown laterally.

Abdomen dark, brownish; sternites often with greenish pruinescence. Hypandrium broad, ventral margin strongly arched. Phallus heavily sclerotized, straight, with expanded pick-like apex; distiphallus lacking. Clasping cercus broad, truncate, apical third of inner surface bearing patch of peg-like setae. Surstylus strongly arched, slightly shorter than clasping cercus (Fig. 12*a*).

Female. Similar to male except lacking ventral setae on forefemur and tibia.

Geographic distribution and seasonal occurrence. This species is found from June to September and is confined to the western Cordillera (Map 4).

Remarks. This species is most closely related to *C. pacifica* on the basis of the long straight phallus, lacking a curved distiphallus.

Etymology. The specific name is from the Spanish *amarillo* ("yellow"), referring to the yellowish legs of this species.

Clinocera arnaudi sp. nov.

Fig. 12*b*; Map 5

Type material. HOLOTYPE, male labelled "CANADA: ALBERTA: Banff/ National Park, Lake/ Louise, meadow at Tea/ House near Six Glaciers/ 12 August 1986/ Paul H. Arnaud, Jr. [dissected]"; "HOLOTYPE/ Clinocera/ arnaudi/ Sinclair [red label]" (CAS).

Paratypes: **CANADA. Alberta:** Banff NP, Lk. Louise Tr., 12.viii.1986, PHA (1 ♀, CAS); Banff, 29.viii.1922, CBDG (1 ♂, USNM); Banff, Cascade Mtn, 7000–8000′, 10,13.viii.1925, OB (2 ♂♂, 1 ♀, CAS); Jasper NP, Mt Whistler, 7500–8085′, 25.vii.1979, A.C. Pont (1 ♂, BMNH). **British Columbia**: 45 km N Revelstoke, 17.ix.1995, AB (1 ♂, 1 ♀, CNC). **USA. Wyoming:** Yellowstone NP Falls, 23.vii.1923, ALM (1 ♂, 4 ♀♀, USNM).

Possible additional material (males are required for confirmation of identification). **CANADA. Alberta:** Laggan, Popes Pass, 8000′, snow, 26.viii.1925, OB (2 ♀♀, CAS). **British Columbia:** Mt Revelstoke NP, Eva Lk., 22.vii.1955, GJS (1 ♀, CNC). **USA. Washington:** Mt Rainier, Paradise Pk, viii.1917, ALM (1 ♀, USNM).

Recognition. This species is similar to *C. borkenti, C. coloradensis*, and *C. cordillerana*, but is distinguished by several long costal setae proximally, apex of phallus spear-shaped, with distiphallus absent, and long, slender clasping cerci arched medially.

Description (wing length 4.2–5.2 mm).

Male. Face with blue pruinescence, lower third flat. Arista-like stylus short and stout. Postocular setae pale.

Scutum bivittate; pleura and notopleural depression with pale-blue pruinescence. Several inconspicuous acrostichal setae anterior to first dc; 5 dc, sometimes 6 or 7; 1 pprn; 1 presut spal; 2 npl, lower short and slender; 1 psut spal; 1 pal; 2 sctl. Laterotergite with patch of long pale setae. Posterior margin of anepisternum lacking setae.

Wing infuscate, with spots or clouding absent; stigma beyond R_1 faint and elongate; 2 long basal costal setae and several costal setae proximal to Sc; erect costal spines present. Auxiliary crossvein lacking between R_4 and R_{2+3}; stump vein absent; R_4 bell-shaped; anal vein indistinct; cell dm slightly produced apically. Halter brown.

Coxae with pale-blue pruinescence, concolorous with pleura. Femora dark brown. Forefemur with anterior preapical comb reduced to transverse row of 4 setae; dense white pile beneath; 6 anteroventral stout setae and 3 or 4 posteroventral stout setae. Foretibia with single row of stout setae beneath. Hind tibia with long erect dorsal setae along entire length. Fifth tarsomere lacking dorsoapical extension; empodium white laterally.

Abdominal sternites and lateral margin of tergites with blue pruinescence, remaining tergites brown. Hypandrium narrow, tapering on apical third. Phallus straight, lacking long distiphallus; arched posteriorly to sharp, spear-shaped apex. Clasping cercus long, parallel-sided, arched medially to opposing cercus, apex round bearing stout, pointed setae. Surstylus short, extending less than half length of clasping cercus; apex with distinct notch (Fig. 12*b*).

Female. Similar to male except as follows: face brown; costal setae

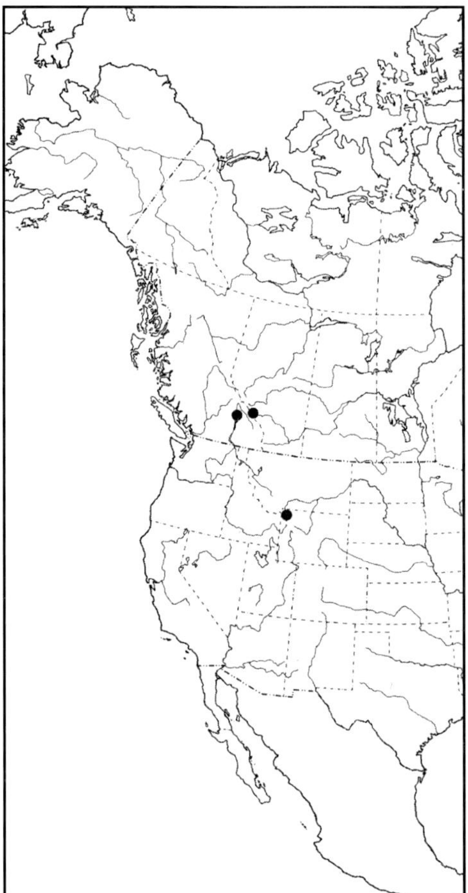

Map 5. Known distribution of *Clinocera arnaudi*.

subequal in length to basal costal setae, extending to near Sc; ventral bristles of forefemur and foretibia lacking.

Geographic distribution and seasonal occurrence. This species is confined primarily to the Rocky Mountains (Map 5), with flight records from July to September.

Etymology. The specific name is a patronym in honour of Paul H. Arnaud, Jr., the collector of the holotype and many other interesting western specimens of Empididae.

Clinocera borkenti sp. nov.

Figs. 12*c*, 29*f*; Map 6

Type material. HOLOTYPE, male labelled "50 mi SW of\ Terrace, B.C.\ 9.VII.1960\ J. G. Chillcott"; "HOLOTYPE/ Clinocera/ borkenti/ Sinclair [red label]" (CNC).

Paratypes: **CANADA. British Columbia:** Same data as holotype (7 ♂♂, 2 ♀♀, CNC); Glacier NP, 3.3 km W Rogers Pass Information Centre, roadcut seep, 1.vii.1989, BJS (2 ♀♀, CNC); Gold Ck nr Revelstoke, 17.viii.1937, HBL (2 ♂♂, 1 ♀, CNC); Mt Revelstoke NP, 4.8 km Summit Rd, roadcut waterfall, 3.vii.1989, BJS (1 ♀, CNC); 13 km N Revelstoke, Hwy 23, roadcut seep, 2.vii.1989, BJS (1 ♂, 3 ♀♀, CNC); 16 km N Revelstoke, 3.vii.1989, AB (10 ♂♂, 8 ♀♀, CNC). **USA. Colorado:** Gothic, 9500′, 19.vii.1934, M.A. Cazier (2 ♂♂, 2 ♀♀, AMNH). **Oregon:** Hood R. Co., Starvation Ck SP, cascading str., 10.vii.1989, BJS (3 ♂♂, 1 ♀, CNC); ditto, at falls, 26.vi.1978, DDW (1 ♂, 1 ♀, CAS); Lane Co., Willamette NF, Proxy Falls, 22.vi.1978, DDW (1 ♂, CAS). **Washington:** Lewis Co., Mt Rainier NP, Stevens Cyn Rd, 4000–4500′, Stevens Ck, 24.viii.1973, WJT (4 ♂♂, 12 ♀♀, WSU); Mt Rainier, Burroughs Mtn (21.viii.1934), White R. (20.vii.1924), van Trump (21.vii.1922), Yakima Pk (19.viii.1934), ALM (2 ♂♂, 3 ♀♀, USNM); Mt Rainier, 31.vii.1966, FCH (1 ♂, USNM); Mt Rainier NP, nr Ohanapecosh R., 720 m, Panther Ck, 2.x.1979, DDW (1 ♀, CAS); Mt Rainier NP, nr Christine Falls, roadcut seep, 8.vii.1989, BJS (1 ♀, CNC); Whatcom Co., Mt Baker NF, Silver Fir Cpgd, 610 m, 10.x.1979, DDW (1 ♂, 1 ♀, CAS). **Wyoming:** Yellowstone NP Falls, 23.vii.1923, ALM (3 ♀♀, USNM).

Additional material examined. CANADA. British Columbia: Bijou Falls, Pine Pass, Hwy 97, 16.viii.1978, PHA (4 ♂♂, 5 ♀♀, USNM); Glacier NP, Illccillewaet, Perley Rock Tr., 22.ix.2005, BJS (1 ♂, 1 ♀, ZFMK).

Possible additional material (male is required for species confirmation). **MEXICO. Sonora:** 22 km N Moetezuma, 25.ix.1980, DDW (1 ♀, CAS).

Recognition. This species is assigned to the subgroup with more than one basal costal bristle; distinguished from among these species primarily by the form of the phallic tip and the short ventral setae of the foretibia.

Description (wing length 5.5–6.5 mm).

Male. Face with dull bluish white pruinescence except just beneath antennae, lower third flat. Gena and posterior region of head with tinge of blue pruinescence; vertex dark. Arista-like stylus short and stout. Upper postocular setae dark, lower pale.

Scutum bivittate. Pleura, notopleuron, and postpronotal lobe with blue pruinescence. Several acrostichals anterior to first dc; 5 dc; 1 or 2 pale pprn setulae; 1 presut spal, fine and short; 2 npl, upper strong, lower weak, pale, and short; 1 psut spal; 1 pal; 2 sctl. Laterotergite with patch of long pale setae; posterior margin of anepisternum with pale setulae.

Wing (Fig. 29*f*) infuscate, margins of veins with clouding; stigma beyond R_1 distinct and narrow; 2 or 3 basal costal setae with series of costal setae half length

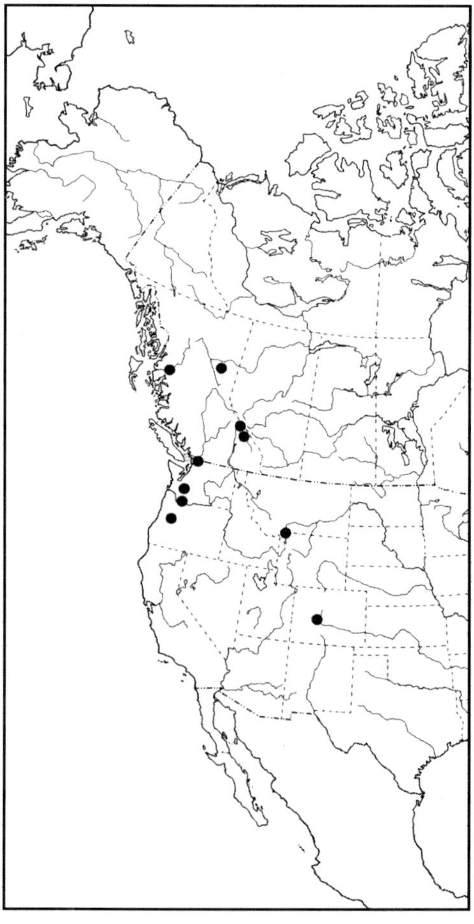

Map 6. Known distribution of *Clinocera borkenti*.

of basal costal setae ending near Sc; costal spines present. Veins straight; auxiliary crossvein between R_{2+3} and R_4 absent; stump vein absent; R_4 bell-shaped; anal vein distinct streak; cell dm produced apically. Halter brown.

Coxae with blue pruinescence, remaining segments dark. Forecoxa with fine pale setae. Forefemur with anterior preapical comb reduced to transverse row of 3–5 setae; dense white pile beneath; antero- and postero-ventral row of short setae extending nearly length of femur. Foretibia with row of short erect setae beneath. Mid and hind femora lacking preapical setae. Hind tibia with erect dorsal seta on apical third. Fifth tarsomere lacking dorsoapical extension; empodium pale brown laterally.

Abdominal sternites and lateral margin of tergites with blue pruinescence, remaining tergites brown. Hypandrium long and narrow. Phallus gently arched; lacking distiphallus; apex divided into oblique pair of narrow converging plates. Clasping cercus short and wide, bearing peg-like spines on inner face. Surstylus arched with rounded apex; nearly subequal in length to clasping cercus (Fig. 12*c*).

Female. Similar to male except as follows: face brown; costal setae subequal in length to basal costal setae, extending to near Sc; ventral bristles of forefemur and foretibia lacking.

Geographic distribution and seasonal occurrence. *Clinocera borkenti* is confined to the western Cordillera (Map 6), with flight records from July to October.

Etymology. The specific name is a patronym in honour of A. Borkent, who has collected many valuable western Nearctic species of Clinocerinae.

Clinocera coloradensis sp. nov.

Fig. 12*d*; Map 7

Type material. HOLOTYPE; male labelled "Corona Pass, Colo[rado; USA]./ Boulder Co., 10,600′/ 11. VII. 1961/ J.G. Chillcott"; "HOLOTYPE/ Clinocera/ coloradensis/ Sinclair [red label]" (CNC).

Paratypes: **USA. Colorado:** Same data as holotype (6 ♂♂, 14 ♀♀, CNC); Gilpin Co., 2 mi. E Rollins Pass, 3048 m, 16.viii.1973, DDW (1 ♂, 1 ♀, CAS); Hoosier Pass, 12 000′, 8.viii.1961, JGC (5 ♂♂, 2 ♀♀, CNC); Larlmer Co., above Trail Ridge Rd, 5.viii.1964, J. Hendrickson (2 ♀♀, SEMC); Mt Evans, 13 200′, 28.vii.1961, WRMM (1 ♂, CNC); Mt Evans, 14 000′, 25.vii.1961, JGC (1 ♀, CNC); Rocky Mtn Pk, Hidden Valley, 12.vii.1934, ALM (3 ♂♂, 1 ♀, USNM); Summit Lk., Mt Evans, along grassy str. bank, 12 800′, 24.vii.1961, JGC (7 ♂♂, 14 ♀♀, CNC); Trail Ridge Rd, 29.vii.1938, 24.vii.1939, MTJ (1 ♂, 1 ♀, CSU).

Recognition. This species is distinguished by the 2–5 basal costal setae and the numerous setae interspersed among the dorsocentral bristles.

Description (wing length 4.8–5.2 mm).

Male. Face flat with silver-blue pruinescence. Gena and posterior region of head with tinge of blue pruinescence; vertex dark. Arista-like stylus short and stout. Postocular setae dark.

Scutum bivittate. Pleura and npl with blue pruinescence. Several acrostichals extend to first dc; 8–18 dc (interspersed setae difficult to distinguish from true dc bristles); 1 pprn and 3 pale short setae; 1 presut spal; 2 npl, upper dark, lower pale and fine; 1 psut spal; 1 pal; 2 sctl. Laterotergite with patch of long pale setae; posterior margin of anepisternum lacking setulae.

Wing infuscate, margins of veins with clouding, especially crossveins;

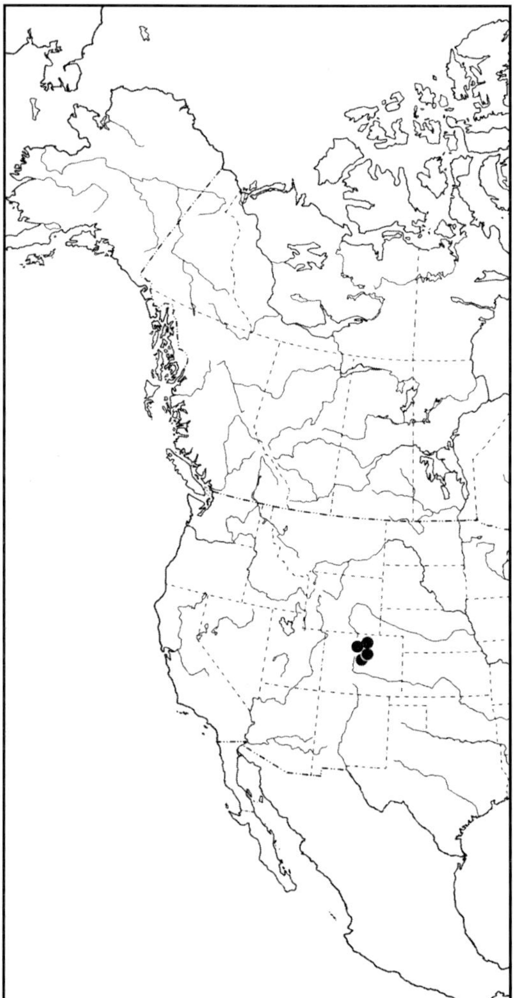

Map 7. Known distribution of *Clinocera coloradensis*.

stigma beyond R_1 faint and narrow; 2–5 basal costal setae; costal spines present. Veins straight; auxiliary crossvein between R_{2+3} and R_4 absent; stump vein absent; R_4 bell-shaped; anal vein distinct streak; cell dm produced apically. Halter brown.

Coxae and femora with pale-blue pruinescence, remaining segments brown. Forecoxa with long pale setae. Forefemur with antero- and postero-ventral row of small black setae on basal three-quarters; anterior, preapical comb reduced to transverse row of 3 or 4 setae; ventral surface on apical third with short dense white pile. Foretibia with row of short stout setae beneath. Mid and hind femora lacking preapical setae. Hind tibiae with erect dorsal setae on apical half. Fifth tarsomere lacking dorsoapical extension; empodium brown laterally.

Abdominal sternites and lateral margin of tergites with blue pruinescence, remaining tergites brown. Hypandrium long and narrow. Phallus strongly arched, lacking distiphallus; apex with pair of small plates and flap of membranous tissue. Clasping cercus with inner basal lobe; bearing scattered peg-like setae on inner face and basal lobe. Surstylus arched, not projecting free from epandrium; shorter than clasping cercus (Fig. 12*d*).

Female. Similar to male except as follows: face brown; foretibia lacking erect setae beneath; forefemur lacking ventral biserial row of small setae.

Geographic distribution and seasonal occurrence. This species is confined to the Rocky Mountains of Colorado (Map 7), with flight records from July and August.

Etymology. The specific name is derived from the State from which all known specimens were collected.

Clinocera conjuncta Loew

Figs. 13*a*, 29*g*, *h*; Map 8

Clinocera conjuncta Loew, 1860: 80.

Clinocera taos Melander, 1902: 242 **syn. nov.**

Clinocera (*Hydrodromia*) *undulata* Melander, 1928: 230 **syn. nov.**

Clinocera conjuncta: Melander, 1902, 1928, 1965: 468.

Type material examined. *C. conjuncta*: HOLOTYPE, male labelled “Type/ 1179”; “62.[blue label]”; “Loew/ Coll.”; “conjuncta/ m. [handwritten]” (MCZ).

C. taos: HOLOTYPE, female labelled “Franconia, N.[ew]H.[ampshire, USA]”; “W.M. Wheeler,/ Collection.”; “TYPE/ No./ A.M.N.H. [red label]”; “Am. Mus. Nat. Hist./ Dept/ Invert. Zool./ No. 813” (AMNH).

C. undulata: HOLOTYPE, male labelled “Moscow Mt./ 6.VII.[19]12 Idaho [USA]”; “TYPE/ Hydrodromia/ undulata/ Mel. [red label] [missing head, forelegs, left tibia and tarsus]” (USNM).

Paratypes: **USA. Idaho:** Moscow Mtn, 16.xi.1917, ALM (1 ♂, USNM); **Washington:** Mt Rainier, above Longmire's, 5000′, 3.viii.[19]05 (1 ♀, USNM).

Notes on synonymy. After a very large number of specimens from across North America were examined, it was evident that this species includes many forms with variable wing patterns, including the length of cell dm. This variation included *C. taos* and *C. undulata*, which are here considered conspecific with *C. conjuncta*. Loew (1860) stated that the type specimen of *C. conjuncta* was from Washington, D.C., collected by O. Stacken.

Additional material examined. Some 2300 specimens from the following provinces and states: **CANADA. Alberta:** Banff, 20.vi.1928 (CAS); Banff NP, Lk. Louise, Six Glaciers, 12.viii.1986, PHA (CAS), 26 km N Lk. Louise, trib. of Mosquito Ck, 28.vi.1989, BJS (CNC); Base Prospect Mtn, Cadomin, 19.vii.1987, SAM (DEBU); Jasper NP, 60 km S Jasper, Bubbling Springs, 28.vi.1989, BJS (CNC); Kananaskis For. Exp. Stn, Seebe, 24.vii.1968, HJT (CNC), Kananaskis, 26.vi.1928, OB (CAS); Laggan (Lk. Louise), 16.vii.1928, OB (CAS); Waterton Lks. NP, 14–20.vii.1980, HJT (CNC). **British Columbia:** 37–52 km NW Anahim Lk., 6–15.vii.1992, AB (CNC); 14 km SW Bella Coola, 10–11.vii.1992, AB (CNC); Bijou Falls, Pine Pass, Hwy 97, 16.viii.1978, PHA (USNM); Bowser, 30.v.1955, R. Coyles (CNC); Capilano, N Vancouver, 17.x.1972, JRV (CNC); Cathedral PP, Quiniscoe Lk., 6.vii.1986, S.G. Cannings (UBC); Cultus Lk., 22–27.x.1938, J.K. Jacob (CNC); 40 km E Enderby, 10.vii.1992, AB (CNC); 5–6 km NE Falkland, 12, 13.vii.1989, 17.vii.1995, AB (CNC); 29 km SE Hope, 14.vii.1994, AB (CNC); Heckmann Pass, Tweedsmuir PP, 18.vii.1978, G.G.E. Scudder (UBC); Hector, 19.vi.1932, OB (CAS); Horseshoe Bay, 30.v.1961, JRV (CNC); Kelsall, 18.v.1948, WRMM (CNC); Ketchum Lk., 58°22′N, 131°35′W, 3600′, 23.viiii.1960, WWM (CNC); Kootenay NP, Radium Hot Springs + Hwy 93, roadcut seep, 17.ix.2005, BJS (ZFMK); Lk. Cowichan, 33 km NW, 9.vii.1991, AB (CNC); Lisadele Lk., 4000′, 58°41′N, 133°04′W, 5–10.viii.1960, WWM, RP (CNC); Manning PP, rd to lookout, 8.x.1979, DDW (CAS); Manning PP, Lightning Lk. Cpgd, 5.viii.1986, PHA (CAS); Moosehorn Lk., 4500′, 58°10′N, 132°07′W, tundra/meadows, 26–28.vii., 1.viii.1960, WWM, RP (CNC); Mt Revelstoke NP, Eva Lk. Tr., 6000′ (14.viii.1952), Jade Lk., 6300′ (19.viii.1952), GJS (CNC); Penticton, Apex Mtn, 6500′, JRV (CNC); Pete Lk, 57°56′N, 131°56′W, 20.viii.1960, RP (CNC); Prince Rupert, Ridley Is., 7.viii.1996, PHA (USNM); Queen Charlotte Is., Graham Is. nr Queen Charlotte City, 10–23.vii.1988, SAM (DEBU); 6 km NE Revelstoke, 19.viii.1990, AB (CNC); 13 km N Revelstoke, Hwy 23, roadcut str., 2.vii.1989, BJS (CNC); Robson, 2.ix.1947, 15.x.1955, 9–30.x., 6.xi.1956, 30.x.1957, 13–27.x.1958, 19.x.1960, 9–15.ix.1961, 15–17.x.1961, 15.x.1962, HRF (CNC, UBC); Silver Star Prov. Rec. Area, 22 km E Vernon, 3.vii.1989, BJS (CNC); Squamish, 17.ix.1975, G.F. Hevel (USNM); Squamish, Diamond Head Tr., 3200′, 25.viii.1953, WRMM (CNC); Terrace, Mt Thornhill, boggy seep, 3500′, 14.vii.1960, JGC (CNC); 23 mi. N Terrace, Mt Allard, 20.vii., 11.viii.1960, CHM

(CNC); 40 mi. W Terrace, marginal sedges, 9.vii.1960, JGC (CNC); 50 mi. SW Terrace, under waterfall, 24.vi., 9.vii.1960, JGC (CNC); Wells Gray PP, Clearwater Valley, Shadow Lk., Mink Ck, 16.viii.1986, PHA (CAS); Yoho NP, Kicking Horse Cpgd, str., 29.vi.1989, BJS (CNC). **Newfoundland and Labrador:** Goose Bay, 13.vi., 5.vii., 12.viii.1950, J.J. Tibbles (CNC); Great Caribou Is., 27.vii.1896, CWJ (MCZ); Nain, 8.viii.1918 (MCZ, USNM). **Nova Scotia:** Cape Breton Highlands NP, Beuach Ban Falls, wet rock face, 1–8.vii.1984, HJT, JRV (CNC), Lone Shieling, str., 1–8.vii.1984, JRV (CNC), nr Pleasant Bay, seepage, 1–8.vii.1984, DMW (CNC). **Quebec:** Gaspé Bay,

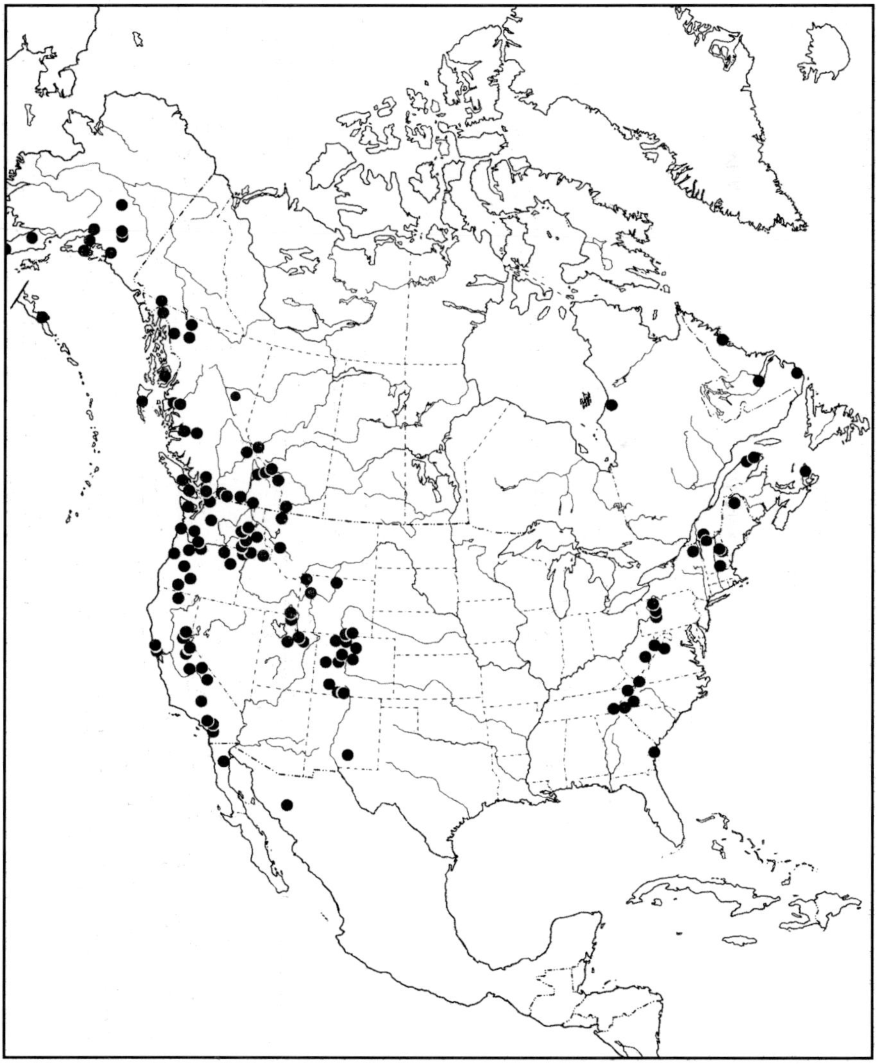

Map 8. Known distribution of *Clinocera conjuncta*.

28.viii.1959, CWS (USNM); Gaspé Bay, Gaspésie Pk, Mt Jacques Cartier, 1000 m, 11.viii.1983, B.M. Bissett (CNC); Grande Vallee, Gaspé, 20.vii.1931, JMA (USNM); Great Whale R., 29.viii.1960, JRV (CNC); Sutton, Round Top Mtn, 5.vi.1963, JGC (CNC); Tabutiere, 11.vii.1929, W.J. Brown (CNC). **MEXICO. Baja California Norte:** Sierra San Pedro Martir, 2900 m, 17 km E Park Entrance, 7.ix.1980, DDW (CAS). **Sonora:** 22 km E Moctezuma, 25.iv.1980, DDW (CAS). **USA. Alaska:** Anchorage, Flat Top Mtn, South Fk, 5, 19.vii.1993, TS (BLKU); Aniakchuk NM, Surprise Ck, 56°56′N, 158°6′W, vii–viii.1988, B. Wisseman (OSU); Cold Bay, 26.vii., 20.viii.1952, WRMM (CAS, CNC); Denali Hwy, mi. 36, Boulder Mtn, 14.vii.1993, TS (BLKU); Fern Mine, 27.vii.1956, W.C.E. (WSU); George Parks Hwy, mi. 244, Bison Gulch, 21.vii.1993, TS (BLKU); Glen Hwy, mi. 55, Moose Ck, 19.vii.1993, 4–17.vi.1994, TS (BLKU); Granite Ck, 12.viii.1958, W.C.E. (WSU); Hainer, rd to Chilkoot Lk. SP, 26.vii.1996, PHA (USNM); Hyder, rd to Rish Ck, 5.viii.1996, PHA (USNM); Independence Mine, 4.viii.1956, W.C.E. (WSU); Ketchikan, 18–27.vii.1951, B. Malkin (CAS); Ketchikan, Deer Mtn, 5.viii.1970, N.L.H. Krauss (USNM); King Salmon, Naknek R., 1–10.viii.1952, J.B. Hartley (CNC); Mt McKinley NP, Teklanika R. Cpgd, 760 m, 15.vii.1978, PHA (USNM); One Mile Ck, nr Paxson, 7–16, 27.vii.1993, TS (BLKU); Palmer, 1.viii.1948, Sailor, Ridenour (USNM); Richardson Hwy, mi. 24, Blueberry Lk. (18.vii.1993), mi. 26, Thompson Pass (18.vii.1993), mi. 29, Worthington Glacier (17.vii.1993), mi. 98, Isabel Pass (8.vii.1993), mi. 184, nr Paxson (7.vii.1993), mi. 204, Phelan Ck (8.vi.1994), mi. 208, Rainbow Ridge (8.vii.1993), TS, OY (BLKU); Richardson Hwy, mi. 206, beaver pond margin, 16.vii.1989, DMW (CNC); Richardson Hwy, mi. 250, 27.vii.1951, WRMM (CNC); Richardson Hwy, mi. 206, Isabel Pass, 13.vii.1962, REL (CNC); Seward, 24.vii.1921, JMA (USNM); Whittier, 60°46.5′N, 148°41′W, 8.viii.2003, W.N. Mathis (USNM). **California:** Alpine Co., Winnemucca Lk., 14.vii.1964, R.A. Belmont, R.M. Bohart (UCD); Big Pines, 8.ii.1946, ALM (USNM); Bishop, 28.vii.1940, DEH (SEMC); Cresson Cyn, 10.vii.1930, JMA (USNM); E Barton Flat, 25.vi.1946, ALM (USNM); Fresno Co., Huntington Lk., Rancheria Cyn, 2600 m, 23.vii.–20.viii.1984, JFM (PERC), nr lakeshore, Bear Lk., 2135 m, 8.viii.1975, PHA (CAS); Fresno Co., Sierra NF, Badger Flat Cpgd, Rancheria Ck, 2500 m, 1.viii.1979, DDW (CAS), Forebay Lk., 2200 m, E and W Fks, 31.vii.1979, DDW (CAS), nr Kaiser Pass, 2700 m, ck, 2.viii.1979, DDW (CAS); Inyo Co., Lone Pine Ck, 25.i.1985, RWB, CRN (BYU); Kern Co., Kernville, 27.v.1986, D.J. Burdick (PERC); Los Angeles Co., 5 mi. E Islip Saddle, Angeles NF, Cortelyou Spring, 25.iv.1977, DDW (CAS); Los Angeles Co., Angeles NF, Windy Spring, 27.vi.1974, DDW (CAS); Marin Co., Golden Lk. Bridge, 12.iv.1953, HBL (CAS); Marin Co., Hicks Valley, Arroyo Sausal, 12.v.1977, PHA (USNM); Mono Co., Leavitt Meadows, 2470 m, 23.vii.1966, PHA (USNM); Mono Co., Rt. 108 nr Sonora Pass, 9.vi.1990, P.H. Adler (CNC); Mt Home Cyn, 20.v.1945 (USNM); Nevada Co., Hwy 20, Drue Canal, 7.xii.1986, BJS (DEBU); Nevada Co., Norden, Cal Lodge, 2130 m, 21.vi.1987, PHA (USNM); Nevada Co., Sagehen Ck, 27.vi.1972, 16.vii.1980, R.A. Belmont,

R.M. Bohart (UCD); Plumas Co., Lassen NF, N Fk Feather R., 28–30.v.1977, DDW (CAS); Riverside Co., Cottonwood Cyn, 21.iii.1967, G.R. Ballmer (UCR); San Bernadino Mtns, 1000 Springs, 24.x.1946, ALM (USNM); Siskiyou Co., McBride Springs [41°21′N, 122°16′W], 5200′, 10–14.vi.1974, J. Doyen (UCB); Siskiyou Co., S Fk Sacramento R., 19.vii.1948 (USNM); Sonoma Co., Stillwater Cove, 23.v.1954, E.I. Schlinger (UCD); Tuolumne Co., Brightman Flat, 11.vii.1977, DDW (CAS); Tuolumne Co., Sonora Pass, 9000′, 22.vii.1951, W.C. Bentinck (UCB). **Colorado:** Big Ck Lks., 5.vii.1965, FCH (FSCA); Boulder Co., Corona Pass, 10 600′, 11.viii.1961, JGC (CNC); Cameron Pass, 1.viii.1959, H. Knutson (USNM), 14.viii.1946, MTJ (CSU), 19–22.viii.1940, CWS (USNM); Cameron Pass nr Gould, 19–21.viii.1940, MTJ (CSU, WSU); Clear Ck Co., Floral Pk, Berthoud Pass, 9.vii.1974, G.W. Byers, C.W. Young (SEMC); Eagle Co., Fulford Cave Cpgd, 22.vi.1996, SF, AF (CSU); Estes Pk, 9.viii.1947, D.L. Denning (UWL); Floral Pk, 11.viii.1973, JRV (CNC); Gilpin Co., 2 mi. E Rollins Pass, 2048 m, 6.viii.1973, DDW (CAS); Grand Co., 2 mi. W Rollins Pass, 238 m, 6.viii.1973, DDW (CAS); Grand Co., FR100, 0.5 mi. off Rt. 40, 19.viii.1993, SF et al. (CSU); Gunnison Co., Gothic, 9800′, 18.vii.1934, CPA (AMNH); Hinsdale Co., Red Cloud Pk, Red Mtn, 10 000′, 22.vii.1934, CPA (AMNH); Independence Pass, 21.vii.1962, FCH (USNM); Jackson Co., Cameron Pass, 16.viii.1966, GFK (USUL), Lk. Agnes, 19.viii.1947, MTJ (WSU); Lake Co., Independence Pass, 12 100′, tundra, 7.viii.1961, JGC, CHM (CNC); Larimer Co., Joe Wright Res., 3 mi. E Cameron Pass, 16.viii.1982, RH (MSUB); Larimer Co., W Br. Larimer R. above Blue Lk., 22.viii.1993, SF (CSU); Loveland Pass, 20.vii.1947, FCH (USNM), summit, 12 000′, 7.viii.1961, JGC (CNC); Mesa Co., Grand Mesa, Atkinson Res., 16.viii.1986, J. Jenkins (WSU); Mineral Co., Wolf Ck Pass, Hwy 160, 24.viii.1993, SF et al. (CSU); Mt Evans, 11 300′, 10.viii.1961, JGC (CNC); Rio Grande Co., Beaver Ck, 10 000′, MT, 21.vi.1972, WWW (USNM); Rocky Mtn NP, 29.vi.1959, 24.viii.1967, FCH (USNM); Routt Co., 8 mi. S Rabbit Ears Pass, 15.viii.1973, GFK, WJH (USUL); San Juan Co., nr Silverton, Molas Elk Ck Tr., 26–29.vii.1985, J.J. Jenkins (WSU); Summit Co., Eagles Nest Wilderness, Wheeler Lks., 11 100′, 15.vii.1998, SF, AF (CSU), Last Lk., 11 700′, SF, AF (CSU); Summit Co., Tenmile Range, 10 500′, 15–17.vii.1998, SF, AF (CSU); Tennessee Pass, 10 240′, 9–10.vii., JMA (USNM); Trail Ridge Rd, 24.vii.1939, MTJ (CSU); Vail Pass, 21.vii.1966, FCH (USUL); Ward, 14.viii.1939, MTJ (CSU). **Georgia:** McIntosh Co., Sapelo Is., FIT (Savannah), 28.iv.–9.v.1987, BRC Hym Team (CNC). **Idaho:** Franklin Co., Cub R. Cyn, 7.viii.–17.ix.1985, WJH (USUL); Gunnison Co., Cottonwood Pass, 29.vii.1961, JGC (CNC); Idaho Co., Legend Lk., 9 air mi. NE Selway Falls, 6800′, 19.vii.1977, WJT (WSU); Latah Co., Moscow Mtn, East Twin Peak, 19.vii.1983, RSZ (WSU), St. Joe NF, 8 mi. NE Harvard, 11.xi.1981, RSZ (WSU), 7 mi. ENE Moscow, nr Moscow Mtn, 15, 16.v.1985, WJT (WSU), 7 mi. N Troy nr Big Meadow Rec. Area, 3000′, 13.vii.1979, WJT (WSU); Lewiston Hill, 25.vi.1923, 10.v.1924, ALM (USNM); Moscow Mtn, 16.ix.1917, 10.ix.1922, ALM (USNM), 11.x.1963, E.J. Allen (UIM); Priest Lk., Cavanaugh, 18.viii.1919, ALM

(USNM); Waha, 9.vi.1918, 30.v.1924, ALM (USNM). **Maine:** Mt Katahdin, Hunt Tr., 1–6.vii.1968, DMW (CNC). **Montana:** Coram, Bad Rock Cyn, Shepard Memorial Fountain, 28.viii.1981, PHA (CAS); Glacier Pk, Avalanche Lk. (14.vii.1935), Logan Pass (18.vii.1935), ALM (USNM); Pipestone Pass, 13 mi. S Butte, 6200′, 19.viii.1961, JGC (CNC). **Nevada:** Washoe Co., Mt Rose, 25.viii.1983, PHA (CAS), Mt Rose, summit, 9.vii.1964, R.M. Bohart (UCD); White Pine Co., above Lehman Caves NM, Lehman Ck, 13.iii.1985, RWB, CRN (BYU). **New Hampshire:** Bretton Woods, 1.vii.1936, ALM (USNM); Carroll Co., 1 mi. N Wonalancet, E Fk Spring Brook, 29.viii.–5.ix.1985, DSC (UNH), 2–2.5 mi. NW Wonalancet, The Bowl, 27.vii.–1.viii., 2–17.ix., 18.ix.–3.x.1984, DSC (UNH); Cheshier Co., Monadnock Mtn, 1.x.1972, A. Slater (SEMC); Coos Co., Mt Washington, 3500′, 23.viii.1988, JMC (CNC), Auto Rd, Alpine Garden, 18, 25.vi.1982, DSC (UNH), Lake of Clouds, 15.vii.1982, DSC (UNH); Grafton Co., Franconia Notch SP, 8 mi. S Franconia, along Pemigewasset R., PT, 2, 3.vii.1992, JMC (CNC); Lost R., 7.vii.1941, ALM (USNM); Mt Monadnock, 26.vii.1926, ALM (USNM); Mt Washington, 3500′ (12.vii.1931), Alpine Garden, 5900′ (2.vii.1936, 30.viii.1937), ALM (USNM); Mt Washington, Alpine Garden, 1650 m (8.vii.1981), Auto Rd, 5000′ (14.viii.1958), JRV (CNC); Mt Washington, Lakes of the Clouds (29.viii.1954), Oaks Gulf (3.viii.1954), Becker, Munroe, Mason (CAS, CNC); Mt Washington, 4200′, 22.viii.1974, R.N. Story (UNH), 3000′, 12.viii.1961, WWW (USNM); White Mtns, Dolly Copp (11–13.vii.1931, 29.viii.1932), The Flume (7.vii.1936), ALM (USNM). **New Mexico:** Lincoln Co., 9.7 km W Alto, 3120 m, Lincoln NF, Cyn of Sky Apache, 33°23′N, 105°47′W, 21–24.vi.1995, S.D. Gaimari (CNC). **New York:** Adirondacks, Avalanche Tr., 30.vii.1929, ALM (USNM); Clinton Co., Chazy Lk., 8.ix.1985, RH (MSUB); Trenton Falls, O. Sacken (MCZ, USNM); Whiteface Mtn, 4000′, 19.viii.1962, JRV (CNC), summit, 14.vii.1938, ALM (USNM); Wilmington (Notch), 21.viii.1941, ALM (USNM). **North Carolina:** Andrews Bald, 9.vii.1941, ALM (USNM); Buncombe Co., Blue Ridge Pkwy, Craggy Gardens Picnic Area, 1580 m, 18.v.1986, 12.vi.1990, BJS (CNC, DEBU); GSMNP, Clingmans Dome (18.vii.1941), Forney Ridge (26.vi.1941), Newfound Ridge (11.vii.1941), ALM (USNM); GSMNP, Clingmans Dome, 6647′, 21–22.v.1965, JGC (CNC); Highlands, 7.v., 1.vi.1957, JRV, WRMM (CNC); Indian Gap, 5200′, 18.v.1957, JRV (CNC); Jackson Co., Glen Falls, Hwy 106, E Fk Overflow Ck, 3800′, 25.v.1984, RWB, CRN (BYU); Swain Co., GSMNP, Clingmans Dome Rd, roadcut seeps (15.vi.1990), Newfound Gap (15.vi.1990), 6 km S Balsam Mtn Cpgd (13.vi.1990), Balsam Mtn, Flat Ck (13.vi.1990), BJS (CNC); Yancey Co., Mt Mitchell SP, 2000 m, Balsam Tr., spr., 11.vi.1990, BJS (CNC). **Oregon:** Benton Co., Siuslaw NF, Marys Peak, waterfall, 20.vi.1977, DDW (CAS), Parkers Ck, 14.vii.1989, BJS (CNC); Clackamas Co., Still Ck Cpgd nr Gov't Camp, 3800′, 22–23.vi.1982, WJT (WSU), Bruin Run Ck, 24.i.1974, PHA (CAS); Crater Lk., 16.ix.1934, ALM (USNM); Deschutes Co., Deschutes NF, Cold Spr. Cpgd, 22.vi.1978, DDW (CAS); Hood R. Co., Mt Hood NF, Still Ck Cpgd, 27.vi.1978, DDW (CAS), nr Sahelie Falls, Mitchell Ck, 27.vi.1978, DDW

(CAS); Jackson Co., Dead Indian Soda Spr., 21.v..1964, H. Schuh (WSU); Lane–Linn Co., H.J. Andrews EF, 10 km NE Blue R., 29.vii.1981, DEH (OSU); Lane Co., Willamette NF, Proxy Falls, 22.vi.1978, DDW (CAS); Lane Co., 8 mi. SE Cougar Dam, 880 m, Grasshopper Ck, 11.v., 10.vi.1982, GWC (OSU); Lane Co., 11 mi. NE Blue R, H.J. Andrews EF, Mack Ck, 20.vi.1982, GWC (OSU); Mt Hood NF, 6–13 km E Gov't Camp, 11.vii.1989, BJS (CNC); Mt Hood NF, 3.5 mi. below Timberline Lodge, 11.vii.1989, BJS (CNC); Umatilla Co., Cold Springs, 11.iv.1954, MTJ, J.J. Davis (WSU); Union Co., 28 mi. SSW La Grande, Whitman NF, 23.vi.1976, WJT (WSU); Wallowa Co., Eagle Cap Wilderness, E Lostine R., 7100′, 14–16.xi.1984, J. Jenkins (WSU); Wallowa Co., Aneroid Lk., 4 mi. SE Lk. Wallowa, 7500′, 9.vii.1991, WJT (WSU); Wallowa Co., 39 mi. N Enterprise, Hwy 3, 3400′, 21.vi.1976, WJT (WSU). **Pennsylvania:** Clearfield Co., 0.5 km N Mahaffery, Hwy 219, 30.v.1990, BJS (CNC); Elk Co., Johnsonburg, Hwy 219, cliff seeps, 30.v.1990, BJS (CNC); McKean Co., Allegeny NF, Red Bridge Cpgd, SR 122, spr., 29.v.1990, BJS (CNC). **Tennessee:** Carter Co., 3.vi.1958, HER (USNM); Sevier Co., GSMNP, Alum Cave Tr. (16.vi.1990), Laurel Falls, seep (12.v.1986), Mt LeConte Tr., str., (16.vi.1990), 10 km E Sugarlands (3–6.vi.1993), 18 km E Sugarlands (5.vi.1993), BJS (CNC, DEBU). **Utah:** Boulder, 23.viii.1957, FCH (USNM); Cache Co., Blacksmith Fk Cyn, 1.vii.1976, WJH (USUL), Green Cyn, 19–23.vii.1985, N.N. Youssef (USUL); Logan Cyn, 20.iv.1968, WJH (USUL), Logan Cyn, Turner, 16–20.viii.1985, WJH (USUL), Logan, 16.ix.1974, GFK (USUL); Cache Co., Spring Hollow Ck, 24.v.1977, RWB, Stewart (BYU); Duchesne Co., Butterfly Lk., 12.viii.1971, GFK (USUL); Logan, Spring Hollow, 23.iii.1959, R. Amoureux (USUL); Mt Timpanogus, 26.viii.1943, GFK, D.R. Maddock (USUL); Summit Co., Beth Lk., Uinta Mtns, 23.vii.1981, RWB (BYU); Summit Co., S Bald Mtn 12.viii.1971, GFK (USUL); Summit Co., Wasatch NF, Butterfly Lk., 28.viii.1975, EJR (ROGC); Tooele Co., Loop Cpgd, S Willow Ck, 4.vi.1976, RWB, C. Hogue (BYU); Utah Co., S Fk American Fk R., Altmont Cpgd, 11.ix.1987, RWB, CRN (BYU); Utah Co., N Fk Ck, Timpanogos Tr., 21.ix.1985, RWB (BYU, in alc.); Utah Co., above S Fk Provo R., 21.vii.1982, S.M. Clark (BYU); Wasatch, 5.viii.1956, ALM (USNM). **Vermont:** Jay Peak, 2600–3000′, 20.vii.1968, JRV (CNC); Jay Peak, 21.vii.1968, JRV (CNC). **Virginia:** Grayson Co., Grayson Highlands SP, Big Wilson Ck, 10.vi.1990, BJS (CNC); Shenandoah Co., Wolf Gap, 6.v.1980, DDW (CAS); Shenandoah, Big Meadows, 15.vi.1941, ALM (USNM); Smyth Co., Mt Rogers, 4700–5300′, 1.vi.1962, JRV (CNC). **Washington:** Asotin Co., Field Spring SP, 4 mi. S Anatone, 7.vi.1973, 12, 13.vi.1974, 31.v., 30.vi.1975, 28.vi.1976, 30.iv., 18.vi.1977, 26.vi.1979, WJT (WSU), 24.v.1975, NEW (WSU), 30.vi.1975, EJR (ROGC); Chelan Co., Peshstin Ck, Rt. 97, 10.x.1983, G.R. Fiala (BYU); Chinook Pass, 27.vii.1949, MTJ (WSU); Clallam Co., Olympic NP, Mt Angeles, 1550 m, 22.vii.1978, DDW (CAS), Lk. Ck, 650 m, 24.vii.1978, DDW (CAS); Gifford Pinchot NF, La Wis Wis, Ohanapocosh R., 1.x.1979, DDW (CAS); Ilwaco, 27.viii.1917, ALM (USNM); Kittitas Co., Manastash Ridge ca. 27.5 mi. NW Selah, Section 29, 3250′, 25.v.1986, NEW (WOOC); Lewis Co., Mt

Rainier NP, Comet Falls Tr., above van Trump Ck (13.viii.1977), Paradise R., 0.5 mi. S Narada Falls (8.viii.1979), Reflection Lks. (24.viii.1973), Stevens Ck, Stevens Cyn Rd (24.viii.1973), van Trump Ck, above Christine Falls (28.viii.1979), WJT (WSU); Lewis Co., Mt Rainier NP, Paradise R., 0.5 mi. S Narada Falls (8.viii.1979), RSZ (WSU); Lewis Co., Mt Rainier NP, Comet Falls Tr., above van Trump Ck, 10.ix.1977, E.J. Davis (WSU); Mt Baker NF, Glacier Ck Rd, 1100 m, Grouse Ck and Thompson Ck, 4.x.1979, DDW (CAS); Mt Rainier, Alta Vista (22.vii.1922), Berkeley Pk (19–27.viii.1934), Burroughs Mtn (21.viii.1934), Frying Pan Tr. (29.viii.1934), Ind. Henry (2.viii.1922), Narada Falls (16.viii.1917), Mazama Ridge (23.vii.1922) Paradise Pk (viii.1917), Sluiskin (28.vii.1922), Snoqualmie Pass (29.vii.1924), Summerland (24.vii.1924, 29.viii.1934), van Trump Ck (1.ix.1917), White R. (20.vii.1924), Yakima Pk (19.viii.1934), ALM (USNM); Mt Rainier, above Longmire's, 5000′, 3.viii.1905, JMA (USNM); Mt Rainier NP, Paradise R. nr Paradise, 20.viii.1977, NEW (WOOC); Mt Rainier NP, E end Yakima Pk, 1830 m (3.x.1979), above Louise Lk., 1520 m (4.viii.1977), Ohanapecosh R., 720 m (2.x.1979), Paradise Pk, 1650 m (2.viii.1977), nr White R. Cpgd, 1220 m (3.x.1979), DDW (CAS); Mt Rainier NP, Carter Falls, seep, 3660′, 7.vii.1989, BJS (CNC); Pierce Co., Frog Pond, Paradise, 20.vii.1977, RSZ (WSU); Pierce Co., Mt Rainier NP, Ghost Lk. and Cayuse Pass, 4700′, 8.viii.1979, RSZ (WSU), 2 mi. NE White R. Cpgd, 2.ix.1978, RSZ (WSU); Skagit Co., Baker Lk., 27.v.1965, F. Schmid (CNC); Skamania Co., Gifford Pinchot NF, 20 mi. N Trout Lk., 9.vii.1989, BJS (CNC); Spokane Co., Bald Knob Cpgd, Mt Spokane SP, 6.vii.1978, WJT (WSU), 1500 m, 25.vi.1978, DDW (CAS); Whatcom Co., Mt Baker NF, 850 m, Bagley Ck, 10.x.1979, DDW (CAS); Whitman Co., Steptoe Cyn, 5 mi. W Colton, 8.iii.1973, J.F. Brunner (WSU); Whitman Co., 6 mi. S Wawawai, Snake R., 800′, 10.v.1987, WJT (WSU). **West Virginia:** Greenbrier Co., Monongahela NF, 18 km W Information Center nr Windy Run, 4.vi.1990, BJS (CNC), Summit Lk. Rd, 3.vi.1990, BJS (CNC); Tucker Co., Blackwater Falls SP, Falls of Elakala, seeps, 2.vi., 3.vii.1986, 31.v.–2.vi.1990, 6.vii.2004, BJS (CNC, ZFMK), Davis, SR 32, roadside spr., 1.vi.1990, BJS (CNC). **Wyoming:** Albany Co., Snowy Range Mtns, 15.viii.1951, D.G. Denning (UWL); Big Horn Mtns, 15.viii.1947, FCH (USNM); Canyon Village, 21.vii.1971, GCS (USNM); Carbon Co., Hwy 130, E Ryan Pk, 2990 m, 1.viii.1973, PHA (CAS); Carbon Co., Lk. Marie, 20 km W Centennial, 3230 m, 1.viii.1973, PHA (CAS); Laramie, 8.viii.1949, P.B. Lawson (SEMC); Sierra Madre Range, Battle Lk. Rd, 8500′, 18.vii.1961, JGC (CNC); Sheridan Co., Granite Pass, 7.vii.1959, G. Byers (SEMC); Teto Co., Togwotee Pass, 16.viii.1961, P.H. Poole (CNC), 9000′, 22.vii.1953, R.R. Dreisbach (CNC); Yellowstone NP, Bridge Bay, Yellowstone Lk., 7800′, 18.vi.1952, E.B. Hayden (SEMC); Yellowstone NP, Mt Washburn, 18.viii.1962, P.J.S. (USNM); Yellowstone NP, Cañon Camp, 7700′, 27.viii.1915 (CUIC); Yellowstone NP, Apollinaris (8.vii.1923), Canyon Camp (12.viii.1918), Continental Divide, 6200′ (8.viii.1918), Dunraven Pass (28.vii.1934, 26.vii.1956), falls (23.vii.1923), Lewis Falls (17.vii.1923), Sylvan Pass (22.vii.1934), Thumb Stn (16.vii.1923), W

Craigs Pass (8.viii.1918), ALM (USNM); 8.7 km W Eastgate Yellowstone NP, str., 20.vii.1989, BJS (CNC).

Recognition. This species is characterized by distinct clouding and spots on the wings, auxiliary crossvein, stump vein in the middle of R_{4+5}, and lower half of the face with blue pruinescence.

Description (wing length 3.5–4.3 mm).

Male. Face with shiny blue pruinescence on lower half, lower third flat. Arista-like stylus short and slender. Postocular setae pale.

Scutum bivittate; lower half of pleura, scutellum, and median stripe with blue pruinescence. Several inconspicuous acrostichal setae anterior to first dc; 5 dc; 1 pprn; 1 presut spal; 2 npl, lower short and slender; 1 psut spal; 1 pal; 2 sctl. Laterotergite with patch of long pale setae. Posterior margin of anepisternum with setae.

Wing (Figs. 29*g*, *h*) infuscate; clouding at base of R_4, apex of cell dm and basal cells, midlength of cell m_1, r-m, and stump vein; stigma beyond R_1 distinct and elongate; 1 long basal costal seta; erect costal spines present. Veins often undulating, R_{2+3} dipped below stigma; auxiliary crossvein between R_4 and R_{2+3} and stump vein at midlength of R_{4+5}; R_4 straight; anal vein indistinct; cell dm strongly produced apically. Halter pale.

Coxae with pale-blue pruinescence, concolorous with pleura. Femora dark brown. Forefemur with anterior preapical comb; dense white pile beneath; antero- and postero-ventral row of short stout setae on apical two-thirds. Foretibia with single row of stout setae beneath. Hind tibia with long erect dorsal setae along apical third. Fifth tarsomere lacking dorsoapical extension; empodium white laterally.

Abdominal sternites and lateral margin of tergites with blue pruinescence, remaining tergites brown. Hypandrium elongate, subtriangular. Phallus straight, distiphallus long and curved, with notch near base. Clasping cercus broad and truncate, anteroapical corner round, inner apical face with short stout setae. Surstylus subequal in length to clasping cercus, arched anteriorly, apex with distinct notch (Fig. 13*a*).

Female. Similar to male except ventral bristles of foreleg less pronounced.

Geographic distribution and seasonal occurrence. *Clinocera conjuncta* is a transcontinental species, yet widely disjunct (Map 8). The western population is fairly common, is confined to the mountain ranges, and extends from northern Mexico to Alaska. There are no records from northwestern Canada. The eastern population is restricted largely to the Appalachians and, surprisingly (at least to the author), no records were found from Ontario. A small series of specimens were collected in the highest (2000 m) spring in the east (Mount Mitchell). In the east it appears that *C. conjuncta* is confined mostly to habitats that are cooler (e.g., alpine and boreal regions) than those it occupies in the west.

Clinocera cordillerana sp. nov.

Figs. 13*b*, 29*i*; Map 9

Type material. HOLOTYPE, male labelled "Corona Pass, Colo[rado,USA]./ Boulder Co. 10,600′/ 11.VIII.1961/ J. G. Chillcott"; "HOLOTYPE/ Clinocera/ cordillerana/ Sinclair [red label]" (CNC).

Paratypes: **CANADA. Alberta:** Banff, 5–7.ix.1922, CBDG (3 ♂♂, 3 ♀♀, CNC, DEBU). **USA. Colorado:** Same data as holotype (3 ♂♂, 3 ♀♀, CNC); Lake Co., Independence Pass, 12 100′, tundra, 7.viii.1961, JGC (1 ♀, CNC). **Utah:** Utah Co., North Fk Ck, Timpanogos Tr., 21.ix.1986, RWB (1 ♂, 1 ♀, BYU). **Wyoming:** Yellowstone NP Falls, 23.vii.1923, ALM (1 ♂, 3 ♀♀, USNM).

Possible additional material (males are required for species confirmation). **CANADA. Alberta:** Laggan, Popes Peak on snow (8000′) and Lk. Agnes on str. rocks (6800′), 26, 30.viii.1925, OB (2 ♀♀, CAS). **British Columbia:** 6 km NE Revelstoke, 19.viii.1990, AB (1 ♀, CNC). **USA. Colorado:** Hinsdale Co., Sherman, Cottonwood Ck, 10.ix.1993, SF, T. Eckberg (1 ♀, CSU); Silverton, 9800′, 11.viii., F.M. Carpenter (1 ♀, MCZ).

Recognition. Males of this species have the stout setae on the forefemur confined to near midlength, the phallus with the apex oblique and not distinctly bent, the clasping cercus long and narrow near the base, and the surstylus with a subapical notch. The female wing has a series of long costal bristles and the face is clothed in bluish pruinescence.

Description (wing length 4.8–6 mm).

Male. Face with bright blue pruinescence, lower third flat. Gena and posterior region of head with tinge of blue pruinescence; vertex dark. Arista-like stylus very short and stout. Upper postocular setae pale.

Scutum bivittate. Pleura, notopleuron, and postpronotal lobe with blue pruinescence; scutellum often with tinge of pruinescence. Several acrostichals anterior to first dc; 5 dc; 1 short, dark pprn and 1 pale setula; 1 presut spal; 2 npl, upper strong, lower weak, pale, and short; 1 psut spal; 1 pal; 2 sctl. Laterotergite with patch of long pale setae; posterior margin of anepisternum lacking setulae.

Wing (Fig. 29*i*) infuscate, with clouding or spots absent; stigma beyond R_1 distinct and narrow; 2 or 3 basal costal setae with series of costal setae one-third length of basal costal setae ending near Sc; costal spines present. Veins straight; auxiliary crossvein between R_{2+3} and R_4 absent; stump vein absent; R_4 bell-shaped; anal vein distinct streak; cell dm produced apically. Halter brown.

Coxae with blue pruinescence, remaining segments dark. Forecoxa with fine pale setae. Forefemur with anterior preapical comb reduced to transverse row of 3–5 setae; dense white pile beneath; antero- and postero-ventral row of 7–9 short setae near midlength, lacking on apical and basal fourth. Foretibia with row of short stout erect setae beneath. Mid and hind femora lacking preapical setae. Hind tibia with erect dorsal seta on apical third. Fifth tarsomere lacking dorsoapical extension; empodium white laterally.

Abdominal sternites and lateral margin of tergites with blue pruinescence, remaining tergites brown. Hypandrium long and narrow, apical half-tapered. Phallus gently arched; lacking whip-like distiphallus; apex reduced to slender, pointed oblique plate. Clasping cercus long and broad apically, narrower near base; bearing peg-like spines on inner face; extending well beyond apex of surstylus. Surstylus arched with subapical notch (Fig. 13*b*).

Female. Similar to male except as follows: face with dull blue pruinescence; costa with several more long bristles; ventral bristles and setae of forefemur and foretibia lacking.

Geographic distribution and seasonal occurrence. This species is confined to the Rocky Mountains (Map 9) and is sympatric with *C. coloradensis* in Colorado. The flight period extends from July to September.

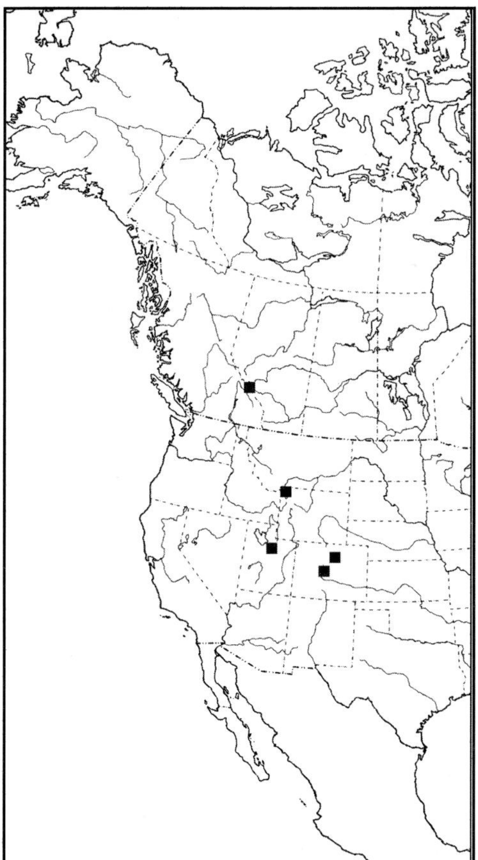

Map 9. Known distribution of *Clinocera cordillerana*.

Etymology. The specific name is from the Spanish *cordillera* ("chain of mountains"), referring to the restriction of this species to mountainous regions.

Clinocera melanderi sp. nov.

Fig. 13*c*; Map 11

Type material. HOLOTYPE, male labelled "[USA: Washington] MT. RAINIER/ Yakima Park/ 19 Aug. 1934/ A.L.Melander"; "ALMelander/ Collection/ 1961"; "HOLOTYPE/ Clinocera/ melanderi/ Sinclair [red label]" (USNM).

Paratypes: **CANADA. British Columbia:** 50 mi. SW Terrace, rock face behind waterfall, 24.vi.1960, JGC (2 ♂♂, CNC). **USA. Washington:** Mt Rainier NP, creek draining E end Yakima Pk, 1830 m, 3.x.1979, DDW (1 ♂, CAS).

Recognition. This species is characterized by a spinose lobe at the base of the clasping cercus, hypandrium long and slender, and distiphallus long and straight.

Description (wing length 3–3.5 mm).

Male. Face dark, with very faint blue pruinescence; lower third flat. Arista-like stylus short and stout. Postocular setae pale.

Scutum bivittae faint; lower half of pleura and notopleural depression with pale-blue pruinescence. Several inconspicuous acrostichal setae anterior to first dc; 5 dc; 1 pprn; 1 or 2 presut spal; 2 npl, lower shorter; 1 psut spal; 1 pal; 2 sctl. Laterotergite with patch of long pale setae. Posterior margin of anepisternum lacking setae.

Wing infuscate, with spots or clouding absent; stigma beyond R_1 faint and elongate; 1 long basal costal seta; erect costal spines present. Auxiliary crossvein between R_4 and R_{2+3} absent; R_4 bell-shaped; anal vein long and indistinct; cell dm slightly produced apically. Halter dark brown.

Coxae with pale-blue pruinescence, concolorous with pleura. Femora dark brown. Forefemur with anterior preapical comb; dense white pile beneath lacking; antero- and postero-ventral row of 7 or 8 long setae. Foretibia with single row of short stout setae beneath. Hind tibia with erect dorsal setae on apical third. Fifth tarsomere lacking dorsoapical extension; empodium dark laterally.

Abdominal sternites and lateral margin of tergites with pale-blue pruinescence, remaining tergites brown. Hypandrium long, narrow, nearly parallel-sided. Phallus gently arched medially, with long straight distiphallus. Clasping cercus broad, apical margin round, with several stout, blunt setae; inner apical surface with scattered blunt, peg-like setae; spinose lobe projecting beyond base. Surstylus narrow, extending to half length of clasping cercus; apex round (Fig. 13*c*).

Female. Unknown.

Geographic distribution and seasonal occurrence. This appears to be a rare species, confined to the Cascade Range of the Pacific Northwest (Map 11). The flight period extends from June to October.

Etymology. The specific name is a patronym in honour of A.L. Melander, collector of the holotype and provider of the framework for all subsequent studies of Nearctic Empidoidea.

Clinocera olivacea Melander

Figs. 13*d*, 29*j*; Map 10

Clinocera olivacea Melander, 1928: 233.

Clinocera olivacea: Melander, 1965: 468.

Type material examined. HOLOTYPE, female labelled "Muir Inlet/ Alaska, [USA] 6.[?].[18]99 [handwritten]"; "Harriman Expedition '99/ T. Kincaid, Collector"; "Type/ Clinocera/ olivacea/ Mel. [red label]" (USNM).

Additional material examined. Some 1345 specimens from the following states and provinces: **CANADA. Alberta:** Banff, 19.viii.1925, OB (CAS), 30.viii.1922, CBDG (CNC), Sunshine Lodge, 7500′, 9.viii.1962, K.C. Herrmann (CNC); Banff NP, Lk. Louise, Six Glaciers, 12.viii.1986, PHA (CAS), Moraine Lk, 11.viii.1986, PHA (CAS); Grissly Ck, Kananaskis Rd, 18 mi. S Hwy 1, 5600′, 17.viii.1969, EMF (CAS); Jasper NP, 100 km S Jasper, trib. of Tangle Ck, 28.vi.1989, BJS (CNC), 60 km S Jasper, Bubbly Springs, 28.vi.1989, BJS (CNC); Laggan, Lk. Agnes, 6800′, 30.viii.1925, OB (CAS); Lk. Louise, 30.vi.1925, E. Hearle (CNC); Moraine Lk., 31.vii.1935, ALM (USNM); Peter Lougheed PP, Hwy 40, 20.vii.1987, SAM (DEBU), Hwy 40, 1.4 km S Lk. Kananaskis, seep, 14.vii.1990, K.M. Fry (CNC), Hwy 40, 20.vii.1987, J. Troubridge (CNC); Plateau Mtn Ecological Res., 19.ix.2003, BJS (ZFMK); Waterton Lks. NP, 14–20.vii.1980, HJT (CNC), Rowe Ck, 24.viii.1981, PHA (CAS). **British Columbia:** Bijou Falls, Pine Pass, Hwy 97, 16.viii.1978, PHA (USNM); 40 km E Enderby, 10.vii.1992, AB (CNC); Glacier NP, 1250 m, 30.vi.1989, BJS (CNC), Hemlock Grove Tr., 11.ix.2003, BJS (ZFMK), Perly Rock Tr., Illecillewaet, 13.ix.2003, 22.ix.2005, BJS (ZFMK), 3.3 km W Rogers Pass Information Centre, roadcut seep, 1.vii.1989, BJS (CNC), 4 km W, str., 1.vii.1989, BJS (CNC); Great Divide, 30.vii.1935, ALM (USNM); Hector, 15.vii.1928, OB (CAS); Klondike Hwy 2, White Pass, 1.viii. 1996, PHA (USNM); Lisadele Lk., 58°41′N, 133°04′W, 6.viii.1960, RP (CNC); Kootenay NP, Radium Hot Springs + Hwy 93, roadcut seep, 17.ix.2005, BJS (ZFMK); Manning PP, Similkameca R., 4.viii.1986, PHA (CAS), Goodfellow Ck, 8.x.1979, DDW (CAS), Buckhorn Ck, Coldspring Cpgd, 7.x.1979, DDW (CAS), Lightning Lk. Cpgd, 5.viii.1986, PHA (CAS); Mt Revelstoke NP, 4.8 km Summit Rd, waterfall, 3.vii.1989, BJS (CNC), 6.3 mi. below summit, 8.viii.1986, PHA (CAS), Jake Lk. Tr., 6000–6300′, 14, 19.viii.1952, GJS (CNC), nr Eva Lk., on snow, 4.ix.1988, AB (CNC), Eva Lk. Tr., str. nr Miller Lk, 1900 m, 12.ix.2003, BJS (ZFMK); Penticton, Apex Mtn, 5800′,

4.viii.1957, JRV (CNC); 30 km N Penticton, Hardy Falls, Deep Ck, 3.vii.1989, BJS (CNC); Pete Lk., 57°56′N, 131°56′W, meadow, 20.viii.1960, RP (CNC); Peterson Ck, Alaska Hwy DC-445, 12.viii.1978, PHA (USNM); 13 km SE Port Renfrew, 5.vii.1991, AB (CNC); 6 km NE Revelstoke, 19.viii.1990, AB (CNC); 13 km N Revelstoke, Hwy 23, 2.vii.1989, BJS (CNC), 16 km N Revelstoke, 3.vii.1989, AB (CNC); 45 km N Revelstoke, 17.ix.1995, AB (CNC); Robson, 29.x., 6.xi.1956, 30.x.1957, 9.ix.1959, HRF (UBC); 7 km SE Rogers Pass, 23.vii.1990, AB (CNC); Silver Star Prov. Rec. Area, 22 km E Vernon, 3.vii.1989, BJS (CNC); Mt Allard, 23 mi. N Terrace, 28.vi.1960, JGC (CNC); Mt Thornhill nr Terrace, 14.vii.1960, JGC, 26.vii.1960, BH, CHM, 14.vii.–8.viii.1960, CHM (CNC); 10 mi. W Terrace, seep, 9.vi.1960, JGC (CNC); 14 mi. E Terrace,

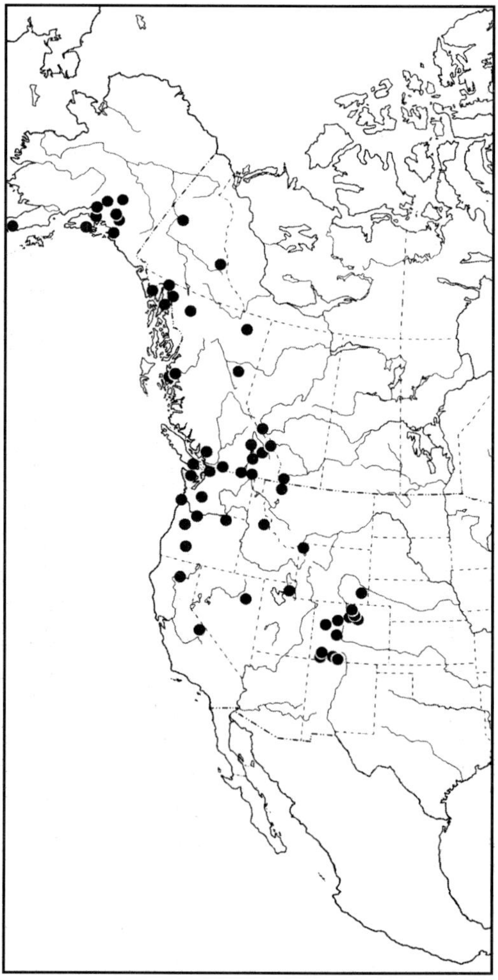

Map 10. Known distribution of *Clinocera olivacea.*

Kleanza Ck, 13.vii.1960, BH (CNC); 49 mi. W Terrace, Slickenslide Ck, 9.vii.1960, JGC (CNC); 50 mi. SW Terrace, wet cliff face, 24.vi.1960, JGC (CNC); Spahats Ck PP, Spahats Ck, 15.viii.1986, PHA (CAS); Squamish, Diamond Head Tr., 3200′, 26.viii.1953, WRMM (CNC);Yoho NP, Kicking Horse Cpgd, 29.vi.1989, BJS (CNC). **Yukon:** Dempster Hwy, km 40.8, Beason Ck, 2.vii.1980, Wood and Lafontaine (CNC), km 45, str., 13.vii.1987, SAM (DEBU); Otter Lk., 62°30′N, 130°25′W, 4000′, 17.vii.1960, J.E.H. Martin (CNC). **USA. Alaska:** Anchorage, Flattop Mtn, South Fk, 19.vii.1993, TS (BLKU); Aniakchuk NM, 56°56′N, 158°6′W, 400 m, Surprise Lk, vii–viii.1988, B. Wisseman (OSU); Denali Hwy, mi. 36, Boulder Mtn, 1305–1450 m, 14.vii.1993, OY, TS (BLKU); Fern Mine, 61°49′N, 149°14′W, 27.vii.1956, W.C.F. (WSU); Glenn Hwy, mi. 55, Moose Ck, 19.vii.1993, 22.vi.1994, TS (BLKU); Girdwood, Alyeska Hwy, 28.vii.1978, PHA (USNM); Hurricane, 18.vi.1921, JMA (USNM); Independence Mine, 4.viii.1956, W.C.F. (WSU, USNM); Kenai Peninsula, McHugh Ck, Seward–Anchorage Hwy A-15, 28.vii.1978, PHA (USNM); Paxson, One Mile Ck, 16.vii.1993, TS (BLKU); Richardson Hwy, Huddleston Falls, 3.viii.1978, PHA (CAS); Richardson Hwy, mi. 14, Horsetail Falls, 18.vii.1993, TS (BLKU), mi. 24, Blueberry Lk., 18.vii.1993, TS (BLKU), mi. 26, Thompson Pass, 18.vii.1993, TS (BLKU), mi. 29, Worthington Glacier, 17.vii.1993, TS (BLKU), mi. 198, Isabel Pass, 8.vii.1993, OY (BLKU), mi. 204, Phelan Ck, 8–27.vii.1993, TS, OY (BLKU), mi. 208, Rainbow Ridge, 8.vii.1993, TS (BLKU); Seward, 24.vii.1921, JMA (USNM); Thane, 58°47′N, 134°19′W, 29.vii.1958 (WSU); Valdez, 15 km N, 26.viii.1948, B.L. Morris (USNM). **California:** Inyo Co., Mosquito Flat, E Mono Pass, 10 000′, 17.vii.1966, PHA (USNM); Shasta Co., McArthur–Burney Falls Memorial SP, 1.viii.1970, PHA (CAS). **Colorado:** Boulder Co., Corona Pass, 10 600′, 11.viii.1961, JGC (CNC); Chaffee Co., Garfield, S Arkansas R., 2990 m, 12.viii.1973, PHA (CAS); Chaffee Co., 3.2 mi. E Monarch Pass, 17.viii.1983, PHA (CAS); Conejo Co., 3 mi. Cumbres, 16.viii.1961, F.P.J. Rindge (AMNH); Gilpin Co., 2 mi. E Rollins Pass, 3048 m, 16.viii.1973, DDW (CAS); Hinsdale Co., Cooper Ck, 8.viii.1993, SF (CSU), Cottonwood Ck, Sherman, 10.ix.1993, T. Eckberg (CSU), Redcloud Pk, Silver Ck Tr., 12 000′, 13.viii.1995, SF, AF (CSU); Hoosier Pass, 12 000′, 8.viii.1961, JGC (CNC); Lake Co., Independence Pass, 11 500′, treeline, 31.vii.1961, JGC (CNC); Larimer Co., Laramie R., W Branch, 22.viii.1993, SF (CSU), Rawah Wilderness, Twin Crater Lks., Outlet Ck, 10 800 – 11 000′, 17.ix.1995, SF, AF (CSU); Loveland Pass, 12 000′, summit, 7.viii.1961, JGC (CNC); Mesa Co., Grand Mesa, Atkinson Res., 26.viii.1986, J. Jenkins (WSU); Mt Evans, 14 000′, 25.vii.1961, JGC (CNC), 10 600′, 24.viii.1961, CHM (CNC); Mt Evans, Doolittle Ranch, 9800′, 22.viii.1961, JGC (CNC); Ouray Co., 5 km S Ouray, Uncompahgre R., 2697 m, 14.viii.1973, PHA (CAS); Red Mtn, 11 000′, 22.vi.1934, CPA (AMNH); Pitkin Co., Ashcroft, Express Ck, 12.viii.1993, SF (CSU); San Juan Co., Red Mtn Pass, 3597 m, 14.viii.1973, PHA (CAS); Shrine Pass, 12.viii.1958, FCH (USNM); Summit Co., Tenmile Ridge, 10 500′, MT, 15–17.vii.1998, SF, AF (CSU). **Idaho:** Idaho Co., Legend Lk., 9 air mi. NE

Selway Falls, 19.vii.1979, WJT (WSU). **Montana:** Glacier NP, Baring Falls, 26.vii.1981, PHA (CAS). **Nevada:** Elko Co., Ruby Mtns, 8000 – 10 000′, 28.vii.1976, J. Doyen (UCB). **Oregon:** Crater Lk., 16.ix.1934, ALM (USNM); Hermiston, 9.vii.1926, ALM (USNM); Hood River Co., Mt Hood NF, E Fk Hood R., Ash Ck, 27.vi.1978, DDW (CAS); Hood River Co., Mt Hood NF, Mitchell Ck, 1400 m, nr Sahalie Falls, 27.vi.1978, DDW (CAS); Linn Co., Monument Pk, Guard Stn, 4000′, 2.viii., EMF (CAS); Mt Hood, 29.vii.1966, FCH (USNM); Mt Hood NF, 3.5 mi. below Timberline Lodge, roadcut seep, 11.vii.1989, BJS (CNC); Viento, 1.viii.1921, ALM (USNM). **Utah:** Cache Co., Logan Cyn, Tony Ck, 27.vii.–6.viii.1987, N. Yousset (USUL). **Washington:** Clallam Co., Olympic

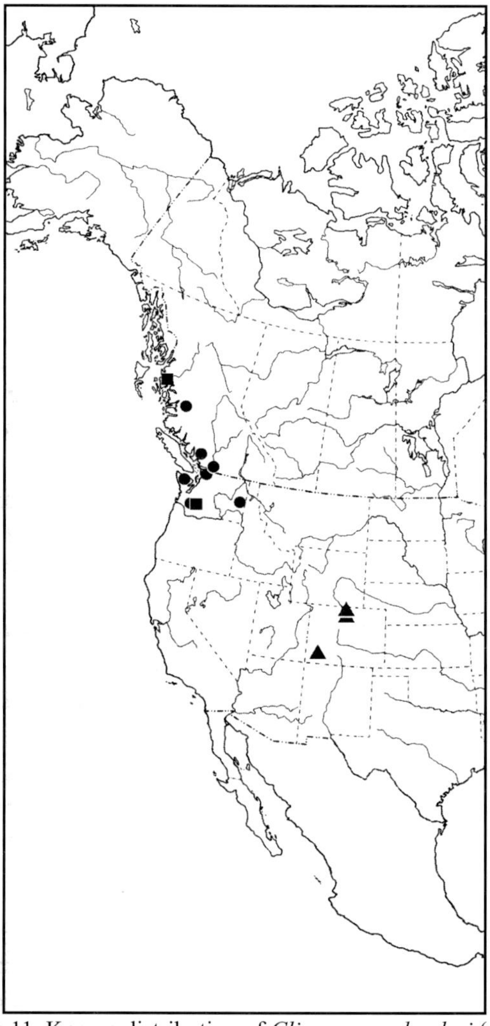

Map 11. Known distribution of *Clinocera melanderi* (■); *C. pacifica* (●); *C. sabroskyi* (▲).

NP, Sunrise Ridge, Mt Angeles, meadow, 16.viii.1973, DDW (CAS); Ilwaco, 28.vi.1925, ALM (USNM); Mt Baker, NF, Glacier Ck Rd, 1100 m, Grouse Ck, 4.x.1979, DDW (CAS), Skyline Tr., 10.viii.1925, ALM (USNM); Mt Rainier, 5500–6500′, 1.vii.1934, OB (CAS), Ind. Henry, 6.viii.1922, ALM (USNM); Mt Rainier NP, 31.vii.1966, FCH (USNM), Cowlitz Divide Tr., Stevens Cyn Rd, 3.viii.1979, PHA (CAS), E end Yakima Pk, 1830 m, 3.x.1979, DDW (CAS), Edith Ck at Paradise Pk, 1650 m, 2.viii.1977, DDW (CAS), Paradise R. nr Paradise, 20.viii.1977, NEW (WOOC), 8.vii.1989, BJS (CNC), nr Christine Falls, roadcut seep, 8.vii.1989, BJS (CNC), Comet Falls above van Trump Ck, 4500′, 13.viii., 10.ix.1977, WJT, E.J. Davis (WSU), Panther Ck, 2.x.1979, DDW (CAS), Reflection Lks., 4500–5000′, 24.viii.1973, WJT (WSU), Sunbeam Ck, Louise Lk., 1520 m, 4.viii.1977, DDW (CAS), van Trump Ck above Christine Falls, 28.viii.1979, WJT (WSU), Yakima Pk, 19.viii.1934, ALM (USNM), Alta Vista, 22.vii.1922, ALM (USNM), Burroughs Mtn, 21.viii.1934, ALM (USNM), Berkley Pk, 19.viii.1934, ALM (USNM), Frying Pan Tr., 29.viii.1934, ALM (USNM), Glacier Stn, 15.viii.1917, ALM (USNM), Mazama Ridge, 23.vii.1922, ALM (USNM), Paradise, viii.1917, viii.1921, ALM (USNM), Sluiskin, 28.vii.1922, ALM (USNM), Summerland, 29.viii.1934, ALM (CAS), van Trump Ck, 1.ix.1917, ALM (USNM), White R., 20.vii.1924, 28.viii.1934, ALM (USNM), White R., White R. Cpgd, 1220 m, 3.x.1979, DDW (CAS); Lewis Co., Mt Rainier NP, Steven Ck at Stevens Cyn Rd, 24.viii.1973, Narada Falls, Paradise R., 24.viii.1973, WJT (WSU); Pierce Co., Mt Rainier NP, 1 mi. W Sunrise Pt, 5500–6000′, 22.viii.1973, WJT (WSU); Pierce Co., Mt Rainier NP, 1 mi. above Christine Falls, 4500′, RSZ (WSU); Whatcom Co., Mt Baker NF, Bagley Ck, 850 m, 10.x.1979, DDW (CAS); Yakima Co., Chinook Pass nr summit, 5400′, 11.viii.1979, NEW (WOOC). **Wyoming:** Platte Co., light trap, 20–30.viii.1949, D.G. Denning (UWL); Yellowstone NP, 8.7 km W Eastgate, str., 20.vii.1989, BJS (CNC); Yellowstone NP, Mt Washburn, 18.viii.1962, P.J.S. (USNM); Yellowstone NP, falls, 23.vii.1923, ALM (USNM), W Craigs Pass, 7900′, 8.viii.1918, ALM (USNM), Canyon Camp, 12.viii.1918, ALM (USNM).

Recognition. This species is distinguished by the golden brown facial iridescence in both sexes, male terminalia with straight phallus and distiphallus arising ventrally, and biserial ventral row of setae on the female forefemur.

Description (wing length 4.6–5.5 mm).

Male. Face flat; entire head, including face, with golden brown iridescence. Arista-like stylus short and slender. Postocular setae pale.

Scutum bivittate; pleura (upper half of anepisternum dark) with blue pruinescence. Several short acrostichals extend to just posterior to first dc; 5 dc; 1 pprn; 1 presut spal; 2 npl, lower short; 1 psut spal; 1 pal; 2 sctl. Laterotergite with patch of long pale setae. Posterior margin of anepisternum with pale setulae.

Wing (Fig. 29*j*) hyaline, with clouding or spots absent; stigma beyond R_1 distinct, elongate and elliptical; 1 long basal costal seta; costal spines present. Veins straight; auxiliary crossvein between R_{2+3} and R_4 absent; stump vein absent; R_4 bell-shaped; anal vein faint streak; cell dm produced apically. Halter dark brown.

Coxae with pale-blue pruinescence; remaining leg segments black. Forecoxa with long pale setae. Forefemur with anterior preapical comb; antero- and postero-ventral row of short bristles, extending to apical fifth of femur; white pile beneath reduced. Foretibia with biserial row of setae beneath. Hind tibia with erect dorsal setae on apical third. Fifth tarsomere lacking dorsoapical extension; empodium light brown laterally.

Abdominal sternites and lateral margin of tergites with blue pruinescence, remaining tergites brown. Hypandrium slender, tapering on apical third. Phallus straight, with short distiphallus arising ventrally. Clasping cercus truncate, short; inner apical face with peg-like setae. Surstylus extending to apex of clasping cercus, arched; apex with shallow notch (Fig. 13*d*).

Female. Similar to male except forefemur with antero- and postero-ventral bristles longer and finer than in male.

Geographic distribution and seasonal occurrence. This species is widely distributed along the western mountain ranges, extending from Alaska to Colorado, but rare in California (Map 10). The flight period extends from June to September.

Remarks. A female specimen from British Columbia was collected while it fed on a chironomid larva.

In several instances, individuals (females) were collected as they flew slowly over parking lots in Glacier National Park (British Columbia), some distance from the nearest running-water habitats. Species of *Clinocera* are rarely observed away from streams and perhaps this occurrence is similar to that observed in other aquatic flies (e.g., simuliids), which apparently mistake the reflection from the pavement for a breeding habitat (i.e., water).

Clinocera pacifica sp. nov.

Figs. 14*a*, 30*a*; Map 11

Type material. HOLOTYPE, male labelled "[USA: Washington] Mt Rainier/ 25 Aug. 21/ Eagle Peak/ ALMelander"; "HOLOTYPE/ Clinocera/ pacifica/ Sinclair [red label]" (USNM).

Paratypes: **CANADA. British Columbia:** 14 km SW Bella Coola, 10–11.vii.1992, AB (1 ♀, CNC); Squamish, Diamond Head Tr., 4000′, 30.viii.1953, GJS (1 ♀, CNC); Manning PP, Dry Ridge Tr, small ck, 4.viii.1986, PHA (1 ♂, 1 ♀, CAS). **USA. Washington:** Clallam Co., Olympic NP, Mt Angeles, meadow below Sunrise Ridge, 1550 m, 23–24.vii.1978, DDW (4 ♂♂, 19 ♀♀, CAS); Lewis Co., Mt Rainier NP, Stevens Cyn Rd, Stevens Ck, 24.viii.1973, WJT (5 ♀♀, WSU); Mt Baker, Skyline Tr., 10.viii.1925, ALM (1 ♀,

USNM); Mt Rainier NP, above Louise Lk, Sunbeam Ck, 1520 m, 4.viii.1977, DDW (1 ♂, CAS); Mt Rainier NP, same data as holotype, 10.ix.1977 (E.J. Davis), van Trump Ck, above Christine Falls, 3900′, 28.viii.1979, WJT (18 ♂♂, 14 ♀♀, WSU); Mt Rainier NP, Whitewater Cpgd, 31.viii.1981, PHA (1 ♂, CAS); Mt Rainier NP, Christine Falls (16.viii.1917), Frying Pan Tr. (29.viii.1934), Paradise Pk (viii.1917, viii.1921), Summerland (24.viii.1924), ALM (4 ♂♂, 3 ♀♀, USNM); Pierce Co., Mt Rainier NP, 4500′, above Christine Falls, 10.ix.1977, RSZ (1 ♀, WSU); Spokane Co., Mt Spokane SP, nr Bald Knob Cpgd, 16.ix.1983, WJT (1 ♂, WSU).

Recognition. This species is most similar to *C. amarilla* on the basis of the male terminalia and pale halteres, but the legs in *C. pacifica* are dull pale yellow to dark brown.

Description (wing length 4.6–5.2 mm).

Male. Lower half of face and gena with blue pruinescence. Arista-like stylus short and slender. Postocular setae pale.

Scutum faintly bivittate. Pleura, notopleuron, and postpronotal lobe with dull bluish green pruinescence. Several short pale acrostichals anterior to first dc; 5 dc; 1 pprn; 1 presut spal; 2 npl, lower weak; 1 psut spal; 1 pal; 2 sctl. Laterotergite with patch of pale setae.

Wing (Fig. 30*a*) hyaline, infuscate along veins; stigma beyond R_1 faint, elliptical; 1 basal costal bristle. R_4 and R_5 strongly divergent; anal vein distinct; cell dm produced apically. Halter with dark shaft and pale knob.

Coxae concolorous with pleura, remaining leg segments pale to dark brown. Forefemur with antero- and postero-ventral row of short setae along entire length; preapical anterior comb present. Foretibia with antero- and antero-dorsal row of stout setae. Hind tibia with erect dorsal bristles on apical half. Basitarsis of all legs with short stout setae beneath; hind basitarsis with 3 long ventral bristles at base. Fifth tarsomere not extended; empodium white, half length of fifth tarsomere.

Abdominal sternites and lateral margin of tergites with pale-blue pruinescence, remaining tergites brown. Hypandrium slender, partially tapered on apical third. Phallus heavily sclerotized, straight; tip with dorsal hook; distiphallus short and straight, arising subapically. Clasping cercus broad, truncate, bearing inner apical patch of peg-like setae. Surstylus with round apex, subequal in length to clasping cercus (Fig. 14*a*).

Female. Similar to male except ventral setae on all leg segments reduced.

Geographic distribution and seasonal occurrence. This species is confined to the Pacific Northwest (Map 11) and the flight period extends from July to September.

Etymology. The specific name is in reference to its western distribution.

Clinocera prasinata Melander

Figs. 14*b*, 30*b*; Map 12

Clinocera prasinata Melander, 1928: 232.

Clinocera prasinata: Melander, 1965: 468.

Type material examined. HOLOTYPE, male labelled "Monolake, Cal[ifornia, USA]./ 23.vii.1911"; "TYPE/ Clinocera/ prasinata/ Mel [red label] [genitalia dissected, stored in microvial beneath specimen]" (USNM). Melander (1928) stated that the holotype was collected by J.M. Aldrich.

Additional material examined. Some 200 specimens from the following province and states: **CANADA. British Columbia:** 45 km N Revelstoke, 17.ix.1995, AB (CNC). **USA. California:** Fall Valley, 10.vii.1940, ALM (USNM); Fresno Co., Sierra NF, Huntington Lk., Rancheria Cyn, Kaiser Cyn,

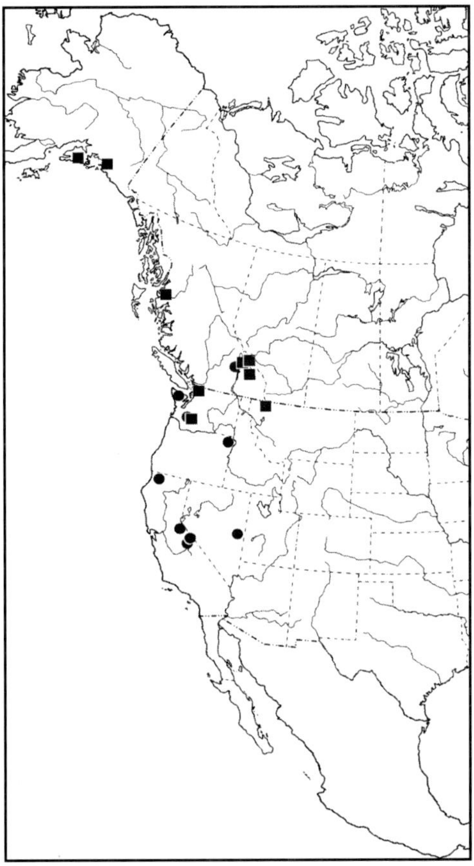

Map 12. Known distribution of *Clinocera prasinata* (●); *C. setitibia* (■).

2700 m, 23.vii.–17.viii.1984, JFM (PERC), Midge Cyn, 2200 m, 13.viii.1984, JFM (PERC), Sample Mead, 2200 m, 16.viii.1984, JFM (PERC), nr Kaiser Pass, 2700 m, 31.vii.–2.viii.1979, DDW (CAS), Sierra NF, E and W Fks Camp Ck, nr Forebay Lk., 31.vii.1979, DDW (CAS); Mono Co., White Mtns, 3 mi. N Inyolo, N Fk Crooked R., 20.viii.1963, HBL (CAS); Siskiyou Co., Klamath NF, S Fk Salmon R., Knownothing Ck, 22.vii.1980, DDW (CAS); Tuolumne Co., Deadman Ck, 12.vii.1979, DDW (CAS). **Nevada:** White Pine Co., Great Basin NP, 3–18.vii.2001, JFM, Brammer (PERC). **Oregon:** Wallawa Co., 4 mi. SE Aneroid Lk., Lk. Wallowa, 7500′, Wallowa Mtns, 9.viii.1991, WJT (WSU). **Washington:** Clallam Co., Olympic NP, 650 m, 24.vii.1978, DDW (CAS); Mt Rainier, Berkeley Pk, 19.viii.1934, ALM, (USNM), Burroughs Mtn, 21.viii.1934, ALM (USNM), Paradise, viii.1921, ALM (USNM), Yakima Pk, 19.viii.1934, ALM (USNM); Mt Rainier NP, Paradise Pk, Edith Ck, 11.x.1979, DDW (CAS), E end Yakima Pk, ck, 2.x.1979, DDW (CAS).

Recognition. This species is distinguished by the following combination of characters: large size, unmarked wings; male face with blue pruinescence; phallus long, slender, gently curved toward apex. Females are distinguished from *C. olivacea* by the blue pruinescent notopleuron.

Description (wing length 4–4.5 mm).

Male. Three-quarters of face and gena with blue pruinescence; region just below antennae and vertex brown; face flat. Arista-like stylus short and slender. Postocular setae pale.

Scutum bivittate; pleura (upper half of anepisternum dark) and notopleuron with dull blue pruinescence. Short acrostichals anterior to first dc; 5 dc; 1 pprn; 1 presut spal; 2 npl, lower weak and short; 1 psut spal; 1 pal; 2 sctl. Laterotergite with patch of long pale setae. Posterior margin of anepisternum with pale setulae. Halter brown.

Wing (Fig. 30*b*) infuscate, with clouding or spots absent; stigma beyond R_1 distinct, elongate, and elliptical; 1 long basal costal seta; costal spines present. Veins straight; auxiliary crossvein between R_{2+3} and R_4 absent; stump vein absent; R_4 bell-shaped; anal vein faint streak; cell dm produced apically.

Coxae with pale-blue pruinescence; remaining leg segments dark. Forecoxa with short pale setae. Forefemur with anterior preapical comb reduced to 2 or 3 bristles; antero- and postero-ventral row of short bristles, extending length of femur; white pile beneath reduced. Foretibia with row of setae beneath. Hind tibia with erect dorsal setae on apical half. Fifth tarsomere lacking dorsoapical extension; empodium dark brown laterally.

Abdominal sternites and lateral margin of tergites with blue pruinescence, remaining tergites brown. Hypandrium narrow and elongate, tapering on apical half. Phallus long, straight, and slender, arched apically to tip of clasping cercus; distiphallus attached abruptly, arising at right angles to shaft. Clasping cercus truncate, with inner apical patch of peg-like setae. Surstylus extending beyond clasping cercus, arched anteriorly (Fig. 14*b*).

Female. Similar to male except as follows: face brown; forefemur and foretibia lacking row of bristles beneath.

Geographic distribution and seasonal occurrence. *Clinocera prasinata* is known only from a single locality in British Columbia, but appears to be more common in the Sierra Nevada National Forest region south of Lake Tahoe (Map 12). The flight period extends from July to October.

Clinocera rainiericola sp. nov.

Fig. 14*c*; Map 17

Type material. HOLOTYPE, male labelled "[USA: Washington] Mt. Rainier/ White River/ 20 July [19]24/ ALMelander"; "HOLOTYPE/ Clinocera/ rainiericola/ Sinclair [red label]" (USNM).

Paratypes: **USA. Washington:** Mt Rainier, Paradise Pk, viii.1917, ALM (1 ♂, USNM).

Recognition. This species is distinguished by its broad clasping cercus and stout phallus with hooked tip (distiphallus absent).

Description (wing length 4.5 mm).

Male. Face brown, lower third flat. Arista-like stylus short and stout. Postocular setae pale.

Scutum bivittate; lower half of pleura with pale-blue pruinescence. Several inconspicuous acrostichal setae anterior to first dc; 5 dc; 1 pprn; 1 presut spal; 2 npl, lower pale; 1 psut spal; 1 pal; 2 sctl. Laterotergite with patch of long pale setae. Posterior margin of anepisternum with pale setae.

Wing infuscate, with spots or clouding absent; stigma beyond R_1 faint and elongate; 1 long basal costal seta; erect costal spines present. Auxiliary crossvein between R_4 and R_{2+3}; R_4 bell-shaped; anal vein short and distinct; cell dm slightly produced apically. Halter dark brown.

Coxae with pale-blue pruinescence, concolorous with pleura. Femora dark brown. Forefemur with anterior preapical comb; dense white pile beneath lacking; antero- and postero-ventral row of short, sharp setae along entire length, except apical fifth, basal setae longer and pale. Foretibia with biserial row of stout setae beneath. Hind tibia with erect dorsal setae on apical third. Fifth tarsomere lacking dorsoapical extension; empodium pale brown laterally.

Abdominal sternites and lateral margin of tergites with blue pruinescence, remaining tergites brown. Hypandrium short, subtriangular. Phallus stout, gently arched apically, with hooked apex; distiphallus absent. Clasping cercus broad, apical margin truncate; inner surface with scattered blunt, peg-like setae. Surstylus narrow, extending to near apex of clasping cercus, with round apex (Fig. 14*c*).

Female. Unknown.

Geographic distribution and seasonal occurrence. This species is known only from two specimens collected in July and August on Mount Rainier (Map 17).

Etymology. The specific name is derived from the type locality, Mount Rainier. This volcanic peak reaches 4392 m and many species of Clinocerinae have been collected on its slopes.

Clinocera sabroskyi sp. nov.

Fig. 14*d*; Map 11

Type material. HOLOTYPE, male labelled "Cameron Pass,/ Colo[rado, USA]. Aug.19–22/ 1940. Sabrosky"; "HOLOTYPE/ Clinocera/ sabroskyi/ Sinclair [red label]" (USNM).

Paratypes: **USA. Colorado:** Same data as holotype (1 ♂, 6 ♀♀, USNM); Latimer Co., 17.ix.1995, Rawah Wilderness, outlet ck, Twin Crater lks., 10 800 – 11 000′, SF, AF (17 ♂♂, 8 ♀♀, CSU); Hinsdale Co., Redcloud Pk, 12 000′, Silver Ck Tr., 13.viii.1995, SF, AF (5 ♂♂, 7 ♀♀, CSU).

Additional material examined. USA. Colorado: San Juan Co, nr Silverton, Vestal Ck, 10 000 – 12 000′, 27.vii.1985, J. and J. Jenkins (1 ♀, WSU).

Recognition. This species is very similar to *C. olivacea*, but is identified by its blue facial pruinescence and distiphallus arising dorsally from the phallus.

Description (wing length 4.1–5.6 mm).

Male. Face flat; lower two-thirds with blue pruinescence. Arista-like stylus short and stout. Postocular setae pale, sometimes upper 1 or 2 black.

Scutum bivittate; pleura and notopleuron with blue pruinescence. Several short acrostichals extend to first dc; 5 dc; 1 pprn; 1 presut spal; 2 npl, lower short; 1 psut spal; 1 pal; 2 sctl. Laterotergite with patch of long pale setae. Posterior margin of anepisternum with pale setulae.

Wing hyaline, with clouding or spots absent; stigma beyond R_1 distinct, elongate and elliptical; 1 long basal costal seta; costal spines present. Veins straight; auxiliary crossvein between R_{2+3} and R_4 absent; stump vein absent; R_4 bell-shaped; anal vein faint streak; cell dm produced apically. Halter dark brown.

Coxae with pale-blue pruinescence; remaining leg segments black. Forecoxa with long pale setae. Forefemur with anterior preapical comb; antero- and postero-ventral row of short bristles, extending to apical fifth of femur; white pile beneath reduced. Foretibia with biserial row of setae beneath. Hind tibia with erect dorsal setae on apical third. Fifth tarsomere lacking dorsoapical extension; empodium light brown laterally.

Abdominal sternites and lateral margin of tergites with blue pruinescence, remaining tergites brown. Hypandrium slender, tapering on apical third. Phallus straight, with short distiphallus arising dorsally. Clasping cercus truncate, short; inner apical face with peg-like setae. Surstylus extended slightly beyond apex of clasping cercus, arched; apex with shallow notch (Fig. 14*d*).

Female. Similar to male except forefemur with antero- and postero-ventral bristles finer than in male

Geographic distribution and seasonal occurrence. This species is currently known only from Colorado (Map 11).

Etymology. The specific name is a patronym in memory of C.W. Sabrosky, a kind and dedicated student of dipterology, who collected the first specimens of this species.

Clinocera setitibia sp. nov.

Figs. 15*a*, 30*c*; Map 12

Type material. HOLOTYPE; male labelled "[USA: Washington] MT RAINIER/ Yakima Park/ 19 Aug. 1934/ A.L. Melander"; "ALMelander/ Collection/ 1961"; "HOLOTYPE/ Clinocera/ setitibia/ Sinclair [red label]" (USNM).

Paratypes: **CANADA. Alberta:** Plateau Mtn Ecological Res., 2200 m, 19.ix.2003, BJS (1 ♂, 1 ♀, ZFMK). **British Columbia:** Mt Revelstoke NP, Eva Lk. Tr., str. nr Miller Lk., 1900 m, 12.ix.2003, BJS (1 ♂, ZFMK); 7 km SE Rogers Pass, 23.vii.1990, AB (1 ♂, 1 ♀, CNC); 6 km E Revelstoke, 23.vii., 19.viii.1990, AB (3 ♂♂, 2 ♀♀, CNC); 45 km N Revelstoke, 17.ix.1995, AB (1 ♀, CNC); 49 mi. W Terrace, Slickenslide Ck, 9.vii.1960, JGC (1 ♂, 1 ♀, CNC). **USA. Alaska:** Seward, 24.vii.1921, JMA (1 ♂, USNM); 5 mi. NW Valdez, 24.viii.1948, B.L. Morris (1 ♀, USNM). **Montana:** Glacier NP, 1 mi. NW Logan Pass, 6500′, 18.viii.1969, EMF (1 ♂, CAS). **Washington:** Same data as holotype (2 ♂♂, 3 ♀♀, USNM); Mt Baker NF, Grouse Ck, 1100 m, 4.x.1979, DDW (1 ♂, CAS).

Additional material examined. USA. Alaska: Keystone Cyn opposite Bridal Veil Falls, Richardson Hwy, v-13.8, 3.viii.1978, PHA (1 ♂, 2 ♀♀, USNM).

Recognition. Males of this species are unique among those of Nearctic *Clinocera* because of the very long erect ventral setae of the foretibiae. Females are more difficult to distinguish from those of other species, but the lower half of the face is clothed in blue pruinescence, unlike that of *C. borkenti* and *C. arnaudi*.

Description (wing length 4.3–6 mm).

Male. Entire face with bright-blue pruinescence, lower third flat. Gena and posterior region of head with tinge of blue pruinescence; vertex dark. Arista-like stylus very short and stout. Upper postocular setae dark and stout.

Scutum bivittate. Pleura, notopleuron, and postpronotal lobe with blue pruinescence. Several acrostichals anterior to first dc; 5 or 6 dc; 1 short dark pprn;

1 presut spal; 2 npl, upper strong, lower weak, pale, and short; 1 psut spal; 1 pal; 2 sctl. Laterotergite with patch of long pale setae; posterior margin of anepisternum lacking setulae.

Wing (Fig. 30*c*) infuscate, with clouding or spots absent; stigma beyond R_1 distinct and narrow; 2 basal costal setae, with series of costal slender, long setae proximal to junction of Sc and C; costal spines present. Veins straight; auxiliary crossvein between R_{2+3} and R_4 absent; stump vein absent; R_4 bell-shaped; anal vein distinct streak; cell dm produced apically. Halter brown.

Coxae with blue pruinescence, remaining segments dark. Forecoxa with fine pale setae. Forefemur with anterior preapical comb reduced to transverse row of 3–5 setae; dense white pile beneath; antero- and postero-ventral row of 5–8 short stout setae beyond midlength, proximal setae slender and pale. Foretibia with row of long stout erect setae beneath, subequal to width of tibia. Mid and hind femora lacking preapical setae. Hind femur with fine erect setae above and beneath. Hind tibia with erect dorsal seta on apical third. Fifth tarsomere lacking dorsoapical extension; empodium white laterally.

Abdominal sternites and lateral margin of tergites with blue pruinescence, remaining tergites brown. Hypandrium long and narrow, apical half-tapered. Phallus nearly straight, arched apically, extended slightly beyond surstylus; distiphallus absent; apex of phallus split into pair of slender sclerites, separated by membrane. Clasping cercus long, with expanded apical half, narrower near base; bearing peg-like spines on inner face. Surstylus arched toward cercus, arched apically, lacking apical notch; much shorter than clasping cercus (Fig. 15*a*).

Female. Similar to male except as follows: pale-blue pruinescence normally confined to lower half of face; lacking ventral setae on foretibia and forefemur.

Geographic distribution and seasonal occurrence. *Clinocera setitibia* is confined to the Pacific Northwest, ranging from southern Alaska to Washington and Montana (Map 12). This species is somewhat infrequently collected and the localities are fairly disjunct in distribution. The flight period extends from July to October.

Etymology. The specific name is from the Latin *seta* ("bristle"), referring to the row of long erect ventral setae on the male foretibia.

Clinocera subtrunca sp. nov.

Figs. 15*b*, 30*d*; Map 13

Type material. HOLOTYPE, male labelled "CAN[ada]: B.C.20km./ N. Squamish/ 30.VII.1990/ A. Borkent/ CD 1240"; "HOLOTYPE/ Clinocera/ subtrunca/ Sinclair [red label]" (CNC).

Paratypes: **CANADA. Alberta:** Banff, 19.vii.1922, CBDG (1 ♀, USNM);

Laggan, 16.vii.1928, OB (1 ♂, CAS); Morraine Lk., 31.vii.1935, ALM (1 ♂, USNM). **British Columbia:** Same data as holotype (2 ♂♂, 4 ♀♀, CNC); 29 km SE Hope, 13.vii.1994, AB (2 ♂♂, 3 ♀♀, CNC); Manning PP, Lightning Lk. Cpgd, 5.viii.1986, PHA (4 ♂♂, 4 ♀♀, CAS); Mt Revelstoke NP, Jade Lk. Tr., 6500′, 19.viii.1952, GJS (9 ♂♂, 10 ♀♀, CNC); Mt Revelstoke, 6000′, 14.viii.1952, GJS (2 ♂♂, 2 ♀♀, CNC); Mt Revelstoke NP, Mt Harry, 11.viii.1952, GJS (1 ♀, CNC); Mt Revelstoke NP, 6.3 mi. below summit, 8.viii.1986, PHA (1 ♀, CAS); 6 km NE Revelstoke, 19.viii.1990, AB (4 ♂♂, 7 ♀♀, CNC); 45 km N Revelstoke, 17.ix.1995, AB (2 ♂♂, 1 ♀, CNC); 7 km SE Rogers Pass, 23.vii.1990, AB (3 ♂♂, 4 ♀♀, CNC); Penticton, Apex Mtn, 5800′, JRV (2 ♂♂, 1 ♀, CNC); Squamish, Diamond Head Tr., 4000′, 11–17.viii.1953, WRMM (2 ♀♀, CNC); Yoho NP, Kicking Horse Cpgd, 29.vi.1989, BJS (1 ♂, 2 ♀♀, CNC).

Additional material examined. CANADA. British Columbia: 56 km SE Hope, 14–15.ix.1990, AB (1 ♀, CNC). **USA. Idaho:** Idaho Co., Legend Lk., 9 air mi. NE Selway Falls, 6800′, 19.vii.1979, WJT (3 ♂♂, 2 ♀♀, WSU); S Lone Lk., 13 air mi. NNE Selway Falls, 6480′, 17.vii.1979, WJT (1 ♂, 4 ♀♀, WSU). **Washington:** Clallam Co., Olympic NP, Mt Angeles, 550 m, meadow, 22–24.vii.1978, DDW (37 ♂♂, 35 ♀♀, CAS); Ilwaco, vii.1917, 27.viii.1917, ALM (2 ♀♀, USNM); Mt Rainier, Berkeley Pk (23.viii.1934), Burroughs Mtn (21.viii.1934), Eagle Peak (19.vii.1922), Ind. Henry (2.viii.1922), Paradise Pk (viii.1917), Sluiskin (28.vii.1922), Summerland (29.viii.1934), van Trump Ck (1.ix.1917, 22.vii.1922), White R. (20.vii.1924), Yakima Pk (22.vii.1924, 19.viii.1934), ALM (38 ♂♂, 33 ♀♀, USNM); Mt Rainier NP, Edith Ck (2.viii.1977), Louise Lk., 1520 m (4.viii.1977), West Side Rd, Fish Ck (1.viii.1977), DDW (14 ♂♂, 13 ♀♀, CAS); Mt Rainier NP, Comet Falls (13.viii.1977), Narada Falls, Paradise R. (24.viii.1973), van Trump Ck, above Christine Falls (28.viii.1979), WJT (6 ♂♂, 6 ♀♀, WSU); Mt Rainier NP, Frog Pond, Paradise (20.vii.1977), Ghost Lk. (8.viii.1979), RSZ and V.L. Zack (1 ♂, 1 ♀, WSU); Mt Rainier NP, Paradise R., 8.vii.1989, BJS (1 ♀, CNC); Mt Rainier NP, Paradise R., 20.viii.1977, NEW (2 ♂♂, 5 ♀♀, WOOC); Narada Falls, 16.viii.1917, ALM (1 ♀, USNM); Pierce Co., Mt Rainier NP, Cayuse Pass, 4700′, wet meadow, 8.viii.1979, RSZ (1 ♀, WSU); Whatcom Co., Heather Meadows, above Mt Baker Ski Lodge, 11.viii.1974, WJT (2 ♀♀, WSU); Whatcom Co., Mt Baker Tr., 29.viii.1973, A.B. Gurney (1 ♂, 1 ♀, USNM). **Wyoming:** Yellowstone NP, W Craigs Pass, 7900′, 8.viii.1918, ALM (2 ♂♂, USNM); 8.7 km W Eastgate, Yellowstone NP, 20.vii.1989, BJS (2 ♂♂, 1 ♀, CNC).

Recognition. The species is distinguished by the brown face with iridescence in both sexes, relatively narrow trunate clasping cercus with stout inner setae, surstylus with notch, wings lacking markings and normally with an auxiliary crossvein.

Description (wing length 4–4.5 mm).

Male. Face golden brown, concolorous with remaining head, lacking blue pruinescence; lower third of face flat. Arista-like stylus short and slender. Postocular setae pale.

Scutum bivittate; pleura with golden iridescence. Several inconspicuous acrostichal setae extending beyond first dc; 5 dc; 1 pprn; 1 presut spal; 2 npl, lower slender; 1 psut psut spal; 1 pal; 2 sctl. Laterotergite with patch of long pale setae. Posterior margin of anepisternum with setulae.

Wing (Fig. 30*d*) infuscate; with clouding or spots absent; stigma beyond R_1 distinct and elongate; 1 long basal costal seta; erect costal spines present. Veins straight; auxiliary crossvein present between R_4 and R_{2+3}, occasionally absent or only present on one wing; stump vein absent; R_{2+3} dipped below stigma; R_4 bell-shaped; anal vein an elongate streak; cell dm strongly produced apically. Halter brown.

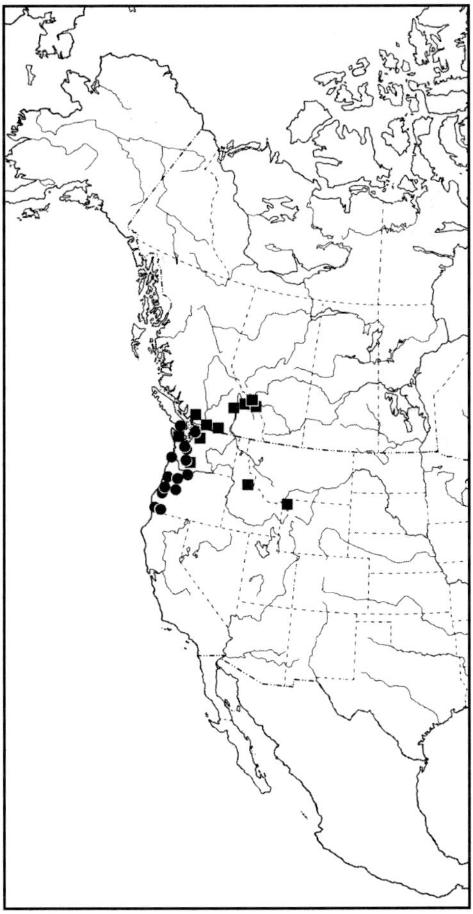

Map 13. Known distribution of *Clinocera subtrunca* (■); *C. trunca* (●).

Coxae with long golden setae, pale bluish pruinescence, concolorous with pleura. Femora dark to reddish brown. Forefemur with anterior preapical comb; dense white pile beneath; antero- and postero-ventral row of short stout setae extending to apical fifth. Foretibia with biserial rows of stout setae beneath. Hind tibia with long erect dorsal setae along apical third. Fifth tarsomere lacking dorsoapical extension; empodium brown laterally.

Abdominal sternites and lateral margin of tergites with blue pruinescence, remaining tergites brown. Hypandrium short, subtriangular. Phallus straight, with roughened dorsal margin; distiphallus long and curved, lacking notch near base; base of distiphallus lacking expansion. Clasping cercus truncate, anteroapical corner round, inner apical face and dorsal margin with short stout setae. Surstylus short, not extending to apex of clasping cercus, apex with distinct notch (Fig. 15*b*).

Female. Similar to male except ventral bristles of foretibia and forefemur present but less pronounced.

Geographic distribution and seasonal occurrence. This species is present farther east than *C. trunca*, with populations known from southern British Columbia and the northern Rocky Mountains (Map 13). The distributions of the two species overlap in Washington State. The flight period ranges from late June to September.

There is an odd Colorado population (Summit County, Tenmile Range) that has the facial coloration of *C. trunca* but the terminalia are similar to those of *C. subtrunca*. Additional collections from this region and neighbouring zones are required to assess the status of this population.

Etymology. The specific name is derived from the Latin *sub* ("from", "somewhat") and *trunca*, alluding to the structural similarity between this species and *C. trunca*.

Clinocera trunca Melander

Figs. 15*c*, 30*e*; Map 13

Clinocera trunca Melander, 1928: 230.

Clinocera trunca: Melander, 1965: 468.

Type material examined. HOLOTYPE, female labelled "Dewatto, Wash[ington, USA]/ June 7, 1906"; "Clinocera/ trunca/ TYPE Mel [red label] [right wing missing]" (USNM).

Paratypes: **USA. Washington:** Friday Harbor (1 ♂, 1 ?, USNM); Olga (2 ♀♀, USNM).

Additional material examined. Some 125 specimens from the following province and states: **CANADA. British Columbia:** 13 km SE Port Renfrew, 5.vii.1991, AB (CNC). **USA. Oregon:** Benton Co., Alsea Falls Cpgd, 21.vi.1977, DDW (CAS), Mary's Peak, summit, 1.vii.1971, GCS (USNM), Mary's Peak, Hwy

34, 4097′, 16.vii.1968, BVP (CNC), Mary's Peak, 4000′, vii.1981, DEH (CSU), Mary's Peak, waterfall nr summit, 20.vi.1977, DDW (CAS), Siuslaw NF, Mary's Peak, 14.vii.1989, BJS (CNC); Curry Co., Humbug Mtn SP, 17.vii.1989, BJS (CNC), Siskiyou NF, 24 km E Pt Orford, Elk R. Rd, 17.vii.1989, BJS (CNC); Dead Man Pass, 30.vii.1966, FCH (USNM); Hood R. Co., Starvation Ck SP, 10.vii.1989, BJS (CNC); Josephine Co., 8.5 km W Galice, 18.vii.1989, BJS (CNC); Lane Co., 8 mi. SE Cougar Dam, 880 m, Grasshopper Ck, 29.vi.1982, GWC (OSU); Lincoln Co., Siuslaw NF, Cape Perpetua, seep, 14.vii.1989, BJS (CNC); Linn Co., Monument Pk, Guard Stn, 4000′, 17.viii.1969, EMF (CAS); Marion Co., Silver Falls SP, behind South Falls, 12.vii.1989, BJS (CNC); Multnomah Co., Troutdale, 23.v.1963, L.S. Miller, FCH (USNM); Neskowin, 10–17.viii.1948, MTJ (WSU). **Washington:** Ilwaco, 27.viii.1917, 28.vi.1925, ALM (USNM); Lewis Co., 23 mi. W Chehalis, Rainbow Falls SP, 300′, 13–14.vii.1977, WJT (WSU); Mt Rainier, Burroughs Mt (21.viii.1934), Paradise Pk (viii.1917), Summerland (29.vii.1934), ALM (USNM); Nasel, 2.vii.1925, ALM (USNM); Seaview, 27.vi.1925, A. Spuler (WSU); Tacoma, 27.v.1917, ALM (USNM).

Recognition. This species is recognized by the absence of wing markings, R_{2+3} normally with an auxiliary crossvein to R_4, base of distiphallus with a sickle-shaped expansion, and notch on the distiphallus absent.

Description (wing length 3.4–4 mm).

Male. Face with shiny blue pruinescence on lower half, lower third flat. Arista-like stylus short and slender. Lower postocular setae pale, upper ones dark and stout.

Scutum bivittate; pleura with pale-blue pruinescence, more pronounced on lower half. Several inconspicuous acrostichal setulae extending beyond first dc; 5 dc; 1 pprn; 1 presut spal; 2 npl, lower short and slender; 1 psut spal; 1 pal; 2 sctl. Laterotergite with patch of long pale setae. Posterior margin of anepisternum with setulae.

Wing (Fig. 30*e*) infuscate, with clouding or spots absent; stigma beyond R_1 inconspicuous and elongate; 1 long basal costal setae; erect costal spines present. Veins straight; auxiliary crossvein present between R_4 and R_{2+3}, occasionally absent on one wing; stump vein absent; R_{2+3} dipped below stigma; R_4 straight; anal vein indistinct; cell dm strongly produced apically. Halter light brown.

Coxae with pale-blue pruinescence, concolorous with pleura. Femora dark to reddish brown. Forefemur with anterior preapical comb; dense white pile beneath; antero- and postero-ventral row of short stout setae on apical two-thirds. Foretibia with single row of stout setae beneath. Hind tibia with long erect dorsal setae along apical third. Fifth tarsomere lacking dorsoapical extension; empodium white laterally.

Abdominal sternites and lateral margin of tergites with blue pruinescence, remaining tergites brown. Hypandrium slender, strongly tapered on apical fourth. Phallus straight, distiphallus long and curved, lacking notch near base;

base of distiphallus with sickle-shaped expansion. Clasping cercus broad and truncate, anteroapical corner round, inner apical face with short stout setae. Surstylus extending beyond clasping cercus, arched anteriorly, apex complete (Fig. 15*c*).

Female. Similar to male except ventral bristles of foretibia less pronounced.

Geographic distribution and seasonal occurrence. See under *C. subtrunca*. This species is known only from the states of Washington and Oregon (Map 13). The flight period extends from May to August.

Remarks. There are only two submarginal cells present in the holotype (i.e., auxiliary crossvein absent between R_4 and R_{2+3}). This was noted by Melander (1928) and the variability of this vein is evident, with the crossvein often absent from one wing.

Clinocera fuscipennis group

Diagnosis

This species-group is recognized by an indistinct to large facial tubercle, 2 postsutual supra-alar bristles, and phallus slender with long flexible distiphallus.

Remarks

Clinocera fuscipennis is one of the most widespread and commonly collected species in North America, and it is also highly variable in size, leg coloration, and thoracic pruinescence pattern.

The flight period of species related to *C. gubernans* (e.g., *C. guttipennis, C. schlinkerti*) is primarily confined to the cooler months, with teneral specimens frequently observed in early autumn in mountainous regions (e.g., Rocky Mountains west of Calgary). These closely related species are characterized by a distinct facial tubercle, reduced costal spines, dense setae on the forecoxae, fifth tarsomere with a short dorsoapical extension, and proepisternum with short lateral setae.

The monophyly of this group is supported on the basis of the following apomorphies: face with a ventral tubercle and nipple-like apical projection of the clasping cercus.

Distribution

Although this species-group is largely confined to North America, specimens of an undescribed species belonging to the *C. fuscipennis* group have been identified from Hokkaido (Japan) and Korea.

Included species

This species-group was recognized by Sinclair (1995) and includes the following species:

NE *Clinocera bulbosa* **sp. nov.**
NE *Clinocera disjuncta* **sp. nov.**
NE *Clinocera femorata* **sp. nov.**
NE *Clinocera fuscipennis* Loew, 1876: 324.
NE *Clinocera gubernans* (Melander, 1928: 234) (*Wiedemannia*).
NE *Clinocera guttipennis* **sp. nov.**
NE *Clinocera lecta* Melander, 1902: 243.
NE *Clinocera schlinkerti* **sp. nov.**
NE *Clinocera stigma* **sp. nov.**
NE *Clinocera wallowa* **sp. nov.**

Key to species of the *Clinocera fuscipennis* group

1 Fifth tarsomere lacking dorsoapical extension; facial swelling variable . . . 2
— Fifth tarsomere with dorsoapical extension (Fig. 9*b*) (very short in *C. wallowa*); facial swelling distinct 6

2 Wing with distinct circular spots (Fig. 31*b*) (western). . . . . **C. lecta** Melander
— Wing hyaline, spots or cloudy regions absent around veins (Fig. 30*g*) 3

3 Forefemur swollen, especially at base, with long basal setae, some longer than width of femur; forefemur with anteroventral row of closely spaced short stout setae (western) . . . . **C. femorata** sp. nov.
— Forefemur slender, not especially swollen at base, lacking basal setae; forefemur with anteroventral row of widely spaced short setae present or absent 4

4 Stigma broad, conspicuous, often dark with proximal margin truncate (Fig. 31d); if faint and narrow proximately, then face with small distinct tubercle above palpus and biserial row of stout ventral setae of male forefemur absent; clasping cercus pointed apically (Fig. 18*a*). . . . . **C. stigma** sp. nov.
— Stigma normally faint, elliptical, with proximal margin pointed; face lacking distinct tubercle above palpus; male forefemur with biserial row of ventral setae; clasping cercus rounded apically (Figs. 16*a*, *c*). 5

5 Wing with cloudy regions around veins, especially crossveins (Fig. 30*g*); R_{2+3} often distinctly dipped beneath stigma; scutellum with blue pruinescence; legs dark; scutum bivittate (northern). . . . . **C. disjuncta** sp. nov.
— Wing hyaline to infuscate with cloudy regions absent around veins (Fig. 30*i*); R_{2+3} gradually arched, not dipped beneath stigma; scutellum usually dark, rarely with blue pruinescence; legs dark to reddish; scutum bivittate in some western populations, especially in pale-legged forms (widespread) . . . . **C. fuscipennis** Loew

6 Scutum with 5 dorsocentral bristles, lacking interspersed shorter setae . . 7
— Scutum with 5 dorsocentral bristles with interspersed long and short setae 8

7 Wing spots usually distinct and circular (Fig. 31*a*); clasping cercus truncate (Fig. 17*b*) (western) . . . . **C. guttipennis** sp. nov.
— Wing lacking distinct spots (Fig. 31*e*); clasping cercus tapered apically (Fig. 18*b*) (western) . . . . **C. wallowa** sp. nov.

8 Lateral setae on proepisternum stout and dark; wing lacking distinct spots (Fig. 31*c*); clasping cercus widely expanded apically (Fig. 17*d*) (western) . . . . **C. schlinkerti** sp. nov.
— Lateral setae on proepisternum slender and pale; wing with or without spots (Figs. 30*f*, *j*); clasping cercus expanded apically or slender (Figs. 15*d*, 17*a*) 9

9 Wing spots indistinct, reduced to clouding along crossveins or lacking (Fig. 30*f*); distiphallus relatively short, not extending beyond base of surstylus (Fig. 15*d*); clasping cercus narrow subapically (western) . . . . **C. bulbosa** sp. nov.
— Wing with distinct circular spots (Fig. 30*j*); distiphallus long, extending well beyond base of surstylus (Fig. 17*a*); clasping cercus expanded subapically (western) . . . . **C. gubernans** (Melander)

Clinocera bulbosa sp. nov.

Figs. 15*d*, 30*f*; Map 14

Type material. HOLOTYPE, male labelled "USA: ID:Cassia Co./ Howell Ck.Bennett/ Springs Cmpg./ 22.IV.1985/ Baumann & Nelson"; "HOLOTYPE/ Clinocera/ bulbosa/ Sinclair [red label]" (BYU).

Paratypes: **USA. Colorado:** Hinsdale Co., 10.ix.1993, Cottonwood Ck at Sherman, SF, T. Eckberg (1 ♂, 1 ♀, CSU). **Idaho:** Same data as holotype (1 ♀, BYU).

Recognition. This species is most similar to *C. gubernans*, from which it differs in having indistinct wing spots and more numerous and stout ventrobasal setae on the forefemur.

Description (wing length 3.6–4.5 mm).

Male. Large dark tubercle on lower half of face, lateral margins with blue pruinescence; remainder of face and head with blue pruinescence, except dark ocellar triangle. Arista-like stylus short and stout. Postocular setae dark.

Scutum with narrow faint pale bivittae. Pleura (upper half of anepisternum dark), notopleuron, postpronotal lobe, and scutellum with faint blue pruinescence; posterior margin of pprn lobe with orange–brown spot. Several short acrostichals anterior to first dc; 5 dc, 1 or 2 short setae sometimes interspersed; 1 pprn; 1 presut spal and several short setae; 2 npl; 2 psut spal and several short setae; 1 pal; 2 sctl. Laterotergite with patch of short pale setae. Proepisternum with several pale lateral setae. Posterior margin of anepisternum lacking setulae.

Wing (Fig. 30*f*) infuscate; faint clouding at base of R_4, apical margin of cell dm, r-m crossvein and base of R_{4+5}; stigma beyond R_1 indistinct and slender; 1 short basal costal seta; costal spines lacking. Veins straight; auxiliary crossvein between R_4 and R_{2+3} absent; stump vein absent; R_4 bell-shaped; anal vein an elongate streak; cell dm truncate. Halter dark, except pale brown at base of stalk.

Coxae with pale-blue pruinescence, remaining leg segments dark. Anterior surface of forecoxa with dark erect setae. Forefemur with anterior subapical comb; ventral surface with dense white pile; anterior surface bare; anteroventral surface with row of short black setae, increasing in length at base; base of forefemur inflated, bearing more than 10 long setae. Foretibia with biserial row of stout

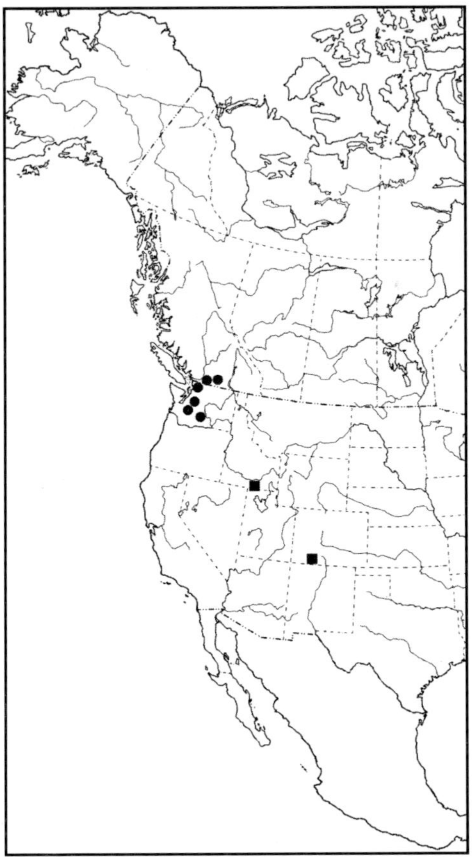

Map 14. Known distribution of *Clinocera bulbosa* (■); *C. guttipennis* (●).

setae beneath. Erect setae of hind tibia poorly differentiated. Fifth tarsomere with dorsoapical extension; empodium brown laterally.

Abdominal sternites and lateral margin of tergites with faint blue pruinescence, remaining tergites brown. Hypandrium subtriangular. Clasping cercus slender, tapered to round apex; inner apical surface half with dark peg-like spines. Phallus long and slightly arched; distiphallus of phallus relatively short, extending back only to slightly beyond tip of surstylus. Surstylus shorter than clasping cercus; ventroapical margin tapered to narrow round apex (Fig. 15*d*).

Female. Very similar to male, including leg chaetotaxy.

Geographic distribution and seasonal occurrence. This species is currently known only from Idaho and Colorado (Map 14). Flight records indicate that this species is present in September and April.

Etymology. The specific name is from the Latin *bulbosus* ("swollen"), referring to the large facial swelling.

Clinocera disjuncta sp. nov.

Figs. 16*a*, 30*g*; Map 15

Type material. HOLOTYPE, male labelled "CANADA: Alberta/ Peter Lougheed P.Pk/ Lk. Rawson, 2020m/ 18.ix.2003/ B.J. Sinclair/ ex. outlet from Lake"; "HOLOTYPE/ Clinocera/ disjuncta/ Sinclair [red label]" (CNC).

Paratypes: **CANADA. Alberta:** Kananaskis Country, Barrier Lk. at Jewells Falls, 1500 m, 16.ix.2003, BJS (1 ♂, 3 ♀♀, ZFMK); same date as holotype (2 ♂♂, 2 ♀♀, CNC, 2 ♂♂, ZFMK). **British Columbia:** Buckinghorse Pk Cpgd, Alaska Hwy, km 281.6, 27.vi.1978, PHA (1 ♀, CAS); Cathedral PP, Quiniscoe Lk., 7.vii.1986, SGC (1 ♀, UBC); Lisadele Lk., 58°41′N, 133°04′W, 4000′, 9.vii.1960, WWM (1 ♂, CNC). **Newfoundland and Labrador:** Squires Memorial, 12.vi.1961, CPA (1 ♀, USNM). **Quebec:** Tasiujaq, riparian vegetation, 25–27.ix.1996, BJS (1 ♀, CNC). **USA. Alaska:** Alaska Peninsula, Aniakchuk NM, 56°56′N, 158°6′W, Surprise Lk., tribs., vii–viii.1988, B. Wisseman (1 ♂, 1 ♀, OSU); Hurricane, 18.vi.1921, JMA (1 ♀, USNM); Flattop Mtn, South Fk, Anchorage, 5.vii.1993 (1 ♀, BLKU), 500–700 m, Anchorage, 2–16.vi.1994, TS (3 ♂♂, 6 ♀♀, BLKU); One Mile Ck, nr Paxson, 8.vi.1994, TS (1 ♀, BLKU). **New Hampshire:** Coos Co., Mt Washington, 3000′, 18–25.vi.1982, DSC (3 ♂♂, 3 ♀♀, UNH), str. bed, 3500′, 23.viii.1988, JMC (1 ♂, CNC); Mt Washington, 4, 21.vii.1914, CWJ (3 ♀♀, USNM).

Additional material examined (status uncertain). **CANADA. British Columbia:** Bijou Falls, Pine Pass, Hwy 97, 16.viii.1978, PHA (4 ♂♂, 5 ♀♀, USNM). **Quebec:** R. aux Castor, station 35, 18.ix.1975, P.A. (1 ♀, UMQ, in alc.). **USA. New Hampshire:** Mt Washington, 4.vii.1914, CWJ (1 ♂, MCZ). **Wyoming:** 8.7 km W Eastgate, Yellowstone NP, 20.vii.1989, BJS (1 ♀, CNC).

Recognition. This species is very similar to *C. fuscipennis*, but is distinguished by blue pruinescence on the scutellum, bivittate scutum, and faint clouding surrounding veins. It may also be confused with *C. stigma*; however, the latter species has large hyaline wings, distinct facial swelling, and broad stigma.

Description (wing length 3.7–4.3 mm).

Male. Face with pale-blue pruinescence, fading toward base of antennae; lower third of face slightly swollen; remainder of head dark. Arista-like stylus long and slender. Postocular setae dark.

Scutum bivittate, separated by pale median stripe; lower two-thirds of pleura,

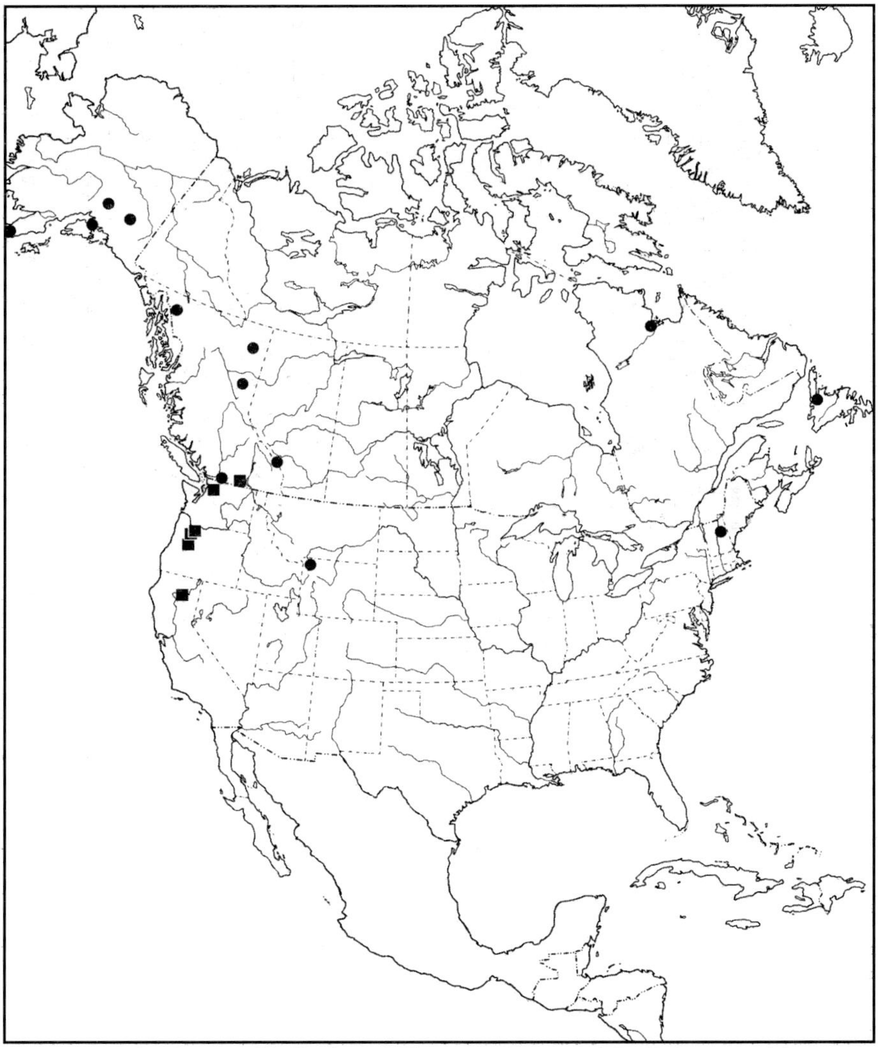

Map 15. Known distribution of *Clinocera disjuncta* (●); *C. femorata* (■).

notopleuron, scutellum, and dorsum of mediotergite with blue pruinescence. Several inconspicuous acrostichal setae anterior to first dc; 5 dc; 1 pprn; 1 presut spal; 2 npl, lower pale and short; 2 psut spal; 1 pal; 2 sctl. Laterotergite with patch of pale setae; posterior margin of anepisternum lacking setulae.

Wing (Fig. 30*g*) infuscate, with faint clouding surrounding crossveins; stigma beyond R_1 faint and elongate; 1 long basal costal seta; costal spines present. All veins straight; auxiliary crossvein between R_4 and R_{2+3} absent; stump vein absent; R_4 bell-shaped; anal vein long and faint; cell dm slightly produced apically. Halter light brown to dark.

Coxae with pale-blue pruinescence, concolorous with pleura; femora reddish brown to dark, remaining segments dark. Forecoxa with pale setae. Forefemur with anterior preapical comb; white ventral pile reduced; antero- and postero-ventral row of short sharp setae extending length of femur. Foretibia with biserial row of stout setae beneath. Mid and hind femora lacking preapical setae. Hind tibia with erect dorsal and ventral setae on apical third. Fifth tarsomere lacking dorsoapical extension; empodium pale brown laterally.

Abdominal sternites and lateral margin of tergites with faint blue pruinescence, remaining tergites brown. Hypandrium short, somewhat tapered. Phallus gently arched, with long, curved distiphallus. Clasping cercus narrow, with expanded apex; anterior apical corner with small round projection, lacking setae; inner surface with fine setae and numerous pointed, peg-like setae. Surstylus shorter than clasping cercus; expanding apically; dorsal and apical margin with matted microtrichia (Fig. 16*a*).

Female. Similar to male except forefemur and foretibia lacking modified setae beneath.

Geographic distribution and seasonal occurrence. This species is widely and disjunctly distributed across northern North America (Map 15). The New Hampshire and Wyoming localities possibly represent glacial relict populations. It appears to be rarely encountered, and flies between June and September.

Remarks. This is a cryptic species, generally difficult to distinguish from the highly variable *C. fuscipennis*. The male terminalia, especially the clasping cercus, are nearly identical in the two species, but generally the peg-like setae of the clasping cercus of *C. disjuncta* are pointed rather than blunt as in *C. fuscipennis*.

Etymology. The specific name is from the Latin *disjunctus* ("disunited"), referring to its widely separated or disjunct populations.

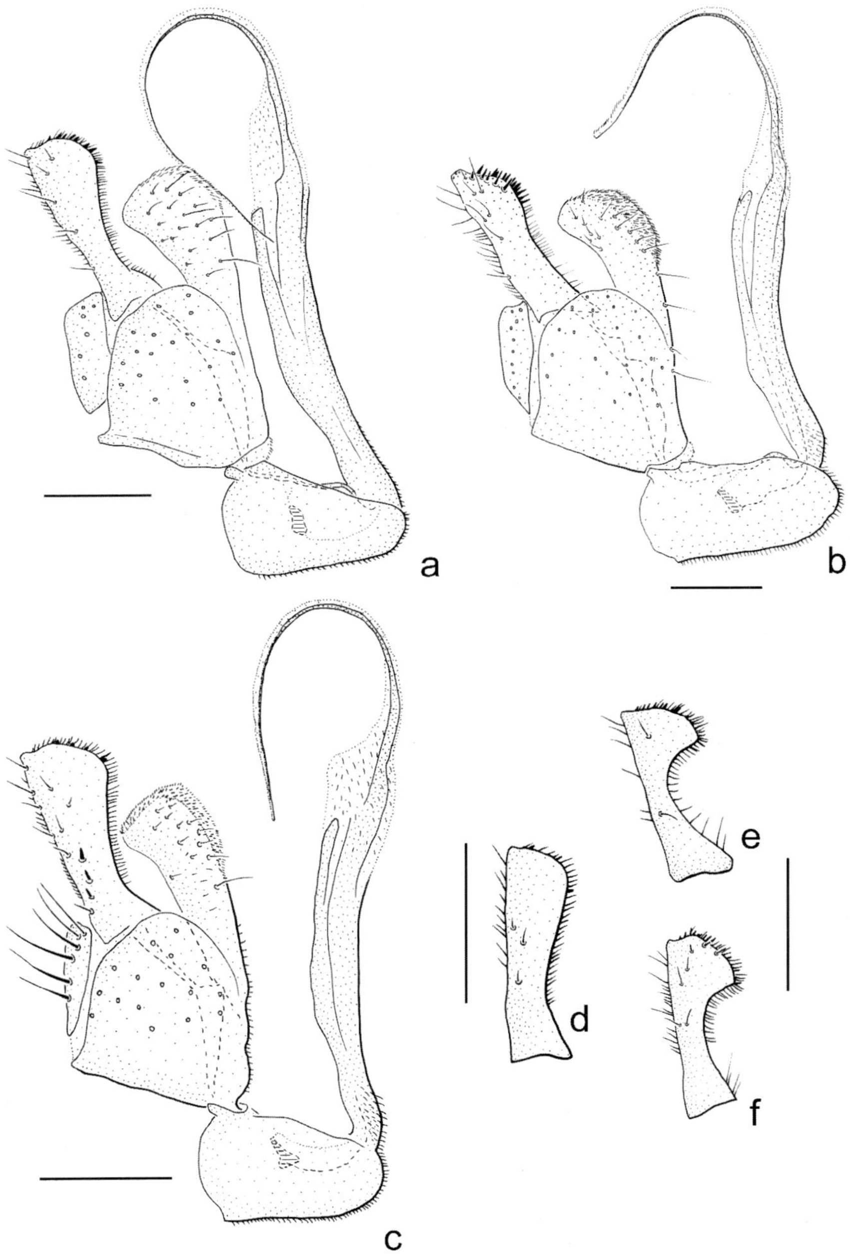

Fig. 16. Male terminalia of *Clinocera*: *a*, *C. disjuncta*; *b*, *C. femorata*; *c*, *C. fuscipennis* (eastern Nearctic); *d*, *C. fuscipennis*, clasping cercus (*C. brevitibia*, paralectotype); *e*, *C. fuscipennis*, clasping cercus (Oregon); *f*, *C. fuscipennis*, clasping cercus (Washington). Scale bars = 0.1 mm.

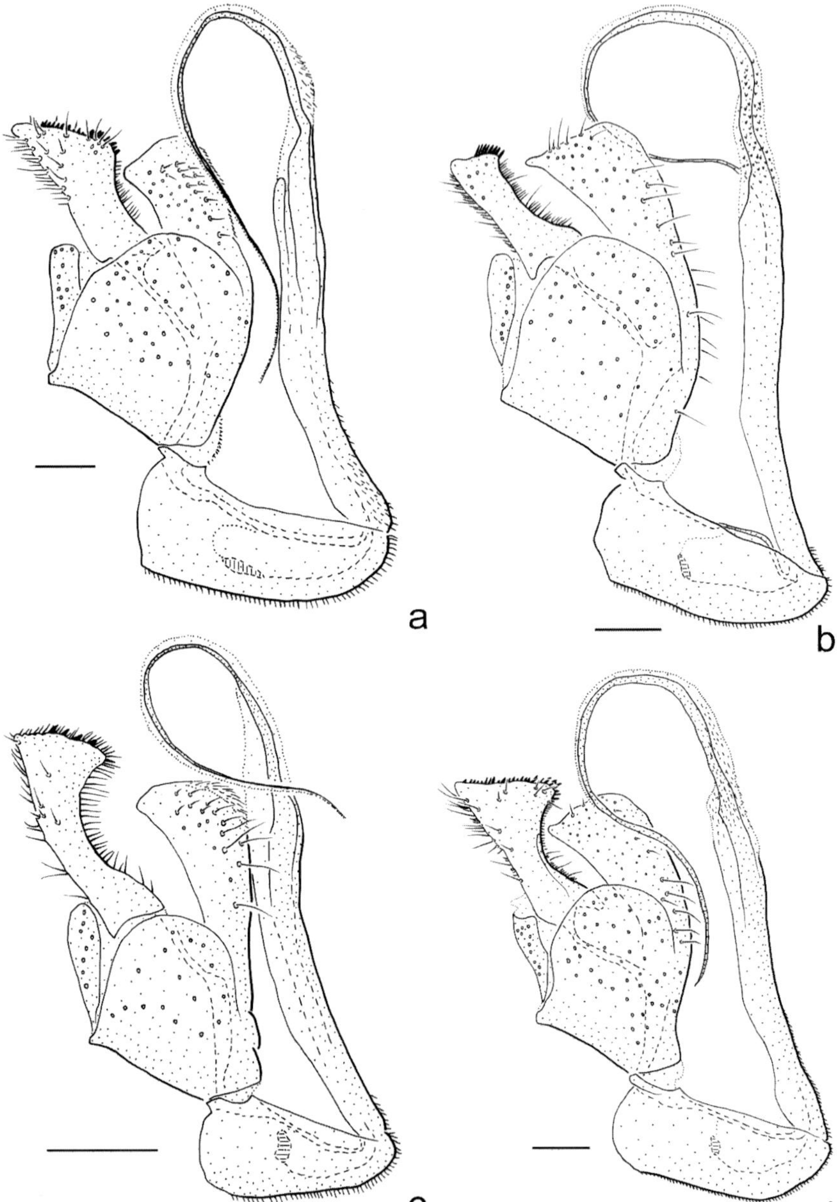

Fig. 17. Male terminalia of *Clinocera*: *a*, *C. gubernans*; *b*, *C. guttipennis*; *c*, *C. lecta*; *d*, *C. schlinkerti*. Scale bars = 0.1 mm.

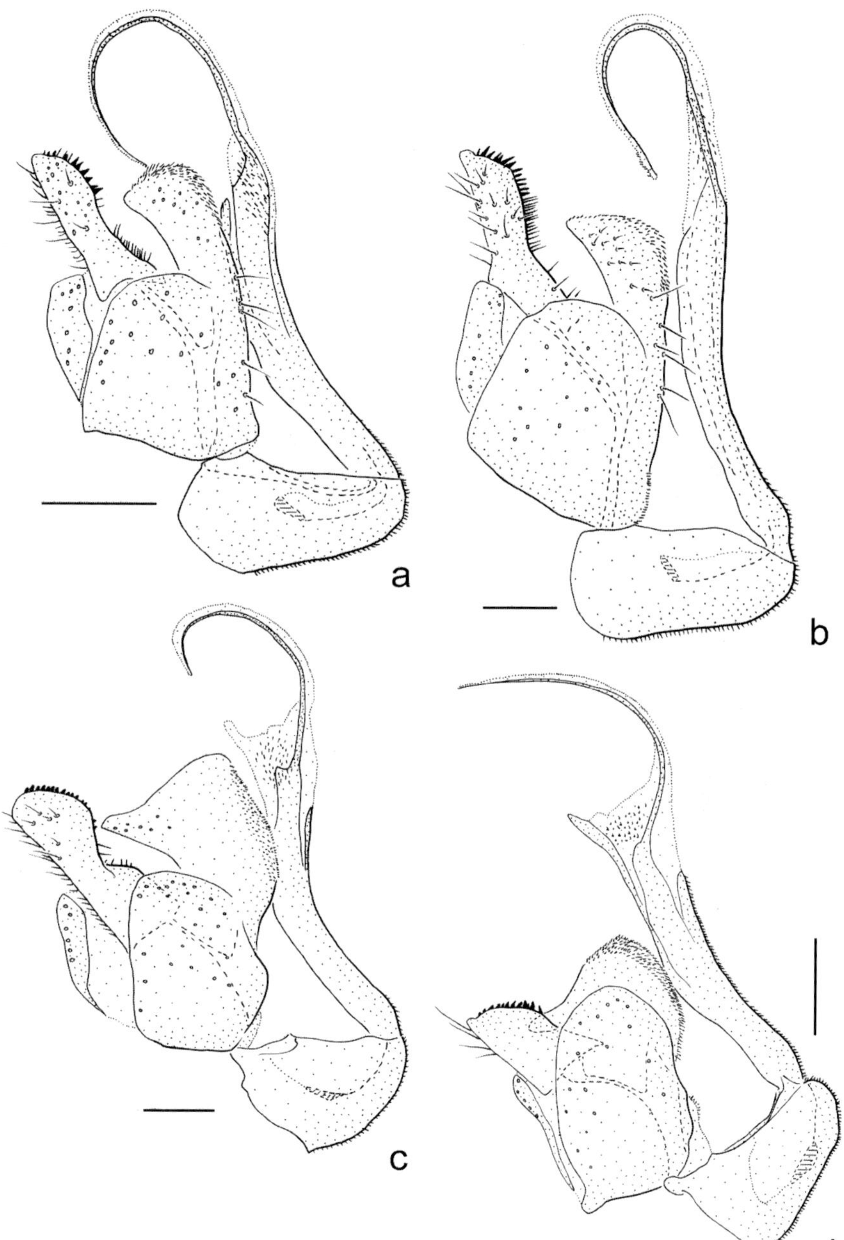

Fig. 18. Male terminalia of *Clinocera*: *a*, *C. stigma*; *b*, *C. wallowa*; *c*, *C. arizonica*; *d*, *C. brevicauda*. Scale bars = 0.1 mm.

Clinocera femorata sp. nov.

Figs. 16*b*, 30*h*; Map 15

Type material. HOLOTYPE, male labelled "CANADA: BRITISH COLUMBIA/ 12kmN Revelstoke-Hwy23/ 2.VII.1989, St. Cyr Ck./ BJ Sinclair"; "HOLOTYPE/ Clinocera/ femorata/ Sinclair [red label]" (CNC).

Paratypes: **CANADA. British Columbia:** Trinity Valley, 29.vi.1937, HBL (2 ♂♂, 2 ♀♀, CNC). **USA. California:** Shasta Co., Burney Falls, 29.vi.1947, R.L. Usinger (1 ♂, 3 ♀♀, UCB). **Oregon:** Lane Co, 11 mi. NE Blue R., H.J. Andrews EF, 800 m, Mack Ck, 4.vi.1982, GWC (1 ♂, 4 ♀♀, OSU); 8 mi. SE Cougar Dam, 880 m, Grasshopper Ck, 8.iv.1983, 4.vi., 2.vii., 6.x.1982, GWC (2 ♂♂, 2 ♀♀, OSU); Mt Hood NF, 12 km E Gov't Camp + Rd 48, cold str., 11.vii.1989, BJS (2 ♂♂, CNC). **Washington:** Whatcom Co., Mt Baker NF, Silver Fir Cpgd, 610 m, 10.x.1979, DDW (1 ♂, 1 ♀, CAS); Whatcom Co., Mt Baker NF, along Bagley Ck, 850 m, 10.x.1979, DDW (1 ♀, CAS); Whatcom Co., Mt Baker NF, along Thompson Ck nr Glacier Ck Rd, 400 m, 4.x.1979, DDW (1 ♂, 2 ♀♀, CAS).

Recognition. This species is similar to *C. stigma* but has long bristles on the base of the swollen forefemur in both sexes.

Description (wing length 4.5–5.6 mm).

Male. Head small in proportion to body. Lower half of face bulbous; lateral margin of facial swelling with blue pruinescence when viewed in front; remainder of head dark. Arista-like stylus short and stout. Postocular setae dark.

Scutum with faint pale vittae. Pleura (upper half of anepisternum dark), notopleuron, scutellum, and prescutellar depression with dull blue pruinescence. Several short acrostichals anterior to first dc; 5 dc; 1 pprn; 1 presut spal; 2 npl, lower pale and short; 2 psut spal; 1 pal; 2 sctl. Laterotergite with patch of short pale setae. Posterior margin of anepisternum with pale setulae.

Wing (Fig. 30*h*) infuscate, with clouding or spots absent; stigma beyond R_1 distinct, elongate and elliptical; 1 short basal costal seta; costal spines present. Veins straight; auxiliary crossvein between R_{2+3} and R_4 absent; stump vein absent; R_4 bell-shaped; anal vein faint streak; cell dm slightly produced apically. Halter light brown.

Coxae with pale-blue pruinescence, remaining leg segments dark brown. Forecoxa with dark erect anterior setae. Forefemur stouter than other femora, with swollen base; anterior preapical comb present; anterior surface bare and flattened; base of femur with 10–15 long stout spines; antero- and postero-ventral row of short setae; dense white pile beneath. Foretibia with short stout setae beneath; 2 long apical ventral bristles present. Hind tibia with short erect dorsal and ventral setae on apical half. Fifth tarsomere lacking dorsoapical extension; empodium brown laterally.

Abdominal sternites and lateral margin of tergites with faint blue pruinescence, remaining tergites brown. Hypandrium parallel-sided. Phallus gently arched; distiphallus relatively short and slender. Clasping cercus tapering to round apex; inner apical face with patch of sharp spines. Surstylus three-quarters length of clasping cercus; narrow, arched apically with matted microtrichia on dorsoapical margin (Fig. 16*b*).

Female. Similar to male except as follows: face dark; ventral bristles of forefemur and foretibia same as in male.

Geographic distribution and seasonal occurrence. Recorded only from the far-western Cascade Ranges (Map 15). Apparently the range of *C. femorata* has not penetrated very far north into previously glaciated regions. The flight period is from June to October.

Remarks. This species is very similar to an undescribed species from South Korea (Samgak-Koji, Macheon-Myeon (BLKU)). Both species have a dark broad stigma and a swollen forefemur with long basal setae.

Etymology. The specific name is in reference to the swollen setose forefemur.

Clinocera fuscipennis Loew

Figs. 1*b*, 4*d*, 6, 7, 8*a*, 10*e*, 16*c*–*f*, 30*i*; Map 16

Clinocera fuscipennis Loew, 1876: 324.

Clinocera (*Phaeobalia*) *brevitibia* Melander, 1928: 235 **syn. nov.**

Clinocera fuscipennis: Melander, 1902, 1928, 1965: 468.

Notes on synonymy. *Clinocera fuscipennis* is transcontinental, with variable leg and facial coloration. Eastern populations generally have dark legs and lack vittae on the scutum. In western populations, vittae are often present and (or) the legs are distinctly pale or reddish brown. Melander (1928) distinguished the latter form and named it *C. brevitibia*. However, leg coloration varies greatly, even within a single population. The form of the clasping cercus of the male terminalia is somewhat variable, with the apex slightly to greatly expanded compared with the midlength width (Figs. 16*c*–*f*). For these reasons, *C. brevitibia* is herein assigned as a junior synonym of *C. fuscipennis*.

Type material examined. *Clinocera brevitibia*: LECTOTYPE (here designated), male labelled "Olga, Wash[ington, USA]/ 26.VII.[190]9"; "TYPE (male sign)/ Phaeobalia/ brevitibia/ Mel [red label]" (USNM). My lectotype label "LECTOTYPE/ of Clinocera/ brevitibia Mel./ des. B.J. Sinclair 1991 [red label]" has been attached to this specimen. The lectotype is designated in order to clarify the application of the name to this taxon.

Paralectotypes: Tacoma, Wash[ington, USA], 27.viii.[19]11 (3 males, USNM); Olga, Wash[ington], 26.vii.[190]9 (1 male, USNM). My paralectotype labels have been additionally attached to these specimens. An additional specimen (Mt. Constitution, Jul.31.8. Wash.) with a red paratype label attached is not considered part of the original syntype series.

Clinocera fuscipennis: HOLOTYPE, female labelled "Type/ 1665 [red label with white upper border]"; "N.H."; "fuscipennis/ Lw [handwritten] [missing left foreleg]" (MCZ).

Additional material examined. Some 1200 specimens from the following provinces and states: **CANADA. Alberta:** Jasper NP, Jasper, Cabin Ck, 27.vi.1989, BJS (CNC); Kananaskis Country, Peter Lougheed PP, Lk. Rawson, 16.ix.2003, BJS (1 ♀, ZFMK). **British Columbia:** 4 km N Canoe, 16.iv.1990, AB (CNC); 33 km NW Lk. Cowichan, 9.vii.1991, AB (CNC); Glacier NP, 1250 m, Illecillewaet Cpgd, ypans, str., 31.vi.–2.vii.1989, BJS (CNC), 1250 m, 30.vi.1989, BJS (CNC); 4 km W Glacier NP, 1.vii.1989, BJS (CNC); 10 km NE Haney, UBC Research For., 3.viii.1988, JMC (CNC); Kitimat, beach, 2.vi.1960, JGC (CNC); Kootenay NP, Radium Hot Springs + Hwy 93, Sinclair Ck, 9.ix.2003, 17.ix.2005, BJS (ZFMK); Manning PP, Coldspring Cpgd, Buckhorn Ck, 7.x.1979, DDW (CAS), Goodfellow Ck, Hwy 3, 8.x.1979, DDW (CAS), rd to lookout, 8.x.1979, DDW (CAS), Three Falls Tr., Shadow Falls, 4.vii.1989, BJS (CNC); Mission City, 28.v.1953, WRMM (CNC); Mt Allard, 23 mi. N Terrace, 3500′, 28.vi., 20.vii.1960, JGC (CNC); Mt Revelstoke NP, Inspiration Woods Tr., 12.ix.2003, BJS (ZFMK), Giant Cedar Tr., 700 m, 2.vii.1989, BJS (CNC), km 4.8 Summit Rd, roadcut str., 3.vii.1989, BJS (CNC); 12–13 km N Revelstoke, Hwy 23, St. Cyr Ck, str., 2.vii.1989, BJS (CNC); Mt Thornhill nr Terrace, seep, 14.vii.1960, JGC (CNC); 30 km N Penticton, Hardy Falls Pk, Deep Ck, 3.vii.1989, BJS (CNC); 13 km SE Port Renfrew, 5.vii.1991, AB (CNC); Tofino, 2 mi. S Hwy 4, str. above beach, 5.vii.1990, JMC (CNC); Trinity Valley, 29.vi.1937, HBL (CNC); Vancouver, Pt Grey, 19.ix.–23.xi.1972, 2.iv., 26.vii.1973, JRV (CNC), North Vancouver, 320 m, base of Grouse Mtn, 14.vii.1999, BJS (ZFMK); Vancouver Is., Goldstream PP, 19 km NW Victoria, Goldstream R., 8.vii.1974, PHA (CAS); Vancouver Is., North Shore Rd, 17 km Lk. Cowichan, 19–28.vii.1985, I.M. Smith (CNC); 22 km W Vernon, Silver Star Prov. Rec. Area, 3.vii.1989, BJS (CNC). **Nova Scotia:** Cape Breton Highlands NP, str., 22.vi.1983, JRV (CNC). **Ontario:** Ancaster, Hwy 403, 29.v.1985, BJS (DEBU); Dundas, Borer's Falls, 24.xii.1984, 17.iii., 8.xii.1985, BJS (CNC); Burke Falls, 13.vii.1926, F.P. Ide (CNC); Collingwood Twp, Black Ash Ck, 12.iv.1955, F.P. Ide (CNC); Dundas, Rock Chapel, 10.viii. (larva), 31.x.1986, BJS (DEBU, reared); Dundas, Webster's Falls, 31.x., 8.xii.1985, 22.i., 12.ii., 29.iii., 12.iv.1986, BJS (DEBU); Halton Co., Rattlesnake Pt Cons. Area, nr buffalo compound, 12.v.1990, BJS (CNC); Hamilton, 15, 29.v., 14, 31.x., 8, 28.xii.1985, 29.iii.1986, BJS (DEBU); Iroquois Falls, 22.vi.1983, 29.vi.1987, JRV (CNC); Kakabeka Falls PP, 19.vi.1989, BJS (CNC); Orangeville, 20.ix.1984, SAM (DEBU); Ouimet Cyn PP, 14.vi.1989, BJS (CNC); Owen Sound, Inglis Falls, 12.v., 4.vi., 2, 26.x.1985, 19.ii.1986, BJS (DEBU); Springmount, Jones Falls, 23.iv.1985, BJS (DEBU); St. Joseph Is., 46°18′N, 83°59′W, seep, 23.x.1988, JES (DEBU); Tobermory, Dunk's Bay, 30.x.1996, SAM (DEBU); Wiarton, 29.xi.1985, 17.iv.1986, BJS (DEBU). **Quebec:**

Beechgrove, 18.x.1956, JRV (CNC); Covey Hill, 28.vi.1923, CHC (CNC); Gatineau Pk, Black Lk., 18.v.1989, BJS, JMC (CNC), Dunlop parking area, 14.vii.1991, BJS (CNC), Luskville Falls, 8.vi.1989, BJS (CNC), Waterfall Tr., 23.v.1990, BJS (CNC); Kingsmere, 16.v.1958, JGC (CNC); Mt Orford, 9.ix.1937, GES (CNC); Old Chelsea, 9.x.1955, 29.iii.1958 (JFM), 15.x.1958, 10.x.1961, 12, 15.x.1964, 17.x.1988, JRV (CNC); R. aux Castor, Baie James, trib. Lac Nathalie, 11.ix.1974, J.L. Frechette (UMQ); St. Hippolyte, tribs. Lk. Achigan, 10, 16.v.1973, 30.iv., 16.v.1974, PPH (UMQ). **MEXICO. Sonora:** 22 km N Moctezuma, 25.ix.1980, DDW (CAS). **USA. Alaska:** Flattop Mtn, 550–600 m, Anchorage, 2.vi.1994, TS, T. Tanaka (BLKU); 15 mi. SE Anchorage, 28.vi.1957, G.W. Byers (SEMC); Glenn Hwy, mi. 54.6, Moose Ck, 4.vi.1994, TS (BLKU); Kenai–Russian R. camp, Sterling Hwy, A-107, 23.vii.1978, PHA (USNM); Muir Inlet, Glacier Bay, 14.vi.1965, D.M. DeLong (USNM); Valdez, 15 mi. N, Alaska Ins Project, 26.viii.1948, B.L. Morris (USNM). **Arizona:** Cochise Co., 12.5 km S Sierra Vista, Ramsey Cyn, MT, 1700 m, 17–24.xii.1986, BVB (CNC). **California:** Alpine Co., Little Teton Ck, 12.vii.1977, DDW (CAS); D.L. Bliss SP, 12.xii.1986, BJS (DEBU); Calaveras Co., Hwy 26, 5 km SW West Pt, 770 m, 27.xi.–27.xii.1987, PHA (USNM); Crestline, 19.vi.1944, ALM (USNM); Fresno Co., Kaiser Cyn, Huntington Lk., 2700 m, 17–23.viii.1984, JFM (PERC), Sierra NF nr Kaiser Pass, 2700 m, 2.viii.1979, DDW (CAS), Sierra NF, E and W Fks Forebay Lk., 31.vii.1979, DDW (CAS), Sierra NF, Pitman Ck nr Big Ck, 1500 m, 2.viii.1979, DDW (CAS); Humboldt Co., Humboldt Redwoods SP, Cabin Ck, 17.v.1978, DDW (CAS), 2 mi. S Orick, 12.v.1966, T.W. Fisher, R.E. Orth (UCR), Six Rivers NF, Ikes Ck at Klamath R., 3.vii.1976, DDW (CAS); Inyo Co., Horton Ck Cpgd, Horton Ck, 15.iii.1995, RWB, CRN (BYU); Madera Co., Boggy Meadows, Big Ck, 6400′, 9.viii.1971, HBL (CAS); Marin Co., 8 mi. SW Fairfax, 23.vi.1951, A.A. Hubert (SEMC), Samuel P. Taylor SP, Lagunitas Ck, 20.v.1977, DDW (CAS), Lagunitas Ck below Alpine Dam, 28.v.1979, DDW (CAS), Mill Valley, 25.vii.1967, PHA (USNM); Monterey Co., 9.7 mi. SE Gorda, Hwy 1, 30.v.1971, PHA (USNM); Placer Co., Shirttail Ck below Sugar Pine Reservoir, 18.v.1983, RWB (BYU); San Mateo Co., Corte Madera Ck, 230 m, 9.ix.1987, PHA (USNM); Santa Cruz Co., Big Basin SP, Opal Ck, 14.v.1978, DDW (CAS); Sequoia NP, Lodge Pole, 3.vii.1984, D.J. Burdick, (PERC); Shasta Co., McArthur Burney Falls Memorial Pk, 2900′, 1.viii.1970, PHA (CAS); Siskiyou Co., Klamath NF, Portugese Pk Tr., 2.vii.1978, DDW (CAS), Klamath NF, 2.5 mi. Wooley Ck, 3.vii.1978, DDW (CAS), McBride Springs, 5200′, 10–14.vi.1974, S. Szerlip (UCB), Klamath NF, Knownothing Ck, S Fks of Salmon R., 22.vii.1980, DDW (CAS); Trinity Co., Shasta NF, Trinity R., 2.vii.1978, DDW (CAS); Tulare Co., Mays Ck, 20.vii.1984 (PERC); Tuolumne Co., Brightman Flat, 11.vii.1978, DDW (CAS). **Colorado:** Larimer Co., Sheep Ck nr jct. Buckhorn Ck, 15.x.1995, SF, AF (CSU). **Georgia:** Dade Co., Cloudland Cyn SP, nr Trenton, seeps, waterfalls, 24.iv.1989, JMC (CNC). **Idaho:** Cassia Co., Howell Ck, mouth of Howell Cyn, 22.iv.1985, CRN, RWB (BYU); Kootenai Co., Hwy 97, 36 mi. S jct. Hwy 90, nr Lk. Coeur d'Alene, 2600′, 25.iv.1992, WJT (WSU). **Indiana:** Crawford Co., Hoosier NF, Hemlock

Cliffs, 19.v.1991, BJS (CNC). **Kentucky:** Meade Co., Twin Cave 1, 17.xii.1957, R.E. Woodruff (FSCA, USNM); Whitely Co., Cumberland Falls SP, 17–18.vi.1990, BJS (CNC). **Maine:** Bar Harbor, 4–13.vi.1921, CWJ (USNM); Mt Katahdin, Hunt Tr., 2400′, 1–6.vii.1968, DMW (CNC). **Maryland:** Plummer Is., 14.xi.1914, RCS (USNM). **Massachusetts:** Franklin Co., Sunderland, Mt Toby SF, 28.v.1984, NEW (WOOC). **Nevada:** White Pine Co., Great Basin NP, Baker Ck and Lehman Ck, 2.vii.2001, JFM (PERC). **New Hampshire:** The Bowl, 2.5 mi. NW Wonalancet, MT, 20.x.–7.xi.1984, 2–17.x.1985, DSC (UNH); Pittsburg, Rt. 3, Connecticut Lks., 22–24.vi.1972, B.J. and F.C. Thompson (AMNH). **New Mexico:** Miguel Co., Santa

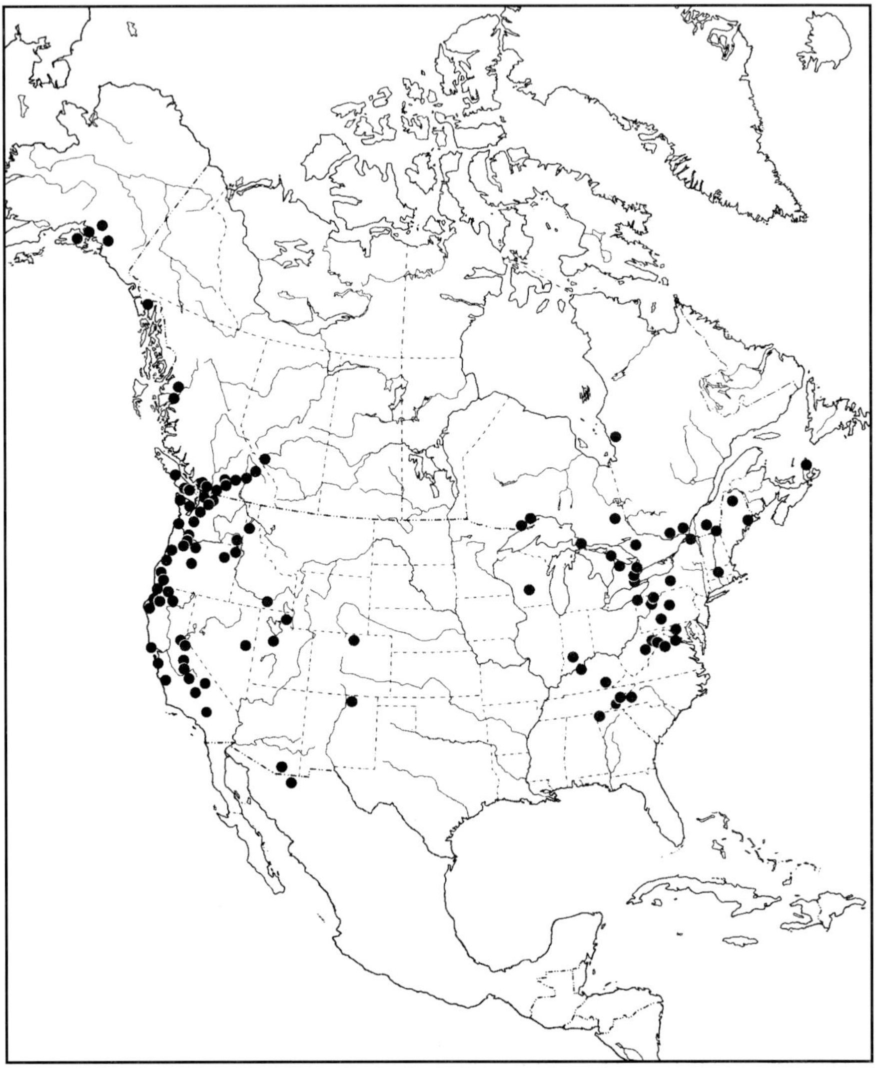

Map 16. Known distribution of *Clinocera fuscipennis*.

Fe NF, nr Cowles, 27.v.1991, BJS (CNC). **New York:** Ithaca, 13.iv.1924, CHC (AMNH), 23.v.1986, BJS (DEBU). **North Carolina:** Buncombe Co., Blue Ridge Pkwy, Craggy Gardens Picnic Area, 1580 m, 18.v.1986, 12.vi.1990, BJS (DEBU); Highlands, Wilson's Gap, 3100′, 12.v.1957, JRV (CNC); Macon Co., Highlands Biological Stn, MT, 27.v.1984, J.K. Liebherr (CUIC); Swain Co., GSMNP, Andrews Bald Tr., 5.vi.1993, BJS (CNC), Clingmans Dome, 21.vi.1941, ALM (USNM), Clingmans Dome Rd, 15.vi.1990, 3–6.vi.1993, 9.vii.2004, BJS (CNC, ZFMK), 20.v.1957, JRV (CNC), 6 km S Balsam Mtn Cpgd, 13.vi.1990, BJS (CNC) Balsam Mtn, Flat Ck, 13.vi.1990, BJS (CNC), Newfound Gap, 15.vi.1990, BJS (CNC); Yancey Co., Mt Mitchell SP, Balsam Tr., spr., 2000 m, 11.vi.1990, BJS (CNC). **Oregon:** Benson Pk, 24.vi.1935, ALM (USNM); Benton Co., Alsea Falls Cpgd, 21.vi.1977, DDW (CAS), Siuslaw NF, Marys Peak, Alder Falls, 14.vii.1989, BJS (CNC); Clackmas Co., Mt Hood NF, Still Ck Cpgd, 23.vi.1977, DDW (CAS), Still Ck Cpgd nr Gov't Camp, 22–23.vi.1982, WJT (WSU); Curry Co, Siskiyou NF, 7 km W Agnes, seep, 18.vii.1989, BJS (CNC), 17.6 and 24 km E Pt Orford, Elk R. Rd, 17.vii.1989, BJS (CNC), Mt Hood NF, 12 km E Gov't Camp, Rd 48, 11.vii.1989, BJS (CNC), Humbug Mtn SP, roadcut Hwy 101, 17.vii.1989, BJS (CNC), 8 km W Sixes, Edson Ck Cpgd, Edson Ck, 16.vii.1989, BJS (CNC); Hood River Co., Starvation Ck SP, falls, 26.vi.1978, DDW (CAS), Starvation Ck SP, below falls, 10.vii.1989, BJS (CNC), Mt Hood NF, Still Ck Cpgd, 1150 m, 27.vi.1978, DDW (CAS); Josephine Co., 8.5 mi. W Galice, 18.vii.1989, BJS (CNC); Lane Co., Willamette NF, Proxy Falls, 22.vi.1978, DDW (CAS); Lane Co., 2 mi. N Blue R., Quartz Ck, 490 m, 26.vi., 3.xi.1982, 6, 27.x.1982, 15.v.1983, GWC (OSU), 8 mi. SE Cougar Dam, Grasshopper Ck, 880 m, 17.v., 2.vii.1982, 18.iv.1983, GWC (OSU), 11 mi. NE Blue R., H.J. Andrews EF, Mack Ck, 880 m, 17.v., 1–20.vi., 2–6.vii., 6–27.x.1982, GWC (OSU), Siuslaw NF, 3 mi. S Yachats, Gwyann Ck, 15.vii.1989, BJS (CNC), Siuslaw NF, MT, 10 mi. S Yachats, Rock Ck, 15–16.vii.1989, BJS (CNC); Marion Co., Silver Falls SP, 23.vi.1974, PHA (CAS), Silver Falls SP, Cougar Tr., 12.vii.1989, BJS (CNC); Multnomah Co., Lower Falls, 25.xi.1984, D.C. McClain (NCSU), ck above falls, 10.vii.1989, BJS (CNC); Union Co., Ladd Cyn, 4280′, 14 mi. S La Grande, MT, 15–21.vi.1975, E.J. Davis (WSU); Wallowa Co., 39 mi. N Enterprise, Hwy 3, 3400′, seep, 21.vi.1976, WJT (WSU); Wasco Co., Warm Springs Indian Res., Beaver Ck, 23.vi.1978, DDW (CAS). **Pennsylvania:** Blair Co., 4 mi. W Duncanville, 23.v.1980, DDW (CAS); Erie Co., Four and Six Mile cks, ECM (PSBC); Centre Co., Pine Grove Mills, 1.v.1972, DDW (CAS); Elk Co., Johnsonburg, Hwy 219, cliff seep, 30.v.1990, BJS (CNC); McKean Co., Allegany NF, Red Bridge Cpgd, str., 29.v.1990, BJS (CNC). **Tennessee:** GSMNP, Alum Cave Tr., 16.vi.1990, BJS (CNC), Chimney Tops Tr., 5.vi.1993, BJS (CNC), Clingmans Dome, 13.v.1986, BJS (DEBU), Grotto Falls, 13.v.1986, BJS (DEBU), Le Conte Tr., 16.vi.1990, BJS (CNC), Little Pigeon R., Ramsay Cascade Tr., 4.vi.1993, BJS (CNC), 10 and 18.5 km E Sugarlands, roadcut seeps, 15.vi.1990, 5.vi.1993, BJS (CNC); GSMNP, 5000′, 9.vi.1939, CPA (AMNH). **Utah:** Cache Co., Blacksmith Fk Cyn, 26.vi.1967, WJH (USUL), Green Cyn, 19–23.vi.1985, N.N. Youssef (USUL), Logan Cyn, White Pine Tr.,

24–31.viii.1984, 1–8.ix.1984, N.N. Youssef (USUL), Logan Cyn, Turner Cpgd, 20–27.vi.1985, WJH (USUL), Logan Cyn, Twin Ck, MT, 21–29.vi.1979 (USUL), Rock Cyn, 9.iii.1963 (WSU), W Hodges Cyn, 9–14, 23–29.ix.1976, GFK, WJH (USUL); Utah Co., Bunnells Fk nr jct. S Fk Prove R., 12.iii.1982, S.M. Clark (BYU). **Virginia:** Fairfax Co., Dead Run, 3.iv.1914, RCS (USNM); Shenandoah Co., Wolf Gap, 6.v.1980, DDW (CAS). **Washington:** Clackamas Co., Still Ck Cpgd, Gov't Camp, 3800′, 22–23.vi.1982, WJT (WSU); Clallam Co., Olympic NP, Mt Angeles, 1550 m, Sunrise Ridge, 24.vii.1978, DDW (CAS); Garfield Co., Pataha Ck, Hwy 128, 8 mi. S Pomeroy, 24.vi.1978, DDW (CAS); Gifford Pinchot NF, 15 mi. S Randle, 1650′, 9.vii.1989, BJS (CNC), La Wis Wis, 1.x.1979, DDW (CAS); Grays Harbor Co., Olympic NF, Willaby Cpgd, 21.vii.1978, DDW (CAS), Quinault Lk., Rainforest Resort, 28.vi.1974, PHA (CAS); Ilwaco, vii.1917, 5.v.1918, 28.vi.1925, ALM (USNM); Mt Baker, 900 m, NE slope, 27.v.1973, JRV (CNC), 1100 m, Grouse Ck, Glacier Ck Rd, 4.x.1979, DDW (CAS), Thompson Ck, 400 m, 4.x.1979, DDW (CAS); Mt Constitution, 31.vii.1908 (USNM); Mt Rainier NP, Comet Falls, str., 8.vii.1989, BJS (CNC), Panther Ck, nr Ohanapecosh R., 2.x.1979, DDW (CAS), White R., 20.vii.1924, ALM (USNM), White R., 1220 m, White R. Cpgd, 3.x.1979, DDW (CAS), E end Yakima Pk, 1460 m, 3.x.1979, DDW (CAS); Pacific Co., Fort Canby SP, nr Ilwaco, 19.vii.1978, DDW (CAS); Pierce Co., Clover Ck, 8.vi.1980, T.L. Whitworth (USUL); Quilcene, 24.vii.1917, ALM (USNM); Skamania Co., Gifford Pinchot NF, 20 mi. N Trout Lk., Big Spring Ck, 9.vii.1989, BJS (CNC); Snoqualmie Pass, 19.vi.1924, ALM (USNM); Stehekin, Lk. Chelan, 30.vii.1919, ALM (USNM); Whatcom Co., Glacier, 25.v.1965, F. Schmid (CNC), Mt Baker NF, Bagley Ck, 850 m, 10.x.1979, DDW (CAS), Silver Fir Cpgd, 610 m, 10.x.1979, DDW (CAS), Razor Hone Ck, 1100 m, Heather Meadows, 9.x.1979, DDW (CAS); Whitman Co., Yakawawa Cyn, 2500′, 7 mi. NW Colton, 25–27.vi.1977, WJT (WSU). **West Virginia:** Greenbrier Co., Monongahela NF, Summit Lk. Rd, roadcut seep, 3.vi.1990, BJS (CNC), 18 km W Information Center, nr Windy Run, 4.vi.1990, BJS (CNC), SR 39, Coats Run, 4.vi.1990, BJS (CNC); Hardy Co., 2 mi. NW Mathias, 3.v.1980, DDW (CAS); Pocahontas Co., Monongahela NF, SR 150, 25.5 km N SR 39, 4.vi.1990, BJS (CNC), Hills Ck Scenic Area, 3, 4.vi.1990, BJS (CNC); Tucker Co., Blackwater Falls SP, 39°11′N, 79°5′W, 2.vii.1986, 6.vii.2004, BJS (DEBU, ZFMK), Falls of Elakala, 31.v., 2.vi.1990, BJS (CNC); Tucker Co., 16.5 km S Davis, SR 32, roadcut seep, 20.v.1986, 1.vi.1990, 7.vii.2004, BJS (CNC, DEBU, ZFMK). **Wisconsin:** Dells, 8.vii.1933, ALM (USNM).

Recognition. Distinguished from other eastern species by its dark coloration, slightly bulbous face, infuscate wings, and general absence of scutal vittae. In western North America this species is more variable, with legs dark to reddish brown and often with distinct scutal vittae. In western regions it is most similar to *C. stigma*, but the proximal end of the broad stigma is narrow, the scutum is bivittate, and the male forefemur has ventral rows of stout setae (lacking in *C. stigma*).

Description (wing length 2.2–4.5 mm).

Male. Face mostly clothed with pale-blue pruinescence, remainder of head dark; lower third slightly swollen. Arista-like stylus long and slender. Postocular setae dark.

Scutum with (western) or lacking (eastern) vittae; pleura and npl with faint blue pruinescence (Figs. 1*b*, 4*d*). Several inconspicuous acrostichal setae anterior to first dc; 5 dc; 1 pprn; 1 presut spal; 2 npl, lower pale and short; 1 or 2 psut spal; 1 pal; 2 sctl. Laterotergite with patch of pale setae. Posterior margin of anepisternum lacking setulae.

Wing (Fig. 30*i*) infuscate, with spots or clouding absent; stigma beyond R_1 faint and elongate, sometimes distinct and broad; 1 long basal costal seta; costal spines present. All veins straight; auxiliary crossvein between R_4 and R_{2+3} absent; stump vein absent; R_4 bell-shaped; anal vein long and faint; cell dm slightly produced apically. Halter yellowish brown to dark brown.

Coxae with pale-blue pruinescence, concolorous with pleura; femora reddish brown, paler on inner ventral margins or dark; remaining segments dark (Fig. 4*d*). Forecoxa with pale setae. Forefemur with anterior preapical comb; dense white pile beneath reduced; antero- and postero-ventral row of short sharp setae on basal two-thirds. Mid and hind femora lacking preapical setae. Foretibia with biserial row of stout setae beneath. Hind tibia with erect dorsal and ventral setae on apical third. Fifth tarsomere lacking dorsoapical extension; empodium pale laterally.

Abdomen dark, brownish. Hypandrium short, somewhat tapered. Phallus gently arched, with relatively short, curved distiphallus. Clasping cercus narrow, with expanded apex; anterior apical corner with small round projection, lacking setae; inner surface with fine setae and scattered blunt peg-like setae. Surstylus shorter than clasping cercus; expanded apically; dorsal and apical margin with matted microtrichia (Figs. 8*a*, 16*c*–*f*).

Female. Similar to male except forefemur and foretibia lacking modified setae beneath.

Geographic distribution and seasonal occurrence. This transcontinental species is generally commonly collected, with the eastern and western populations widely disjunct (Map 16).

Specimens from emergence traps monitored for a number of years were made available through the generosity of E.C. Masteller. This material was collected from small spring-fed streams with soft sandstone–shale substrates in the St. Lawrence River drainage of Lake Erie (Erie County, Pennsylvania, 42°7′17′N, 79°59′25′′W). See Masteller and Flint (1984) and McCreadie et al. (1994) for a description of the site and methodology. At all three sites, specimens emerged primarily during September–May, with no records from July and August (Tables 4, 5). A similar pattern was observed from a number of sites monitored in southern Ontario (see Sinclair and Marshall 1987).

Table 4. Emergence pattern for *Clinocera fuscipennis* in Fourmile Creek, Erie, Pennsylvania, USA (1981–1992).

	Upper site (site 1)		Lower site (site 2)	
	No. of males	No. of females	No. of males	No. of females
January	1	2		
February	3	14		
March	6	17	1	
April	1	11		2
May	5			
June				
July				
August				
September	3	1		
October	6	10	1	7
November	14	30	7	17
December	1	3	1	4
Total	40	88	10	30

Note: Site 2 was downstream from a sewage outflow (see Masteller and Flint 1984).

This species was also collected in emergence traps set on a small tributary of a subarctic watershed in Quebec (Landry and Harper 1985). However, a female specimen was misidentified and is in fact *C. brunnea*. *Clinocera fuscipennis* was also collected in May from emergence traps set in small streams of the Laurentian foothills, Quebec (Harper 1980). An additional species was identified from Harper's (1980) study (*C. fuscipennis*), but this represents a mixed series of *C. fuscipennis* and *Trichoclinocera falcata* Sinclair.

Remarks. An adult of this species was collected with a female chironomid midge held its mouthparts.

Clinocera gubernans (Melander)

Figs. 4*b*, 17*a*, 30*j*; Map 17

Wiedemannia (*Chamaedipsia*) *gubernans* Melander, 1928: 234.

Wiedemannia gubernans: Melander, 1965: 469.

Clinocera gubernans (Melander): Sinclair, 1995: 693.

Type material examined. HOLOTYPE, male labelled "Nelson/ B.C.[Canada]/ 17.VII.[19]10"; "Type/ Chamaedipsia/ gubernans/ Mel. [red label]" (USNM).

Additional material examined. Some 300 specimens from the following provinces and states: **CANADA. Alberta:** Banff, on snow, 28.ix., xii.1925, OB

Table 5. Emergence pattern for *Clinocera fuscipennis* in Sixmile Creek, Erie, Pennsylvania, USA (1977–1986) (see Masteller and Flint 1984).

	No. of males	No. of females
January		
February	1	2
March	7	3
April	10	109
May	3	1
June	4	2
July		
August		
September	4	2
October	53	46
November	76	107
December	3	18
Total	162	191

(CAS, USNM), 16.vii.1912, x.1921, N.B. Sanson (USNM); Jasper NP, Maligne Lk, 27.xi.1988, E. Fuller (CNC); Kananaskis Country, Hwy 40, Evan–Thomas Ck, 15.ix.2003, BJS (ZFMK); Waterton Pk, 4200′, 1.x.1957, JRV (CNC). **Yukon:** Dempster Hwy, km 10, Klondike R., 25.ix.1982, R. Frisch (CNC); North Fk Pass, Tombstone Cpgd, 4.vii.1985, E. Bijdemast (UBC). **USA. Alaska:** Anchorage, Flat Top Mtn, 550–600 m, 2.vi.1994, TS (BLKU). **Idaho:** Blaine Co., 6 mi. NE Sun Valley, Corral Ck, 22.iv.1985, RWB, CRN (BYU), Cyn Ck jct. S Fk Payette R., 23.iv.1985, RWB, CRN (BYU), Trail Ck, 3 mi. above Sun Valley, 22.iv.1985, RWB, CRM (BYU), Hwy 75, 4 mi. W Galona Summit, Frenchman Ck, 23.iv.1985, RWB, CRN (BYU), Big Wood R., Hwy 75, 8 mi. SE Galona Summit, 22.iv.1985, RWB, CRN (BYU), E Fk Wood R., nr jct. Bigwood R. 22.iv.1985, RWB, CRN (BYU), Salmon R., Hwy 75 above jct. Frenchmans Ck, 23.iv.1985, RWB, CRN (BYU); Boise Co., S Fk Payette R. jct. Canyon Ck, 23.iv.1985, RWB, CRN (BYU), Warm Spring Ck jct. S Fk Payette R. and Hwy 21, 23.iv.1985, RWB, CRN (BYU); Custer Co., Hwy 75 above Obsidian, Fourth of July Ck, 23.iv.1985, RWB, CRN (BYU), border Blaire and Custer Cos., Salmon R., 23.iv.1985, RWB, CRN (BYU), Little Boulder Ck nr jct. E Fk Salmon R., 8.iv.1977, RWB, Sakaguchi (BYU), Elk Ck, Elk Ck Cpgd, 23.iv.1985, RWB, CRN (BYU), Salmon R., Hwy 75, 8 mi. N Stanley, 23.iv.1985, RWB, CRN (BYU). **Montana:** Flathead Co., N Fk Flathead R., 31.x.1982, J. Gustin (WSU). **Nevada:** White Pine Co., Baker Ck, Baker Cpgd, 8.v.1976, RWB (BYU). **Oregon:** Wallowa Co., Lostine R., Pole Bridge Picnic Area, 19.v.1977, RWB, Dunster (BYU). **Utah:** Cache Co., Logan Cyn, 27.iii.1938, GFK, FCH (USUL), 20.x.1953, R.P. Olson (USUL), 20.iv.1937, 8.xi.1973, GFK (USUL), 3.vi.1970, WJH (USUL), 26.iv.1970, T.L. Whitworth (USUL), 11.iv.1981, H.V. Egoscue (USUL), 15.ii.1966, J. Karlsson (USUL); Duchesne Co.,

Yellowstone R., Yellowstone Cpgd, 23.iii.1984, RWB, CRN (BYU), Hyde Pk, 31.x.1953, R.P. Olson (USUL); Salt Lk Co., Hwy 210, 3 mi. E Temple Granite Quarry, Little Cottonwood Cyn, 14.iii.1981, S.M. Clark (BYU). **Wyoming:** Lincoln Co., Bridge Cpgd, 2 mi. SE Alpine jct., Greys R., 10.iv.1985, RWB, CRN (BYU); Strawberry Ck above Bedford, 11.iv.1985, RWB, CRN (BYU), Little Greys R., jct. Greys R., 10.iv.1985, RWB, CRN (BYU); Sublette Co., jct. Hoback R., Hwy 189, Bull and Cliff Cks, 10.iv.1985, RWB, CRN (BYU).

Recognition. This species is distinguished by the dark tubercle on the lower third of the face, spotted wings, and short dorsocentral bristles with numerous setulae interspersed.

Description (wing length 4–4.5 mm).

Male. Large dark tubercle on lower third of face, lateral margins with blue pruinescence; remainder of face and head with blue pruinescence, except dark ocellar triangle. Arista-like stylus short and stout. Postocular setae dark.

Scutum with narrow faint pale bivittae. Pleura (upper half of anepisternum dark), notopleuron, postpronotal lobe, and scutellum with faint blue pruinescence; posterior margin of postpronotal lobe with orange–brown spot (Fig. 4*b*). Several short acrostichals anterior to first dc; 5 dc, 2–4 short setae interspersed between each dc bristle; 1 pprn; 1 presut spal and several short setae; 2 or 3 npl; 2 psut spal and several short setae; 1 pal; 2 sctl. Proepisternum with several lateral setae. Laterotergite with patch of short pale setae. Posterior margin of anepisternum lacking setulae.

Wing (Fig. 30*j*) infuscate; faint round spots at base of R_4, apical margin of cell dm, tip of cell bm, and r-m crossvein; stigma beyond R_1 indistinct and slender; 1 short basal costal seta; costal spines lacking. Veins straight; auxiliary crossvein between R_4 and R_{2+3} absent; stump vein absent; R_4 bell-shaped; anal vein an elongate streak; cell dm truncate. Halter light brown.

Coxae with pale-blue pruinescence, remaining leg segments dark. Anterior surface of forecoxa with dark erect setae. Forefemur with anterior subapical comb; ventral surface with dense white pile; anterior surface bare; anteroventral surface with row of short black setae, increasing in length at base; basal fifth with posteroventral row of 6–8 slender pale setae. Base of forefemur inflated. Foretibia with biserial row of stout setae beneath. Erect setae of hind tibia poorly differentiated. Fifth tarsomere with dorsoapical extension; empodium brown laterally.

Abdominal sternites and lateral margin of tergites with blue pruinescence, remaining tergites brown. Hypandrium long, narrow, and slightly arched. Clasping cercus with narrow, tapered apex; inner apical surface with dark spines. Phallus long and slightly arched; distiphallus of phallus very long, extending back to near base of epandrium. Surstylus truncate, shorter than clasping cercus; ventroapical margin slightly projecting (Fig. 17*a*).

Female. Very similar to male, including leg chaetotaxy.

Geographic distribution and seasonal occurrence. This species is widespread along the ranges in western North America, extending from Alaska to Nevada and Utah (Map 17). The records are somewhat scattered, likely owing to the apparent primary flight period occurring during colder months. For example, in Alberta (Banff) specimens have been collected on the snow in November and December and teneral specimens were collected in broad creeks in mid-September (Kananaskis, Alberta; Fig. 2*c*).

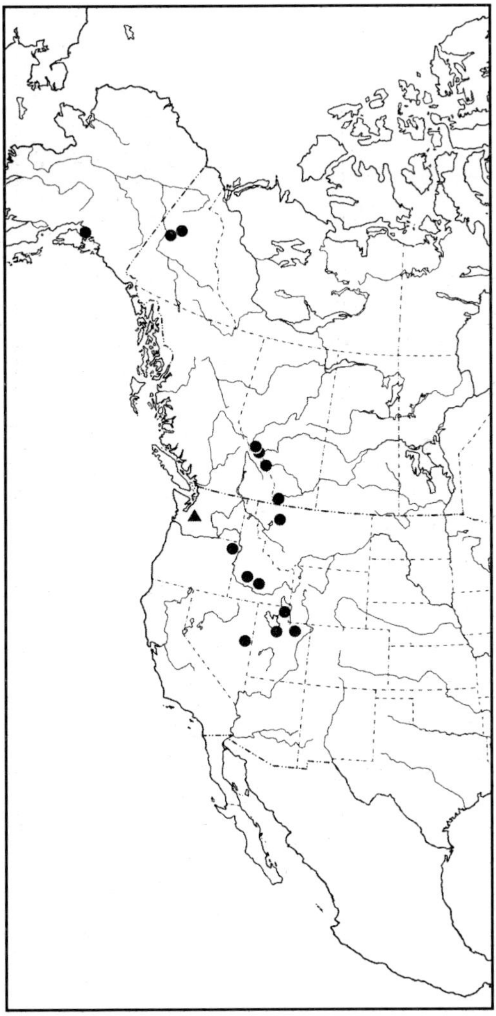

Map 17. Known distribution of *Clinocera gubernans* (●); *C. rainericola* (▲).

Clinocera guttipennis sp. nov.

Figs. 17*b*, 31*a*; Map 14

Type material. HOLOTYPE, male labelled "U.S.A.: WASHINGTON:/ Mt. Baker N.F. along/ Thompson Creek near/ Glacier Cr. Rd. 400 m/ 4-x-1979 D.Dee Wilder"; "D. Dee Wilder/ Collection [yellow label]; Collection of the/ California Academy/ of Sciences, San/ Francisco, Calif."; "HOLOTYPE/ Clinocera/ guttipennis/ Sinclair [red label]" (CAS).

Paratypes: **CANADA. British Columbia:** Keremeos, 10.x.1982, SGC (1 ♂, UBC); Manning PP, Coldspring Cpgd, Buckhorn Ck, 7.x.1979, DDW (12 ♂♂, 15 ♀♀, CAS); Manning PP, Goodfellow Ck, Hwy 3, 8.x.1979, DDW (10 ♂♂, 11 ♀♀, CAS). **USA. Washington:** Gifford Pinchot NF, La Wis Wis along Ohanapecosh R., 390 m, 1.x.1979, DDW (33 ♂♂, 12 ♀♀, CAS); Lewis Co., Gifford Pinchot NF, Miller Ck nr Randle, 290 m, 1.x.1979, DDW (2 ♂♂, 8 ♀♀, CAS); Mt Baker NF, Glacier Ck Rd, Thompson Ck, 400 m, 4.x.1979, DDW (12 ♂♂, 3 ♀♀, CAS); Mt Rainier NP nr White R. Cpgd, White R., 3.x.1979, DDW (4 ♂♂, 1 ♀, CAS).

Recognition. This species is distinguished by the long dorsocentral bristles without interspersed setulae, and wings with large circular spots.

Description (wing length 4.4–5 mm).

Male. Large dark tubercle on lower half of face, lateral margins with blue pruinescence; remainder of face and head with blue pruinescence, except dark ocellar triangle. Arista-like stylus short and stout. Postocular setae dark.

Scutum with faint narrow dark vitta bordered by a pair of bluish vittae, the latter more distinct near scutellum. Pleura (upper half of anepisternum dark), notopleuron, postpronotal lobe, and scutellum with faint blue pruinescence; posterior margin of postpronotal lobe with orange–brown spot. Several short acrostichals anterior to first dc; 5 dc; 1 pprn; 1 or 2 presut spal; 2 or 3 npl; 2 psut spal; 1 pal; 2 sctl. Laterotergite with patch of short dark setae. Proepisternum with several pale lateral setae. Posterior margin of anepisternum lacking setulae.

Wing (Fig. 31*a*) infuscate; faint round spots at base of R_4, apical margin of cell dm, tip of cell bm, and r-m crossvein; stigma beyond R_1 distinct and elongate; 1 short basal costal seta; costal spines lacking. Veins straight; auxiliary crossvein between R_4 and R_{2+3} absent; stump vein absent; R_4 bell-shaped; anal vein an elongate streak; cell dm truncate. Halter brown, paler near base.

Coxae with pale-blue pruinescence, remaining leg segments dark. Anterior surface of forecoxa with dark erect setae. Forefemur with anterior subapical comb; ventral surface with dense white pile; anterior surface bare; anteroventral surface with row of short black setae, increasing in length at base; basal fourth with posteroventral row of short slender setae; base of forefemur inflated. Foretibia with biserial row of stout setae beneath. Erect setae of hind tibia poorly differentiated. Fifth tarsomere with dorsoapical extension; empodium brown laterally.

Abdominal sternites and lateral margin of tergites with blue pruinescence, remaining tergites brown. Hypandrium long and narrow. Clasping cercus finger-like, with long spines on inner dorsoapical tip. Phallus long and straight, apex with scale-like spines; distiphallus very long, ending near base of surstylus. Surstylus truncate, shorter than clasping cercus; ventroapical corner attenuated (Fig. 17*b*).

Female. Very similar to male, including leg chaetotaxy.

Geographic distribution and seasonal occurrence. This species is presently known only from southern British Columbia and Washington State (Map 14). Apparently the range of *C. guttipennis* has not penetrated very far north into previously ice-covered land. All material was collected in October.

Etymology. The specific name is from the Latin *gutta* ("spot") and *penna* ("wing"), referring to the spotted wings.

Clinocera lecta Melander

Figs. 17*c*, 31*b*; Map 18

Clinocera lecta Melander, 1902: 243.

Clinocera lecta: Melander, 1928, 1965: 469.

Type material examined. LECTOTYPE (here designated), male labelled "Lewiston/ Idaho[USA]"; "W.M. Wheeler/ Collection."; "TYPE/ No./ A.M.N.H. [red label]"; "Am. Must. Nat. Hist./ Dept. Inverte. Zool./ No. 811 [head and right midleg missing]" (AMNH). My lectotype label "LECTOTYPE/ of Clinocera/ lecta Mel./ des. B.J. Sinclair 1991 [red label]" has been attached to this specimen. The lectotype is designated in order to clarify the application of the name to this taxon.

Paralectotypes: Kendrick, Idaho (1 ♂, AMNH; 1 ♀, USNM). My paralectotype labels have been additionally attached to these specimens.

Additional material examined. Some 200 specimens from the following provinces and states: **CANADA. Alberta:** Banff, 7.ix.1922, CBDG (CNC); Banff NP, 14 mi. W, 4500′, GES (CNC); Canmore, Grassi Lks., lower tr., 17.ix.2003, BJS (ZFMK); Peter Lougheed PP, Lk. Rawson, 2020 m, 18.ix.2003, BJS (ZFMK); Peter Lougheed PP, Hwy 40, roadcut seep, 20.vii.1987, SAM (DEBU); 5 km W Lk. Louise, seep nr Lk. Agnes, 23.vi.1989, AB (CNC); Plateau Mtn Ecological Res., 2200 m, 19.ix.2003, BJS (ZFMK). **British Columbia:** Glacier NP, 3.3 km W Rogers Pass Information Centre, roadcut seep, 1.vii.1989, BJS (CNC), Illecillewaet Cpgd, Asulkan Ck, 1.vii.1989, BJS (CNC); 1250 m, str., 30.vi.1989, BJS (CNC); 4 km W Glacier NP, str., 1.vii.1989, BJS (CNC); Kootenay NP, Radium Hot Springs + Hwy 93, roadcut seep and Sinclair Ck, 17.ix.2005, BJS (ZFMK); Manning PP, Hwy 3, Goodfellow Ck, 8.x.1979, DDW (CAS); Mt Revelstoke NP, Eva Lk. Tr., 1900 m, str. nr Miller Lk., 12.ix.2003, BJS (ZFMK), Inspiration Woods Tr., 620 m, 12.ix.2003, BJS (ZFMK); 12 km N Revelstoke, Hwy 23, St. Cyr Ck, 2.vii.1989,

BJS (CNC); Robson, 21.x.1965, HRF (UBC); 7 km SE Rogers Pass, 23.vii.1990, AB (CNC); Squamish, Diamond Head Tr., 3300′, 13.viii.1953, WRMM (CNC); Yoho NP, Yoho Lk. Tr., 10.ix.2003, BJS (ZFMK). **USA. Alaska:** Richardson Hwy, mi. 14, Horsetail Falls, 18.vii.1993, TS (BLKU). **Arizona:** Cochise Co., 12 km S Sierra Vista, Ramsey Cyn, MT, 1700 m, 3–17.xii.1986, BVB (CNC), Coronado Co., Huachuca Mtns, Miller Cyn, Miller Ck, 3.vi.1991, BJS (CNC); Graham Co., Hospital Flat, Pinolena Mtns, 8950′, 3.viii.1965, HBL (CAS); Grand Cyn NP, CR3, Saddle Cyn, 10–21.v.1998, JFM (CNC, PERC), CR2 Bert's Cyn, 10–21.v.1998, JFM (CNC, PERC), Elves Cyn, 10–21.v.1998, JFM (CNC, PERC). **California:** Fresno Co., Huntington Lk., 2200 m, Midge Ck, 13.viii.1984, JFM (PERC), Sierra NF, Kaiser Pass, 2700 m, 1.viii.1979, DDW (CAS), E and W Fks Camp 61, Forebay Lk., 31.vii.1979, DDW (CAS); Inyo Co., Lone Pine Ck, Lone Pine Cpgd, 25.i.1985, RWB, CRN (BYU); Los Angeles Co., Hwy 39, 3 mi. S Coldbrook Stn, N Fk San Gabriel R., 1.v.1978, DDW (CAS), 5 mi. E Islip Saddle, Cortelyou Spr., 25.iv.1977, DDW (CAS); Nevada Co., Sagehen Ck, 27.vi.1972, R.A. Belmont (UCD); Plumas Co., Lassen NF, Domingo Springs, 31.v.1976, DDW (CAS), Lassen NF, N Fk Feather R., 28.v.1977, DDW (CAS); Tulare Co., Three Rivers, 29.iii.1986, D.J. Burdick (PERC); Tuolumne Co., Sonora Pass, 9000′, 22.vii.1951 W.C. Bentinck (UCB). **Colorado:** Gothic, 9500′, 19.vi.1934, CPA (AMNH); Hinsdale Co., Sherman, Cottonwood Ck, 10.x.1993, SF, AF (CSU); Jackson Co., 2 mi. W Cameron Pass, 20.ix.1984, RLH (MSUB); Larimer Co., Buckhorn Ck jct. Sheep Ck, 5.x.1995, SF, AF (CSU); Monarch Pass, 21.vi.1940, ALM (USNM); Rio Grande Co., Race Ck, MT, 10 500′, 23.vi.1972, WWW (USNM); San Juan Co., nr jct. Hwy 550, Cascade Ck, 12.x.1996, SF, AF (CSU). **Idaho:** Boise Co., Cyn Ck, Hwy 21, N jct. S Fk Payette R., 23.iv.1985, RWB, CRN (BYU, in alc.); Bonner Co., 10 mi. NW Samuels, 26.ix.1969, A.G. Grittins (UIM); Cassia Co., Howell Ck, Bennett Springs Cpgd, 22.iv.1985, RWB, CRN (BYU, in alc); Idaho Co., Hwy 12, 33 mi. NE Lowell, Holly Ck, 2600′, 19.viii.1969, EMF (CAS); Latah Co., 4.1 mi. S Juliaetta, 28.iii.1975, D.F. Veirs (WSU). **Montana:** Coram, Bad Rock Cyn, Shepard Memorial Fountain, 28.viii.1981, PHA (CAS). **Nevada:** Clark Co., Spring Mtns, Mt Charleston, Big Falls, 2600 m, 8.vi.1997, PHA (USNM); Humboldt Co., Pine For. Range, Big Ck, 27.iv.1977, A.L. Sheldon (BYU); White Pine Co., Great Basin NP, Wheeler Cpgd, Lehman and Backer Cks, 3–18.vii.2001, JFM (PERC). **New Mexico:** Lincoln Co., 9.7 km W Alto, Lincoln NF, cyn nr Ski Apachi, 3120 m, 33°23′N, 105°47′W, 21–24.vi.1995, S.D. Gaimari (CNC). **Oregon:** Hood R. Co., Starvation Ck SP, at falls, 26.vi.1978, DDW (CAS), Mt Hood NF, Ash Ck, E Fk Hood R., 27.vi.1878, DDW (CAS), Mt Hood NF, 3.5 mi. below Timberline Lodge, roadcut seep, 11.vii.1989, BJS (CNC); Lane Co., 8 mi. SE Cougar Dam, 880 m, Grasshopper Ck, 11.vii.1982, GWC (OSU); Malheur Co., 13 mi. W Harper, 30.v.1983, RLH (MSUB); Union Co., 28 mi. SSW La Grande, Wallowa–Whitman NF, spr., 23.vi.1976, WJT (WSU); Wallowa Co., 39 mi. N Enterprise, Hwy 3, 3400′, seep area, 21.vi.1976, WJT (WSU). **Utah:** Box Elder Co., One Mile Ck above Standrod, 28.iii.1979, RWB, D. Webb (BYU), N Fk Ck, Timpanogos Tr., 21.ix.1985, RWB (BYU); Cache Co., H. Hodges Cyn, 23–27.ix.1976, GFK, WJH

(UIM); Utah Co., Rock Cyn, 9.iii.1963 (WSU), S Fk American Fk R., 7.viii.1978, RWB, Unzicker (BYU, in alc.). **Washington:** Asotin Co., Fields Spring SP, 4 mi. S Anatome, 24.v.1975, NEW (WSU), 3500–4000′, 31.v., 30.vi.1975, WJT (WSU); Lewis Co., Mt Rainier NP, Stevens Cyn Rd, Stevens Ck, 24.viii.1973, WJT (WSU); Mt Baker NF, Glacier Ck Rd, Grouse Ck, 1100 m, 4.x.1979, DDW (CAS), nr Glacier Ck Rd, 400 m, Thompson Ck, 4.x.1979, DDW (CAS); Mt Rainier NP, Comet Falls Tr., above van Traump Ck, 4500′, E.J. Davis (WSU), Christine Falls, roadcut seep, 8.vii.1989, BJS (CNC), White R. Cpgd, 1220 m, White R., 3.x.1979, DDW (WSU); Whatcom Co., Mt Baker NF, Bagley Ck, 850 m, 10.x.1979, DDW (CAS), Razor Hone Ck, 1100 m, 9.x.1979, DDW (CAS). **Wyoming:** Albany Co., Snowy Range, 25.viii.1956, P. Freytog, (UWL); Carbon Co., Lk. Marie, 20 km W Centennial, 3230 m, 1.viii.1973, PHA (CAS); Yellowstone NP, Cyn Camp, 12.viii.1918, ALM (USNM); Yellowstone NP, 8.7 km W Eastgate, roadcut seep, 20.vii.1989, BJS (CNC).

Recognition. The round-spotted wings of this species are unique in North

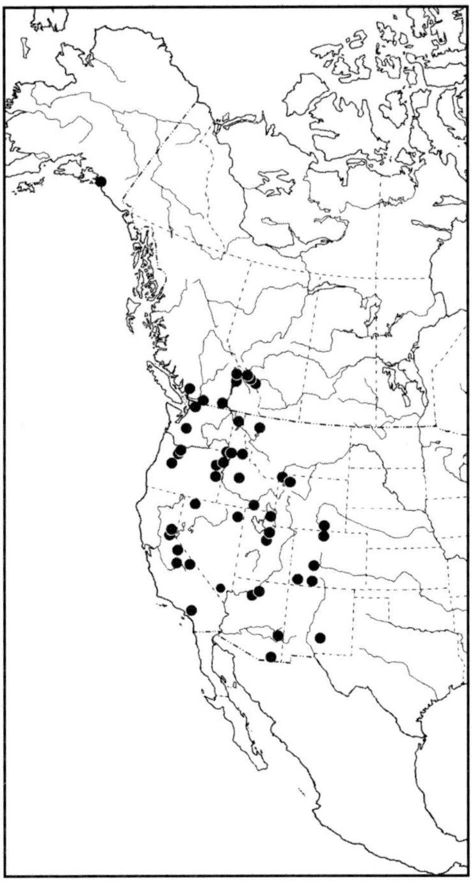

Map 18. Known distribution of *Clinocera lecta*.

America, similar to the patterned wings of the European genus *Phaeobalia* Mik, which led to its assignment to this former subgenus of *Clinocera* (e.g., Melander 1928, 1965).

Description (wing length 4–4.8 mm).

Male. Lower half of face with blue pruinescence; lower half bulbous. Gena and posterior region of head with tinge of blue pruinescence; vertex dark. Arista-like stylus short and slender. Postocular setae mostly pale.

Scutum with two faint black vittae. Pleura (except lower half of anepisternum), notopleuron, and midregion of scutellum with pale-blue pruinescence. Several inconspicuous acrostichals near antepronotum; 5 dc; 1 pprn; 1 presut spal; 2 npl, lower short and slender; 2 psut spal; 1 pal; 2 sctl. Laterotergite with patch of long pale setae; posterior margin of anepisternum lacking setulae.

Wing (Fig. 31*b*) hyaline, infuscate along margin of veins; 1 long basal costal seta; stigma beyond R_1, distinct, elongate, elliptical, slightly deflecting R_{2+3}; costal spines lacking. Three distinct wing spots: circular spot at base of R_4; transverse clouding on distal cell dm crossveins, often extending to R_{4+5} and onto bases of M_1 and M_2 (distal crossveins); clouding on r-m vein, often extending to cell cu*p* (proximal crossveins). Auxiliary crossvein between R_4 and R_{2+3} absent; stump vein absent; R_4 bell-shaped; anal vein faint; cell dm slightly produced apically. Halter dark.

Coxae with blue pruinescence, remaining segments dark. Forecoxa with short pale setae. Forefemur with preapical anterior comb present; antero- and postero-ventral row of short strong setae along entire length; ventral surface with dense white pile. Foretibia with short strong setae beneath. Mid and hind femora lacking preapical setae. Hind tibia with erect dorsal and ventral bristles on apical half. Fifth tarsomere lacking dorsoapical extension; empodium white laterally.

Abdominal sternites and lateral margin of tergites with faint blue pruinescence, remaining tergites brown. Hypandrium gradually tapered. Phallus gently arched, lacking split at base of distiphallus; distiphallus long and curved. Clasping cercus narrow, arched dorsally to expanded, curved tip; peg-like setae on dorsoapical margin and apex of inner face. Surstylus two-thirds length of clasping cercus, apex produced ventrally (Fig. 17*c*).

Female. Similar to male except as follows: face dark; forefemur lacking biserial row of bristles beneath.

Geographic distribution and seasonal occurrence. This species is widespread and fairly common along the western Cordillera from Alaska to Arizona (Map 18). The flight period extends from June to October.

Remarks. Round-spotted wings are rare in *Clinocera*, although they are commonly observed in species of *Phaeobalia* and an undescribed species of *Clinocera* from Japan.

A pale form of *C. lecta* was found in isolated populations in Grand Canyon

National Park. This population is distinguished by shorter wings, slightly pale halteres, and slightly yellowish ventral and inner surfaces of the femora. However, there appears to be a southward grade in these features and no clear separation. The male terminalia of the Grand Canyon population show no significant differences from those of other populations.

Clinocera schlinkerti sp. nov.

Figs. 5*a*, 17*d*, 31*c*; Map 19

Type material. HOLOTYPE, male labelled "CANADA:BRIT[ish]. COLUMBIA:/ Manning Prov.Pk. Cold-/ spring Cmpgrd. along/ Buckhorn Creek 7 Oct./ 1979 D.D.Wilder coll."; "D. Dee Wilder/ Collection [yellow label]; Collection of the/ California Academy/ of Sciences, San/ Francisco, Calif."; "HOLOTYPE/ Clinocera/ schlinkerti/ Sinclair [red label]" (CAS).

Paratypes: **CANADA. Alberta:** Waterton Pk, 4200′, 1.x.1957, JRV (2 ♀♀, CNC). **British Columbia:** Manning PP, Coldspring Cpgd, Buckhorn Ck, 7.x.1979, DDW (9 ♂♂, 8 ♀♀, CAS); Manning PP, Goodfellow Ck, Hwy 3, 8.x.1979, DDW (28 ♂♂, 25 ♀♀, CAS).

Recognition. This species is distinguished by the scattered dark stiff setae of the forecoxa, hyaline wings, cloudy regions absent, and dark stiff setae of the proepisternum.

Description (wing length 4–4.8 mm).

Male. Large tubercle on lower third of face, lateral margins with blue pruinescence; remainder of face and head with blue pruinescence, except dark ocellar triangle. Arista-like stylus long and stout. Postocular setae dark.

Scutum with pale median blue pruinescent vitta extending to prescutellar depression and pair of lateral vittae between dc and spal. Pleura (upper half of anepisternum dark), notopleuron, postpronotal lobe, and scutellum with faint blue pruinescence; posterior margin of pprn lobe with orange–brown spot (Fig. 5*a*). Several short acrostichals anterior to first dc; 5–6 dc, 2–4 short or finer setae between each dc bristle; 1 pprn; 1 presut spal and several short stout setae; 2 or 3 npl; 2 psut spal and several short stout setae; 1 pal; 2 sctl. Laterotergite with patch of short pale setae. Proepisternum with some 6 stout dark lateral setae. Posterior margin of anepisternum lacking setulae.

Wing (Fig. 31*c*) hyaline, with faint clouding or spots absent; stigma beyond R_1 indistinct and slender; 1 short basal costal seta; costal spines lacking. Veins straight; auxiliary crossvein between R_4 and R_{2+3} absent; stump vein absent; R_4 bell-shaped; anal vein an elongate streak; cell dm truncate to extended. Halter light brown at base.

Coxae with pale-blue pruinescence, remaining leg segments dark. Anterior surface of forecoxa with dark erect setae. Forefemur with reduced anterior subapical comb; ventral surface with dense white pile; anterior surface bare; anteroventral surface with row of short black setae, increasing in length at base; base of forefemur inflated with 4–6 long setae. Foretibia with biserial row of stout setae beneath; posteroventral apex of tibia with long setae. Erect setae of hind tibia poorly differentiated. Fifth tarsomere with dorsoapical extension; empodium pale brown laterally.

Abdominal sternites and lateral margin of tergites with faint blue pruinescence, remaining tergites brown. Hypandrium long and narrow. Clasping cercus with narrow base, expanded distally into subtriangular tip; inner apical surface with dark spines. Phallus long and slightly arched; distiphallus very long, longer

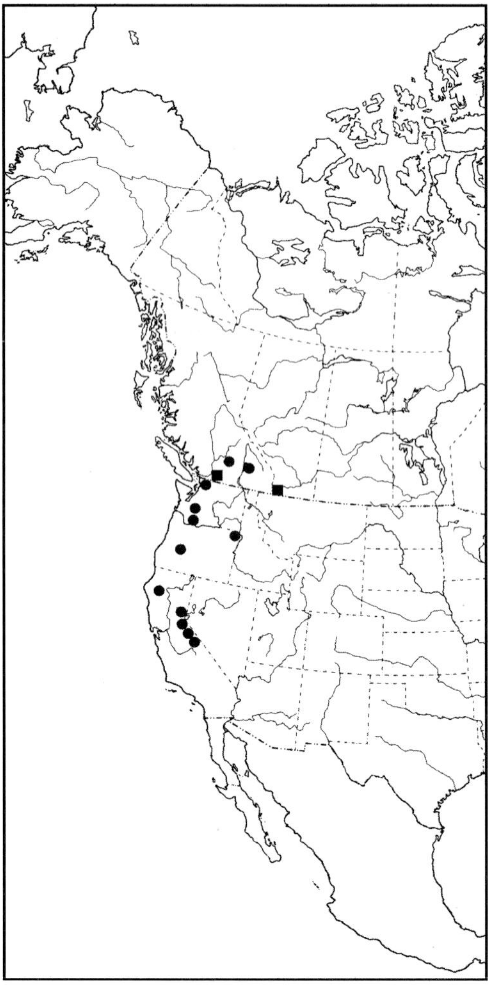

Map 19. Known distribution of *Clinocera schlinkerti* (■); *C. wallowa* (●).

than shaft of phallus. Surstylus truncate, shorter than clasping cercus; ventroapical margin slightly projecting (Fig. 17*d*).

Female. Similar to male, including leg chaetotaxy.

Geographic distribution and seasonal occurrence. This species is known only from two localities in the Cascade Range and Rocky Mountains (Map 19), collected in October.

Etymology. The specific name is in honour of the Schlinkert family of Bonn, Germany, whose generosity has greatly assisted the education of students at Bonn International School.

Clinocera stigma sp. nov.

Figs. 4*c*, 18*a*, 31*d*; Map 20

Type material. HOLOTYPE, male labelled "USA: WASHINGTON/ Mt. Rainier Nat.Pk./ nr.Paradise 8.vii.1989/ Edith Ck. BJ Sinclair"; "HOLOTYPE/ Clinocera/ stigma/ Sinclair [red label]" (CNC).

Paratypes: **CANADA. Alberta:** Jasper NP, 63 km S Jasper, 28.vi.1989, BJS (3 ♀♀, CNC). **British Columbia:** Glacier NP, Illecillewaet Cpgd, 1250 m, ypans, str., 31.vi.–2.vii.1989, BJS (8 ♂♂, 3 ♀♀, CNC), Asulkan Brook, 1.vii.1989, BJS (1 ♂, CNC), 30.vi.1989, cascading str., BJS (4 ♂♂, 2 ♀♀, CNC); Manning PP, Three Falls Tr., mossy str., 4.vii.1989, BJS (10 ♂♂, 3 ♀♀, CNC). **USA. Oregon:** Mt Hood NF, 12 km E Gov't Camp, Rd 48, cold str., 11.vii.1989, BJS (3 ♂♂, 3 ♀♀, CNC), 3.5 mi. below Timberline Lodge, cold str., 11.vii.1989, BJS (2 ♂♂, 1 ♀, CNC). **Washington:** Mt Rainier NP, nr Paradise, Paradise R., 8.vii.1989, BJS (4 ♂♂, 3 ♀♀, CNC), same data as holotype (6 ♂♂, 4 ♀♀, CNC).

Additional material examined. CANADA. Alberta: Banff, on snow, 8.iv.1928, OB (1 ♂, CAS); Kananaskis Country, Peter Lougheed PP, Lk. Rawson, 16.ix.2003, BJS (1 ♀, ZFMK); Lk. Louise, 5 km W, waterfalls at Lk. Agnes, 23.v.1989, AB (23 ♂♂, 3 ♀♀, CNC). **British Columbia:** Glacier NP, Illecillewaet Cpgd, 20.ii.1980, J.G. Woods (1 ♂, 3 ♀♀, CNC); Glacier NP, Perly Rock Tr., Illecillewaet, 22.ix.2005, BJS (3 ♂♂, 3 ♀♀, ZFMK), 10 km NE Nakusp, CD1327, 24.ii.1991, AB (2 ♂♂, 4 ♀♀, CNC); Manning PP, Goodfellow Ck, Hwy 3, 8.x.1979, DDW (65 ♂♂, 63 ♀♀, CAS), King Rd to Lookout, unnamed ck, 8.x.1979, DDW (67 ♂♂, 59 ♀♀, CAS). **USA. Alaska:** Anchorage, Alaska Insect Project, 29.iv., 5.v.1948, R.I. Sailer (2 ?, 8 ♂♂, 10 ♀♀, USNM); Anchorage, Flat Top Mtn, 550–600 m, 2.vi.1994, TS (28 ♂♂, 4 ♀♀, BLKU); Aklutna, Alaska Ins Project, 29.iv.1948, S. Lienk (3 ♀♀, USNM). **California:** Echo Lk., Tahoe, 2.vi.1940, T.G.H. Aitken (1 ♀, CAS). **Idaho:** Blaine Co., 6 mi. NE Sun Valley, Corral Ck, 22.iv.1985, RWB, CRN (2 ♂♂, 14 ♀♀, BYU). **Washington:** Mt Baker, NE slope, 900 m, 27.v.1973, JRV (10 ♂♂, 9 ♀♀, CNC); Mt Baker NF, 720 m, Grouse Ck nr Glacier Ck Rd, 4.x.1979, DDW (130 ♂♂, 100 ♀♀, CAS), Bagley Ck, 850 m, 10.x.1979, DDW (125 ♂♂, 118 ♀♀, CAS),

Heather Meadows, 1100 m, Razor Hone Ck, 9.x.1979, DDW (2 ♂♂, 2 ♀♀, CAS), ck nr Silver Fir Cpgd, 610 m, 10.x.1979, DDW (3 ♂♂, 3 ♀♀, CAS); Mt Rainier NP, 720 m, Panther Ck nr Ohanapecosh R., 2.x.1979, DDW (1 ♂, 6 ♀♀, CAS), 1050 m, Edith Ck, Paradise, 11.x.1979, DDW (164 ♂♂, 197 ♀♀, CAS), 1220 m, White R. nr White R. Cpgd, 3.x.1979, DDW (2 ♀♀, CAS), ck E White R. Cpgd, 1340 m, 3.viii.1977, DDW (1 ♀, CAS), ck E end Yakima Pk, 1460 m, 3.x.1979, DDW (1 ♂, 4 ♀♀, CAS); Whatcom Co., Glacier, 25.v.1965, F. Schmid (1 ♂, CNC).

Recognition. The species is distinguished from related species by its small head with small facial swelling, long slender legs, forefemur lacking distinct ventral setae, and hyaline wings, usually with a broad dark stigma. In addition, the costa lacks short spines, the pleural sclerites are completely clothed in blue pruinescence, and the scutum lacks vittae.

Description (wing length 4.2–5.8 mm).

Male. Head small in proportion to body. Face with light-blue pruinescence; lower half bulbous. Gena and posterior region of head with tinge of blue pruinescence; vertex dark. Arista-like stylus long and slender. Postocular setae mostly pale.

Scutum lacking distinct markings. Pleura, notopleuron, and posterior half of prescutellar depression with blue pruinescence. Several pale acrostichals near collar; 5 dc; 1 pprn; 1 presut spal; 2 npl, lower shorter; 2 psut spal, anterior bristle short; 1 or 2 pal; 2 sctl. Laterotergite with pale setae; posterior margin of anepisternum lacking setulae. Halter dark.

Wing (Fig. 31*d*) hyaline, wide and elongate; clouding around veins absent; stigma beyond R_1 dark, nearly as wide as cell r_1, slightly deflecting R_{2+3}, truncate proximally; 1 short basal costal seta; costa spines lacking. Auxiliary crossvein between R_4 and R_{2+3} absent; stump vein absent; R_4 bell-shaped; anal vein faint; cell dm slightly produced apically.

Coxae with blue pruinescence, remaining segments dark brown. Forecoxa with short dark setae. Forefemur with preapical anterior comb; ventral surface with dense white pile on basal two-thirds; apical third with anteroventral row of fine setae. Foretibia with short stout ventral setae. Mid and hind femora lacking preapical setae. Hind tibia with erect dorsal and ventral setae on apical third. Fifth tarsomere lacking dorsoapical extension; empodium pale brown laterally.

Abdominal sternites and lateral margin of tergites with faint blue pruinescence, remaining tergites brown. Hypandrium large, gradually tapering. Phallus gently curved; distiphallus long and curved. Clasping cercus narrow, tapered to nipple-like tip; dorsomedial spines on apical third. Surstylus shorter than clasping cercus; apex round and produced ventrally (Fig. 18*a*).

Female. Similar to male except as follows: face light brown; forefemur lacking anteroventral row of setae.

Geographic distribution and seasonal occurrence. This species is widespread in western North America, known from widely disjunct localities from Alaska to California (Map 20). Most flight records are from cooler or cold months (e.g., February, October, April) or early July in the mountains (Fig. 2*d*). There is only one record from August and perhaps this is from a small cold microclimate. The broad gap between Alaskan and southern British Columbian populations is likely a collecting artefact resulting from the lack of insect surveys during colder months, when this species is most common.

Etymology. The specific name is in reference to the easily observed large, broad dark stigma.

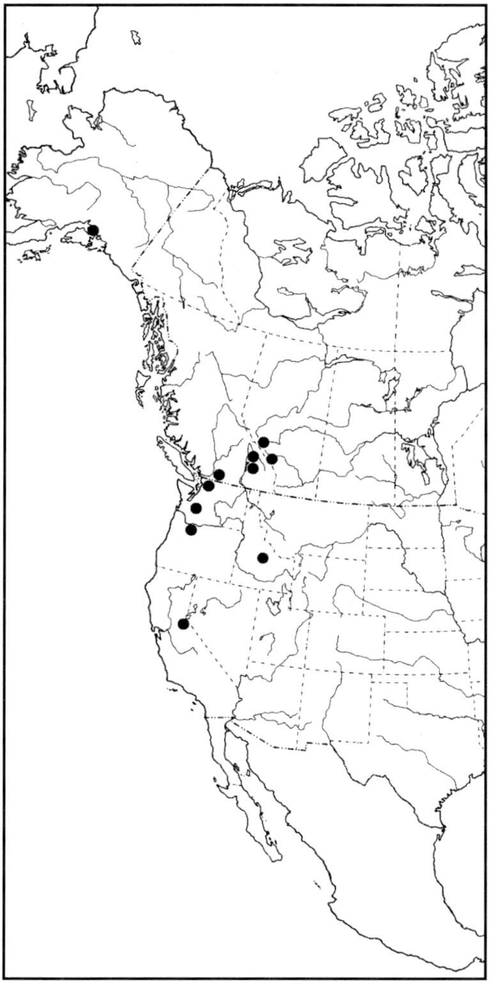

Map 20. Known distribution of *Clinocera stigma*.

Clinocera wallowa sp. nov.

Figs. 18*b*, 31*e*; Map 19

Type material. HOLOTYPE, male labelled "[USA] Oregon: Wallowa Co./ 5.5 miles W. Troy,/ along Wenaha River/ April 7, 1984, J./ Jenkins collector"; "HOLOTYPE/ Clinocera/ wallowa/ Sinclair [red label]" (WSU).

Paratypes: **CANADA. British Columbia:** Chase, waterfall, 13.vii.1974, PHA (1 ♂, CAS); Halfway R., 11 km E Hwy 23 and Upper Arrow Lk., 21.ix.2005, 50°30.17′N, 117°47.21′W, BJS (4 ♂♂, 8 ♀♀, CNC, ZFMK). **USA. California:** El Dorado Co., Middle Ck, China Flats Cpgd, 21.iii.1985, RWB, CRN (2 ♀♀, BYU); Fresno Co., 3.4 mi. W Garnet Dike, Kings R. (main fk), 12.xi.1986, J.A. Halstead (1 ♀, PERC); Plumas Co., Lassen NF, N Fk Feather R., 28.v.1977, DDW (10 ♂♂, 15 ♀♀, CAS); Tehama Co., Mill Ck, Hwy 36, Childs Meadows, 15.ii.1983, RWB, CRN (1 ♂, BYU); Yosemite NP, Merced R., Bridal Veil Ck, 10.iv.1977, DDW (10 ♂♂, 27 ♀♀, CAS). **Oregon:** same data as holotype (5 ♂♂, 5 ♀♀, WSU); Lane Co., Quartz Ck, 2 mi. N Blue R., 15.v., 6, 21, 27.x.1982, 25.iii., 15.v.1983, GWC (6 ♂♂, 7 ♀♀, OSU).

Additional material examined. USA. Washington: Gifford Pinchot NF, La Wis Wis along Ohanapecosh R., 390 m, 1.x.1979, DDW (53 ♂♂, 27 ♀♀, CAS); Lewis Co., Gifford Pinchot NF, Miller Ck nr Randle, 290 m, 1.x.1979, DDW (14 ♂♂, 15 ♀♀, CAS); Mt Baker NF, Glacier Rd, Thompson Ck, 400 m, 4.x.1979, DDW (43 ♂♂, 41 ♀♀, CAS); Mt Rainier NP, nr White R. Cpgd, 1220 m, White R., 3.x.1979, DDW (20 ♂♂, 15 ♀♀, CAS).

Recognition. Males of *C. wallowa* are characterized by a bluish facial tubercle, five dorsocentral bristles and wing infuscation primarily surrounding veins. The facial tubercle in females is dark.

Description (wing length 4–4.6 mm).

Male. Large blue pruinescent tubercle on lower half of face, lateral margins with small dark patch within blue pruinescence; remainder of face and head dark brown, except gena and postgena with blue pruinescence. Arista-like stylus short and stout. Postocular setae dark.

Scutum with bivittae. Pleura (upper and posterior margin of anepisternum and laterotergite dark) and notopleuron with blue pruinescence; posterior margin of postpronotal lobe with orange–brown spot. Several inconspicuous acrostichals anterior to first dc; 5 dc, 1 pprn; 1 presut spal; 2 npl, lower slender; 2 psut spal; 1 pal; 2 sctl. Laterotergite with patch of short pale setae. Proepisternum with several lateral setae. Posterior margin of anepisternum lacking setulae.

Wing (Fig. 31*e*) infuscate; faint clouding along veins; stigma beyond R_1 distinct and broad; 1 short basal costal seta; costal spines present. Veins straight; auxiliary crossvein between R_4 and R_{2+3} absent; stump vein absent; R_4 bell-shaped; anal vein forming long, distinct streak; cell dm produced apically. Halter dark, base of shaft pale.

Coxae with bright-blue pruinescence, remaining leg segments dark. Anterior surface of forecoxa with dark erect setae. Forefemur with anterior subapical comb; ventral surface with short white pile; anterior surface bare; anteroventral surface with row of short black setae, increasing in length at base; basal fourth with posteroventral row of slender setae. Base of forefemur inflated. Foretibia with biserial row of stout setae beneath. Erect dorsal and ventral setae on apical half of hind tibia. Fifth tarsomere with dorsoapical extension weakly developed; empodium brown laterally.

Abdominal sternites and lateral margin of tergites with faint blue pruinescence, remaining tergites brown. Hypandrium not greatly lengthened, nearly parallel-sided. Clasping cercus narrow, tapered to nipple-like apex; inner apical surface with dark spines. Phallus long and slightly arched; distiphallus short, not extending back to surstylus. Surstylus tapering to narrow tip, much shorter than clasping cercus; ventroapical margin slightly projecting (Fig. 18*b*).

Female. Similar to male, including leg chaetotaxy, except facial tubercle dark.

Geographic distribution and seasonal occurrence. This species is confined primarily to the far-western ranges, extending from southern British Columbia to California (Map 19). The range of *C. wallowa* has not penetrated very far north into previously glaciated regions. The flight period is primarily during the cooler months from September to April; a single adult has been recorded in July.

Remarks. This species is very similar to *C. stigma* and links the latter species with the *C. gubernans* complex on the basis of a number of synapomorphies, including facial coloration, apically narrowed clasping cercus, absence of costal spines, and fifth tarsomere with a weakly developed dorsoapical extension.

At the Halfway River locality in British Columbia, adults of *C. wallowa* inhabit a typical *Wiedemannia/Trichoclinocera* habitat (see Sinclair 1994). This site is a wide rocky creek or small river with no overhead canopy and with adults resting on emergent rocks (Figs. 1*d*, 2*e*).

Etymology. The specific name is derived from the type locality.

Clinocera lineata group

Diagnosis

This species-group is recognized by the following characters: absence of the stigma; basally strongly geniculate distiphallus; membranous region at base of distiphallus dorsoventrally expanded; hypandrium small in comparison with epandrium; surstylus subrectangular, often with slender apex; scutum often with dark median vitta; fifth tarsomere normally lacking dorsoapical extension; and small narrow wings (cell dm prolonged apically, clouding usually absent, auxiliary crossvein between R_4 and R_{2+3} absent).

Remarks

This species-group includes the smallest adults of *Clinocera*. They are thin-legged and generally more delicate than some of the other groups, especially the *C. conjuncta* group. The head is somewhat more delicate than that of the other species-groups, being rather small and ventrally pointed.

The monophyly of this group is supported on the basis of the following apomorphies: distiphallus stiff and strongly recurved apically and apex of phallus with wide membranous expansion.

Distribution

This is the most widely distributed species-group, known from the Nearctic, Neotropical, and Oriental regions. It is known also from Australia but not from New Zealand or Africa.

Several undescribed species are known from the Oriental Region and Japan, some of which may have been previously assigned to *Kowarzia* on the basis of facial setulae (see Sinclair 1995).

Included species

This species-group was first recognized by Sinclair (1995) and includes the following species:

NE	*Clinocera arizonica* **sp. nov.**
AU	*Clinocera australiana* Sinclair, 2000*b*: 349.
AU	*Clinocera bickeli* Sinclair, 2000*b*: 350.
NE	*Clinocera brevicauda* **sp. nov.**
OR	*Clinocera chilamche* Smith, 1965: 107.
OR	*Clinocera cuspidata* Wagner & Leese, 2004: 16 (holotype examined) (see Wagner et al. 2004).
OR	*Clinocera guangdongensis* Yang, Grootaert, & Horvat, 2005: 2.
OR	*Clinocera guizhouensis* Yang, Zhu, & An, 2006: 306.
AU	*Clinocera irrorata* Sinclair, 2000*b*: 353.
NE	*Clinocera lineata* Loew, 1862: 207.
OR	*Clinocera longicercus* Wagner & Leese, 2004: 16 (holotype examined) (see Wagner et al. 2004).
NE + NT	*Clinocera macdonaldi* **sp. nov.**
NE + NT	*Clinocera madera* **sp. nov.**
OR	*Clinocera marginesetosa* Wagner & Leese, 2004: 16 (holotype examined) (see Wagner et al. 2004).

NE	*Clinocera mexicana* **sp. nov.**
PA + OR	*Clinocera minutissima* (Vaillant, 1960: 172) (*Atalanta* (*Atalanta*)).
AU	*Clinocera monticola* Sinclair, 2000*b*: 355.
OR	*Clinocera nadi* Smith, 1965: 108 (holotype examined).
NT	*Clinocera nigriscutata* **sp. nov.**
NE + NT	*Clinocera obunca* **sp. nov.**
AU	*Clinocera queenslandica* Sinclair, 2000*b*: 355.
NE	*Clinocera robinsoni* **sp. nov.**
AU	*Clinocera rubriventris* Sinclair, 2000*b*: 356.
OR	*Clinocera sinensis* Yang & Yang, 1995: 235.
OR	*Clinocera sinica* (Yang & Yang, 1995: 237) (*Aclinocera*) **comb. nov.**
AU	*Clinocera spinosa* Sinclair, 2000*b*: 358.
PA + OR	*Clinocera stackelbergi* (Vaillant, 1960: 174) (*Atalanta* (*Kowarzia*)).
NE	*Clinocera sublineata* **sp. nov.**
OR	*Clinocera velutina* (Engel, 1931: 488) (*Kowarzia*) (see Sinclair and Saigusa 2005).
NT	*Clinocera woodi* **sp. nov.**
OR	*Clinocera wui* Yang & Yang, 1995: 236.

Key to species of the *Clinocera lineata* group north of Mexico

1 Face entirely clothed in blue pruinescence; scutum with narrow dark median stripe, flanked by wide shiny stripes 2

— Face with blue pruinescence on lower three-quarters, brown below antenna, or pruinescence reduced to narrow lateral stripes on lower half; scutum with wide dull dark median stripe, flanked by narrower pale brownish stripes (females generally difficult to distinguish) 3

2 Veins wavy (Fig. 31*g*); R_{2+3} often with auxiliary or stump crossvein to costa; lateral stripes of scutum blue (widespread) . . . . ***C. lineata*** Loew

— Veins straight (Fig. 32*a*); R_{2+3} lacking auxiliary crossvein; lateral stripes of scutum grey (eastern) . . . . ***C. sublineata*** sp. nov.

3 Face with lateral pale stripes on lower half (sometimes partially fused), upper half with median stripe; clasping cercus bearing setulae, with slightly hooked apex (Fig. 19*c*); if face mostly blue, clasping cercus narrow to somewhat expanded at midlength (southwestern) . . . . . ***C. madera*** sp. nov.

— Face with blue pruinescence; clasping cercus broad, bearing stout setae, apex straight to tapered (Figs. 18*c*, *d*, 19*b*) 4

4 Thoracic chaetotaxy greatly reduced; ocellar bristle very short and slender, shorter than postocular bristles (southwestern) . . . ***C. macdonaldi*** sp. nov.

— Thoracic chaetotaxy not reduced; ocellar bristle long and slender, longer than postocular bristles 5

5 Clasping cercus extending beyond surstylus, expanded apically (Fig. 18*c*) (southwestern) (female unknown) . . . . ***C. arizonica*** sp. nov.
— Clasping cercus short, extending slightly beyond surstylus, apex tapered (Fig. 18*d*) (southwestern) . . . . ***C. brevicauda*** sp. nov.

Key to species of the *Clinocera lineata* group of the Americas south of the United States

1 Scutum with dark spot at base of wing; pleura yellowish brown, lacking pruinescence 2
— Scutum lacking spot at base of wing; pleura dark brown with blue pruinescence 4

2 Scutum yellowish brown, with brown median stripe bordered by bright yellowish brown stripes (Mexico, Costa Rica) . . . . ***C. woodi*** sp. nov.
— Scutum mostly dark, except yellowish brown postpronotal lobe and notopleuron, stripe pattern not distinct 3

3 Clasping cercus straight, finger-like with broadly rounded or truncate apex (Fig. 20*a*) (Costa Rica) . . . . ***C. nigriscutata*** sp. nov.
— Clasping cercus tapered throughout to narrow tip, long and arched medially (Fig. 20*c*) (Mexico) (female unknown). . . . . ***C. robinsoni*** sp. nov.

4 Face largely clothed in pale pruinescence 5
— Lower half of face with faint lateral pruinescence, upper half with median pale stripe (females difficult to distinguish further). 7

5 Notopleuron brown, lacking pruinescence (Arizona, Mexico) . . . .
. . . . ***C. brevicauda*** sp. nov.
— Notopleuron with bluish pruinescence 6

6 Thoracic chaetotaxy not reduced; clasping cercus finger-like, gradually curved near apex (Fig. 20*b*) (Mexico, Costa Rica). . . . . ***C. obunca*** sp. nov.
— Thoracic chaetotaxy greatly reduced, short and slender; clasping cercus short and broad, tapered apically (Fig. 19*b*) (Arizona, Utah, Mexico) . . . .
. . . . ***C. macdonaldi*** sp. nov.

7 Clasping cercus in shape of arrowhead, apex strongly tapered to point (Fig. 19*d*) (Mexico) . . . . ***C. mexicana*** sp. nov.
— Clasping cercus finger-like, with apex curved (Fig. 19*c*) (Arizona, California, Utah, Mexico). . . . . ***C. madera*** sp. nov.

Clinocera arizonica sp. nov.

Fig. 18*c*; Map 21

Type material. HOLOTYPE, male labelled "ARIZONA [USA] Cochise Co/ Chir. Mtns. E. Turkey/ Creek 15 Sep 1980/ Hanson & Knowlton"; "HOLOTYPE/ Clinocera/ arizonica/ Sinclair [red label]" (USUL).

Recognition. This species is distinguished by the blue pruinescence on the lower two-thirds of the face, with the upper edge pointed, forefemur with two rows of stout setae beneath, and slender wings.

Description (wing length 4.1 mm).

Male. Lower two-thirds of face flat, with blue pruinescence, upper margin of pruinescence attenuated; gena and occiput with tinge of pruinescence; vertex brown. Arista-like stylus long and slender. Postocular bristles dark.

Lower two-thirds of pleura with blue pruinescence. Scutum greasy, vittae not visible. Several short pale acrostichals scattered anterior to first dc; 5 dc; 1 pprn; ? presut spal; ? npl, 1 psut spal; 1 pal; 2 sctl. Laterotergite with patch of long pale bristles; posterior margin of anepisternum with several short pale setae.

Wing hyaline, with spots or clouding absent; stigma absent; 1 long basal costal seta. Veins straight; auxiliary crossvein between R_4 and R_{2+3} absent;

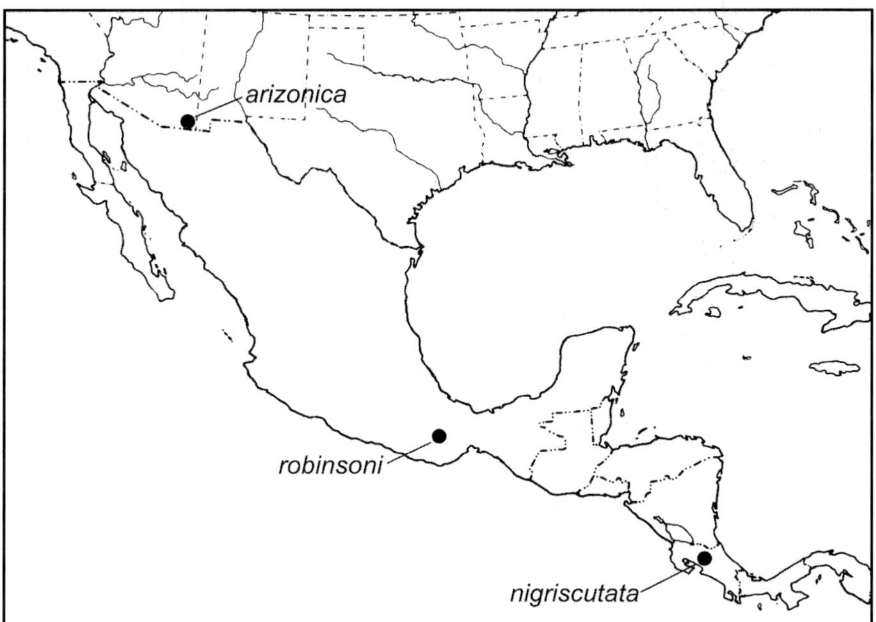

Map 21. Known distribution of *Clinocera arizonica*, *C. nigriscutata*, and *C. robinsoni*.

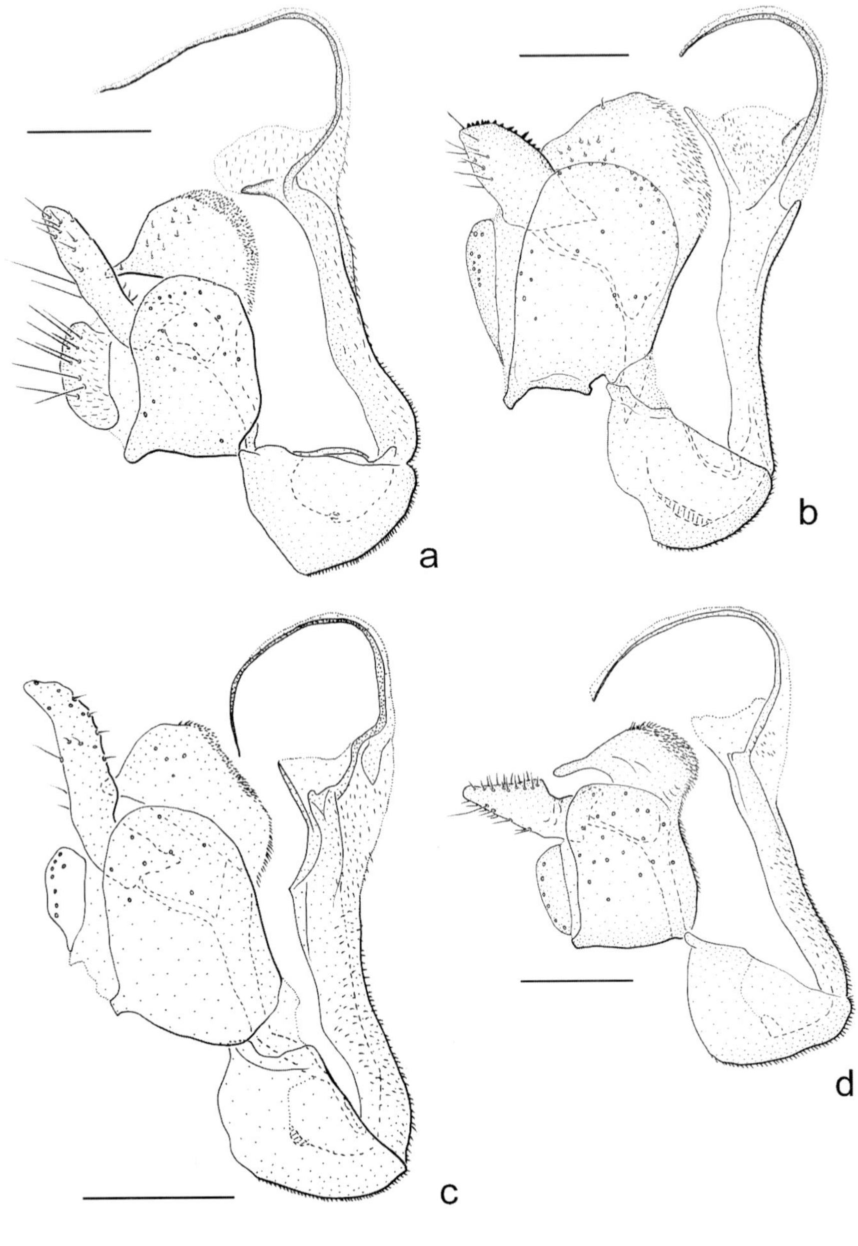

Fig. 19. Male terminalia of *Clinocera*: *a*, *C. lineata*; *b*, *C. macdonaldi*; *c*, *C. madera*; *d*, *C. mexicana*. Scale bars = 0.1 mm.

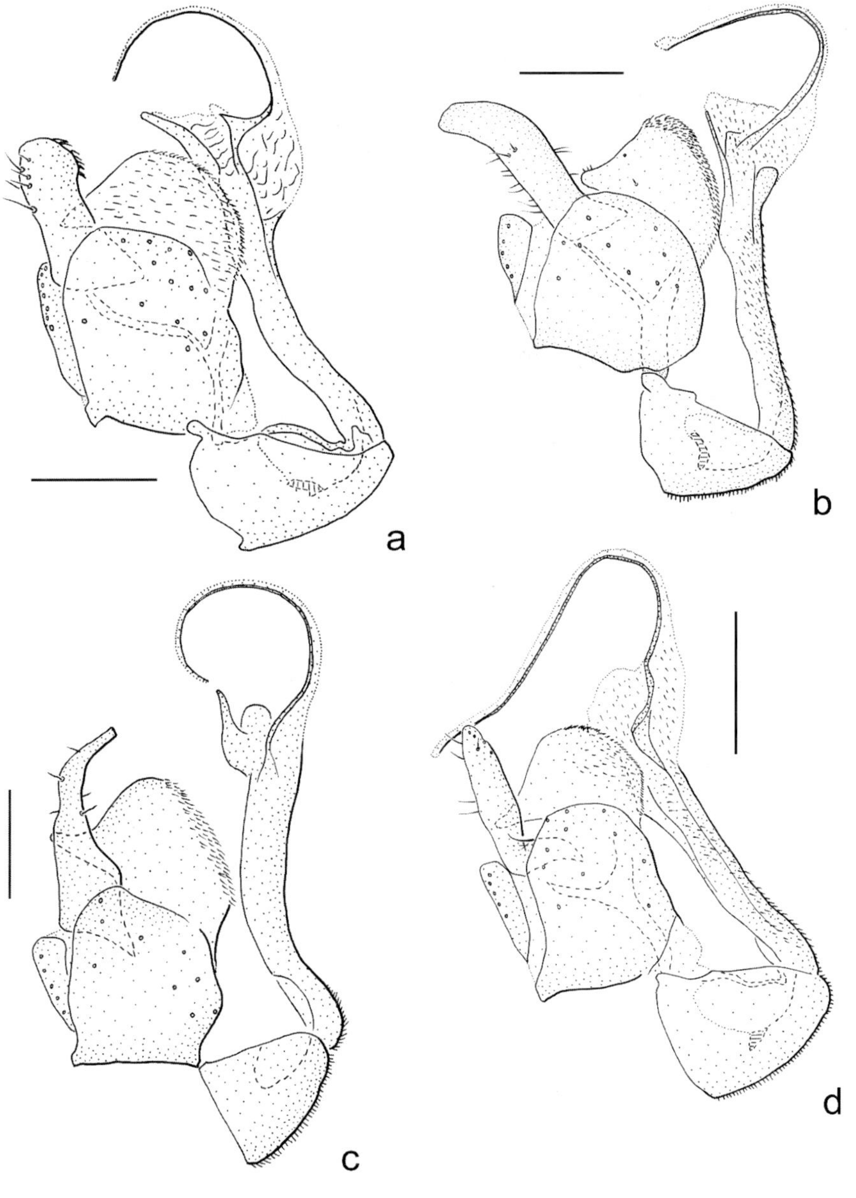

Fig. 20. Male terminalia of *Clinocera*: *a*, *C. nigriscutata*; *b*, *C. obunca*; *c*, *C. robinsoni*; *d*, *C. sublineata*. Scale bars = 0.1 mm.

stump vein absent; R_4 straight; anal vein indistinct or lacking; cell dm produced apically. Halter orange–brown.

Coxae and ventral margin of femora orange–brown; remaining segments dark. Forecoxa with pale setulae. Forefemur with antero- and postero-ventral row of short bristles on apical two-thirds; anterior preapical comb present; dense white ventral pile reduced. Foretibia with biserial row of short setae beneath. Hind tibia with erect dorsal and ventral bristles on apical third. Fifth tarsomere lacking dorsoapical extension; empodium pale laterally.

Hypandrium broadly triangular. Phallus strongly arched; distiphallus relatively short and recurved, less than half length of phallus. Clasping cercus thumb-like, with broadly rounded tip; inner face with stout pointed setae. Dorsal margin of surstylus with matted microtrichia; tip of surstylus produced, narrow and pointed (Fig. 18*c*).

Female. Unknown.

Geographic distribution and seasonal occurrence. This species is known only from the holotype, collected in southeastern Arizona (Map 21).

Etymology. The specific name is derived from the type locality.

Clinocera brevicauda sp. nov.

Figs. 18*d*, 31*f*; Map 22

Type material. HOLOTYPE, male labelled "USA:AZ: Santa Cruz Co./ Coronado Nat. For./ Santa Rita Mtns./ Madera Canyon/ B.J. Sinclair"; "ex. Madera Creek/ 4.VI.1991"; "HOLOTYPE/ Clinocera/ brevicauda/ Sinclair [red label]" (CNC).

Paratypes: **USA. Arizona:** Same data as holotype (5 ♂♂, 11 ♀♀, CNC); same locality as holotype, 3–4.vi.1991, BJS (2 ♂♂, 4 ♀♀, USNM); same locality as holotype, ck, Old Baldy Tr., 4.vi.1991, BJS (1 ♂, 2 ♀♀, CNC); Cochise Co., Coronado NF, Huachuca Mtns, Miller Cyn, Miller Ck, 3.vi.1991, BJS (3 ♂♂, 3 ♀♀, CNC); Cochise Co., Sierra Vista, Ramsey Cyn, 12 km S, 1700 m, oak/juniper, 3–26.xii.1986, BVB (3 ♂♂, 2 ♀♀, CNC). **MEXICO. Puebla:** nr San Martin, 27.v.1963, HER (2 ♂♂, 1 ♀, USNM).

Recognition. This species is very similar to *C. macdonaldi* but the chaetotaxy is not reduced. Females differ also from those of the former species by possessing ventral setae on the forefemur.

Description (wing length 3.3–3.8 mm).

Male. Face with blue pruinescence on lower three-quarters, lower third flat; gena and occiput with tinge of pruinescence; vertex brown. Arista-like stylus short and stout; postocular bristles dark on upper half of eye.

Scutum with pair of pale, opaque vittae; lower two-thirds of pleura and laterotergite with pale blue pruinescence. Chaetotaxy long and slender; several inconspicuous acrostichal setae anterior to first dc; 5 dc; 1 pprn; 1 presut spal; 2 npl, lower fine and short; 1 psut spal; 1 pal; 2 sctl. Laterotergite with patch of short pale bristles. Posterior margin of anepisternum lacking setae.

Wing (Fig. 31*f*) lightly infuscate, with spots or clouding absent; stigma absent; 1 basal costal seta. Auxiliary crossvein between R_4 and R_{2+3} absent; stump vein absent; R_4 nearly straight; anal vein indistinct; cell dm produced apically. Base of halter shaft pale, knob dark.

Coxae with pale-blue pruinescence, yellowish brown. Femora brown dor-

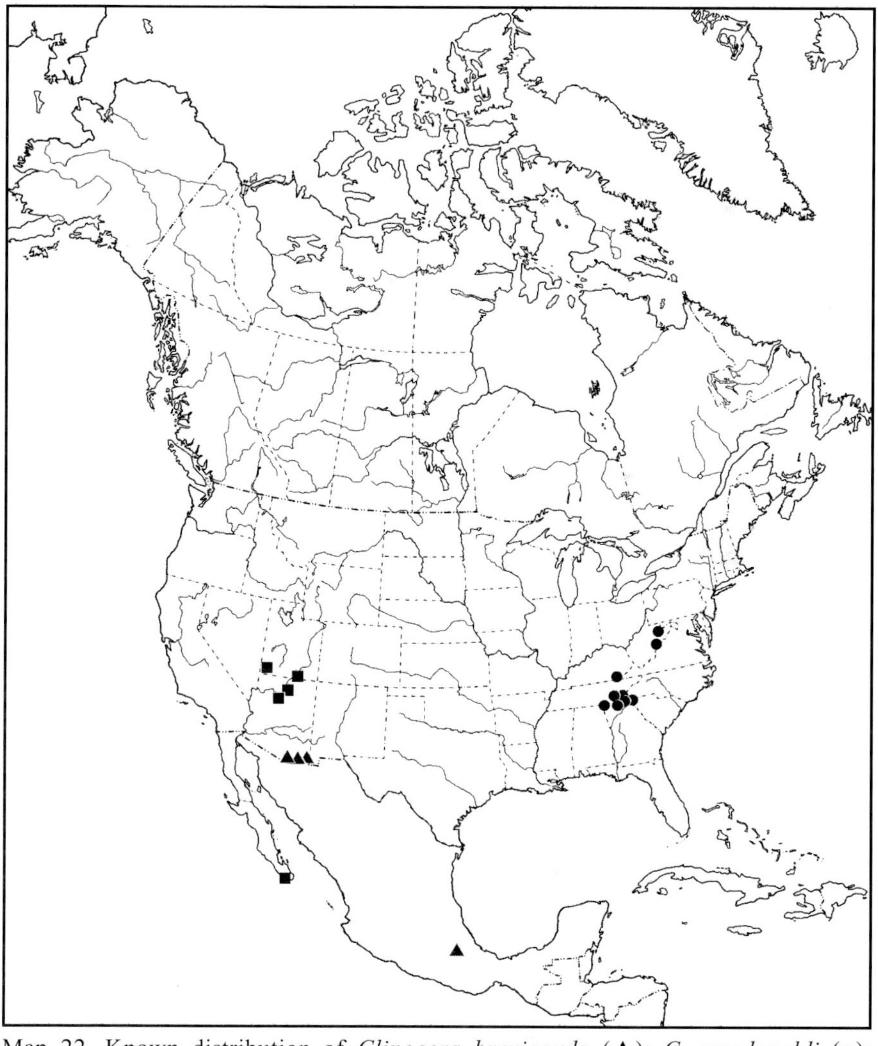

Map 22. Known distribution of *Clinocera brevicauda* (▲); *C. macdonaldi* (■); *C. sublineata* (●).

sally, pale yellowish brown beneath, remaining leg segments dark. Forefemur with anterior preapical comb; dense white pile beneath reduced; antero- and postero-ventral row of short stout bristles nearly along entire length. Foretibia with biserial row of short stout ventral setae. Hind tibia with long erect dorsal and ventral bristles on apical half. Fifth tarsomere lacking dorsoapical extension; empodium brownish laterally.

Abdominal sternites and lateral margin of tergites with faint blue pruinescence, remaining tergites brown. Hypandrium small, subrectangular. Phallus slightly arched, with relatively short distiphallus (subequal to half length of phallus). Clasping cercus very short, broad, somewhat pointed anteriorly; inner margin with short, spine-like bristles. Surstylus short, dorsoapical margin rounded, produced apically into beak-like projection (Fig. 18*d*).

Female. Similar to male except as follows: face with blue pruinescence confined to large central patch; ventral setae of forefemur well developed, especially on apical two-thirds.

Geographic distribution and seasonal occurrence. This species is recorded from southeastern Arizona and Mexico (Map 22).

Etymology. The specific name is derived from the Latin *brevis* ("short") and *cauda* ("tail"), referring to the short male clasping cercus.

Clinocera lineata Loew

Figs. 10*d*, 19*a*, 31*g*; Map 23

Clinocera lineata Loew, 1862: 207.

Clinocera lineata: Melander 1902, 1928, 1965: 468.

Type material examined. LECTOTYPE (here designated), female labelled "Type/ 1182 [red "label with upper white border]"; "[USA] Penn[sylvania; typed on white label]."; "Loew/ Coll."; "lineata [handwritten] [missing left hind tibia and antennae]" (MCZ). My lectotype label "LECTOTYPE/ of Clinocera/ lineata Loew/ des. B.J. Sinclair 1991 [red label]" has been attached to this specimen. The lectotype is designated in order to clarify the application of the name to this taxon.

Paralectotypes: [USA] Penn[sylvania] (2 ♀♀, MCZ). My paralectotype labels have additionally been attached to these specimens.

A male specimen with the same green handwritten label as on the paralectotypes, "Penns, O. Sacken", was identified by various authors as *C. lineata* and was listed as such by Farr (1953). However, this specimen is *C. sublineata* sp. nov., as it lacks wavy wing veins and lacks blue vittae on the scutum.

Additional material examined. Some 700 specimens from the following states, provinces, and territories: **CANADA. Alberta:** Banff, 28.ix.1932, OB (CAS); Cypress Hills PP, Elkwater, str., 5.ix.2003, BJS (ZFMK); Elkwater,

49°42′N, 110°16′W, 2.vi.1955, JRV (CNC); Medicine Hat, 25.xii.1928, F.S. Carr (CAS). **British Columbia:** Aiyansh, Nass R., 25.vi.1960, JGC (CNC); Buckinghorse Cpgd, Alaska Hwy, 281.6 km, 27.vi.1978, PHA (USNM); Cultus Lk., 27.x.1938, J.R. Jacob (CNC); 6 km NW Enderby, 17.viii.1990, AB (CNC); 10 km NE Harey Research For., UBC, 3.vii.1988, JMC (CNC); Mission City, 28.vi.1953, WRMM (CNC); Mt Revelstoke, seep, 5000′, 1.vii.1968, WWW (USNM); Nelson, 17.x.1910, ALM (USNM); Prince Rupert, 17.vi.1919, H.G. Dyar (USNM); Queen Charlotte Is., Graham Is. nr Queen Charlotte City, 10–23.vii.1988, SAM (DEBU), Graham Is., Slatechukk Mtn, 12.vii.1988, T.A. Wheeler (DEBU); Robson, 17–29.iii.1938, 29.x.1960, 3.v.1964, 9.x.1966, 16.vi.1968, 27.ix.1969, HRF (UBC); Salmon Arm, 24.iii., 1.iv.1992, AB (CNC), 6 km E, 4.v.1990, AB (CNC); Terrace, 10 km SW, cliff seep, 9.vii.1960, JGC (CNC); Vancouver, 8.viii.1917, ALM (USNM), Point Grey, 10.xi.1972, 25.vii.1973, JRV (CNC); Vancouver Is., Goldstream, 12.vii.1924, ALM (USNM); Vancouver Is., 13 km Lk. Cowichan, cold spr., 19–28.vii.1985, I.M. Smith (CNC); Vernon, 26.ii.1902, E.P. Venobles (CNC); Vernon, 21 km SE, 5.v.1990, AB (CNC). **Manitoba:** Aweme, 12.xii.1924, E. Criddle (CNC); Fort Churchill, 20.vi.1952, JGC (CNC). **New Brunswick:** Frederiction, Douglas, greenhouse sticky trap, 24.viii.2007 (in alc., CNC). **Northwest Territories:** Aklavik, 18.v.1931, OB (CAS); Yellowknife, 13.v.1953, JGC (CAS). **Nova Scotia:** Cape Breton, 46°48′N, 60°41′W, 400 m, 30.v.1983, H. Goulet (CNC). **Ontario:** Algoma District, Old Mill Bay, Batchawana Bay, 21.x.1985, RLH (MSUB); Ancaster, Heaslips, cold spr., 10–17.vi.1985, Martin, Smith (CNC); Crieff Bog, 3 km W Puslinch, 18–24.vii.1987, 4–17.v.1988, D. Blades (DEBU); Dundas, 24.xii.1984, 3.vii.1985, 29.vi.1986, BJS (DEBU); Guelph, 5.viii.1971, 28.xi.1978, K.S.G. Deacon, B.K. Akey (DEBU); Halton Co., Hilton Falls Cons. Area, seep nr falls, 27.vi.1991, BJS (CNC); Hamilton, Jolly Cut, 15.v., 17.vii., 12.ix., 4–31.x., 14, 21.xi., 28.xii.1985, 22.i., 12, 22.ii., 9.iii.1986, 17.ix.1989, BJS (DEBU); Kingston, 20.x.1985, BJS (DEBU); Lake on the Mtn PP, seep, 7.x.1985, BJS (DEBU); Lambton Co., Rock Glen Cons. Area, 2 km N Arkona, 26.vii.1991, BJS (CNC); Marmora, 20.viii.1952, JFMC (CNC); Orangeville, 20.ii.1977, SAM (ROGC); Ottawa, 15.x.1960, JRV (CNC); Ottawa, Rockcliffe Pk, roadcut seep, 18.ix.1988, 31.vii.2004, BJS (CNC); Owen Sound, Inglis Falls, spring 1984, 17.xii.1984, BJS, SAM (DEBU); St. Joseph Is., 46°18′N, 83°59′W, seep, 23.x.1988, JES (DEBU); Whitby, 28.xii.1973, G.J. Umphrey (DEBU); Wyebridge, 31.iii.1953, JGC (CNC). **Quebec:** Gatineau Pk, 15.iv.1982, Nelson, Roussel (CNC); Knowlton Lodge, 18.vii.1968, JRV (CNC); Lac Nathalie, 8.ix.1975 (UMQ); Lac Philippe, 45°37′N, 76°W, 31.vii.1955, 8.viii.1987, JRV (CNC); Old Chelsea, 14.v.1938, P. Peck (CNC); Old Chelsea, 10.x.1961, 15.x.1964, JRV (CNC); Old Chelsea, Camp Fortune, 28.iii.1950, JFMC (CNC); R. du Castor, 1.vii., 14.viii.1975 (UMQ); Thunder R., 11.vi.1929, WJB (CNC). **Yukon:** Morley River, 25.vi.1989, PHA (USNM). **FRANCE.** St. Pierre, D. Abraham (CNC). **USA. Alabama:** DeSoto SP, Lookout Mtn, Indian Falls, Laurel Ck, 15.v.1986, BJS (DEBU). **Alaska:** Anchorage, Flat Top Mtn,

500–600 m, 2.vi.1994, TS (BLKU); Aniakchak NM, 56°56′N, 158°6′W, Surprise Lk., vii–viii.1988, B. Wisseman (OSU); Cold Bay, 163°W, 26.vii.1952, WRMM (CNC); King Salmon, Naknek R., 10.vii.1952, WRMM (CNC); Middleton Is., 9.vi.1956, J.H. Thomas (USNM); Muir Inlet, Glacier Bay, 4.vi.1965, D.M. DeLong (USNM). **Arkansas:** Johnson Co., Ozark NF, Haw Ck Falls, str., 8.vi.1991, BJS (CNC); Stone Co., Ozark NF, Blanchard Springs, roadcut seep, 21.v.1991, BJS (CNC), Hwy 5, roadcut seep, 21.v.1991, BJS (CNC). **California:** Del Norte Co., Smith R., 23.vii.1932, J.M. Aldrich (USNM); Eldorado Co., Emerald Bay, Eagle Falls, 1880 m, 27.vi.1971, PHA (USNM); Fresno Co., Sierra NF, E and W Fks Camp, 2200 m, 31.vii.1979, DDW (CAS); Fresno Co., Huntington Lk., Kaiser Ck, 17.viii.1984, JFM (PERC); Humboldt Co., Humboldt Redwoods SP, Cabin Ck, 17.v.1978, DDW (CAS); Lake Co., E Fk Middle Ck, 26.x.1967, P.A. Rude (UCB); Plumas Co., Lassen NF, N Fk Feather R., 29–30.v.1977, DDW (CAS); Sequoia Pk, 6200′, 6.vi.1935, ALM (USNM); Siskiyou Co., Klamath NF, Portugese Peak, 2.vii.1978, DDS (CAS); Tuolumne Co., Lost Claim Cpgd, 975 m, 6.vii.1967, PHA (USNM). **Colorado:** Eldora, 3.vii.1961, JGC (CNC); Larimer Co., 17 mi. NE Cameron Pass, 7900′, 20.ix.1984, RLH (MSUB); Rio Grande Co., Beaver Ck, MT, 10 000′, 21.vi.1972, WWW (USNM). **Georgia:** Clarke Co., Athens, 9–16.iv., 20, 30.v.1976, JFM (CNC, PERC); Forsyth, 2.vi.1970 (CNC); Lumpkin Co., Chattahoochee NF, Desoto Falls Scenic Area, Frogtown Ck, 20.vi.1990, BJS (CNC); White Co., Chattahoochee NF, Raven Cliff Falls Tr., Dodd Ck, seep, 21.vi.1990, BJS (CNC). **Idaho:** Craig Mtns, 8.vi.1918, ALM (USNM); Echo Bay, 3.viii.1924, ALM (USNM); Kootenai Co., 10 mi. E Harrison nr Carlin Bay, 2100′, 14–29.vii.1980, 18–24.vii.1982, WJT (WSU); Latah Co., St. Joe NF, 8 mi. NE Harvard, 11.xi.1981, RSZ (WSU); Latch Co., 7 mi. N Troy nr Big Meadows, 13, 30.vii.1979, WJT (WSU); Lewiston Mill, 25.vi.1925, ALM (USNM); New Perce Co., Juliaetta Falls, 2.v.1975, D. Veirs (WSU). **Indiana:** Crawford Co., Hoosier NF, Hemlock Cliffs, str., 19.v.1991, BJS (CNC); Lawrence Co., 16.ix.1982, R.W. Meyer (PERC); Lawrence Co., Spring Mill SP, Donaldson Cave Ck, 19.v.1991, BJS (CNC); Tippe Co., W Lafayette, MT, 14.v.–16.x.1975, JFM (CNC, PERC). **Kentucky:** Mammoth Cave NP, Echo R. spr., 9.vi.1991 BJS (CNC); McCreary Co., Cumberland Falls SP, str., 18.vi.1990, BJS (CNC); Whitley Co., Cumberland Falls SP, 17–18.vi.1990, BJS (CNC). **Maine:** Katahdin, 3.v.1959 (USNM). **Massachusetts:** Middlesex Co., Lincoln, MT, 21.ix.1982, E.T. Armstrong (WOOC); W Springfield, 7.iv.1916, H.E. Smith (MCZ). **Minnesota:** Hennepin, Nine Mile Ck, 19.ii.1935, 29.iv.1939, G. Fredine, H.T. Peters (UMSP). **New Hampshire:** Essex Co., Mt Monadnock, 26.vii.1926, ALM (USNM); Rock Co., 1 mi. W Odiorne Pt, 23–26.vii.1983, DSC (UNH); White Mtns, Dolly Gap, 29.viii.1937, ALM (USNM). **New York:** Clinton Co., Chazy Lk., 8.ix.1985, RLH (MSUB); Long Is., Cold Spring Harbor, 12.viii.1930, C.H. Curran (AMNH), Cold Spring Harbor, 7.vii.1927, ALM (USNM); Rensselaer Co., Brainard, 11–22.vii.1966, P.B. Wygodzinsky (AMNH); Washington Co.,

Eldridge Swamp, 46°48′N, 60°41′W, 23.vi.–6.vii.1987, 5–12.vii.1988, J.K. Barnes (BARC). **North Carolina:** 5 mi. N Arrow Springs, 31.v.1965, JGC (CNC); GSMNP, Clingmans Dome, 21.vi.1941, ALM (USNM), Farney Ridge, ALM (USNM); Haywood Co., GSMNP, Cataloochee Ck, MT, 12.vi.1990, JFM (CNC); Highlands, 3800′, 7.v.1957, JRV (CNC); Jackson Co., 7 mi. S Cashiers, Dularey Bog, MT, 14.vii.1987, WWW (USNM); Swain Co., GSMNP, Newfound Gap, str., 15.vi.1990, BJS (CNC), Balsam Mtn, Flat Ck, spr. str., 13.vi.1990, BJS (CNC); nr Tennessee Bald, Bubbling Spring Ck, 5100′, 17.vii.1957, JGC (CNC). **Oregon:** Benton Co., Marys Peak, summit,

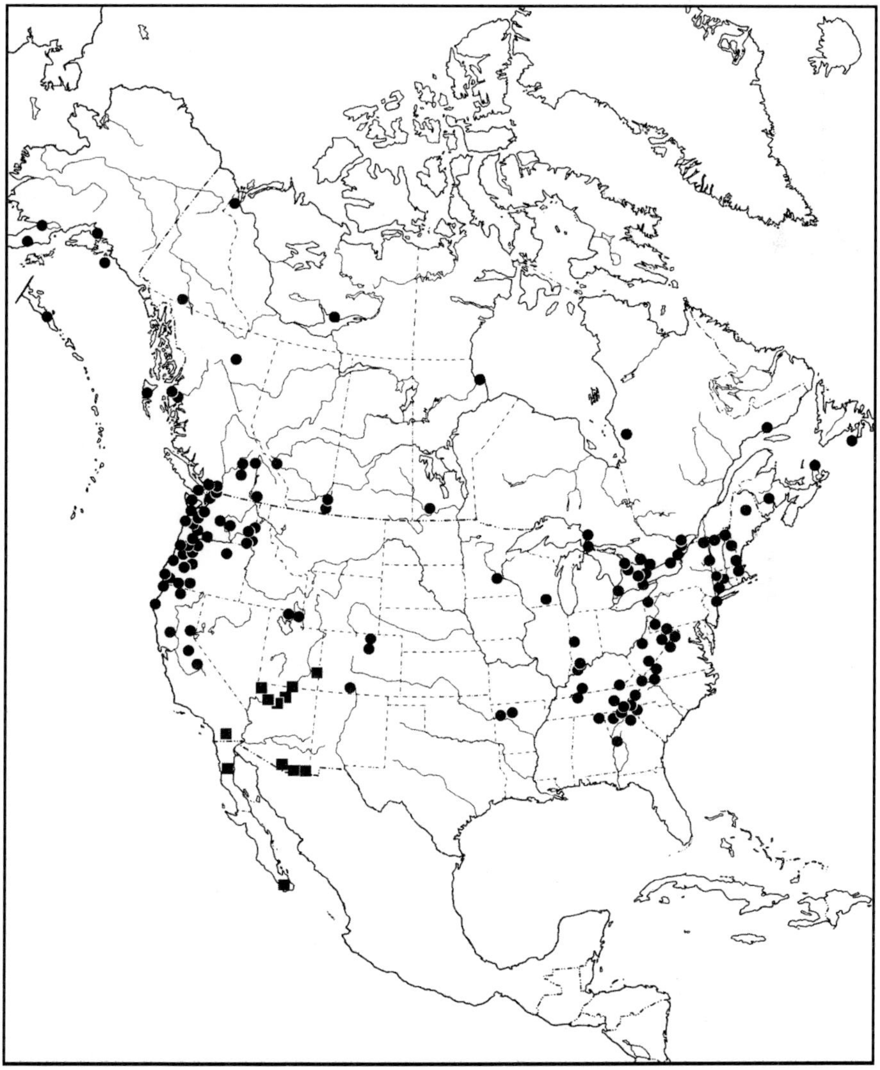

Map 23. Known distribution of *Clinocera lineata* (•); *C. madera* (■).

1.vii.1971, GCS (USNM), waterfall, 20.vi.1977, DDW (CAS); Clackamas Co., Hwy 26 and Timberline Lodge Rd, 4200′, waterfall, 23.vi.1982, WJT (WSU); Clackamas Co., Mt Hood NF, Tollgate Cpgd, 22.vi.1977, DDW (CAS); Clackamas Co., Still Ck Cpgd, nr Gov't Camp, 3800′, 22–23.vi.1982, WJT (WSU); Corvallis, 21.vi.1975, ALM (USNM); Curry Co., Siskiyou NF, 24 km E Pt Orford, Elk R. Rd, 17.vii.1989, BJS (CNC), 8 km W Sixes, Edson Ck Cpgd, Edson Ck, 16.vii.1989, BJS (CNC), Elk R. Rd, Sunshine Bar Cpgd, spr., 17.vii.1989, BJS (CNC), 7 km W Agness, roadcut seep, 18.vii.1989, BJS (CNC); Deschutes Co., Deschutes NF, Cold Spring Cpgd, 22.vi.1978, DDW (CAS); Hood R. Co., Starvation Ck SP, str., 10.vii.1989, BJS (CNC), 26.vi.1978, DDW (CAS), Viento, 1.vii.1917, ALM (USNM); Indian Ford, 3200′, 6 mi. NW Sisters, 10.vii.1965, Deschutes (CNC); Jackson Co., 20 mi. SE Ruch, French Gulch Rd, 22.v.1964, J. Schuh (WSU); Josephine Co., 8.5 km W Galice, str., 18.vii.1989, BJS (CNC); Klamath Co., 5 km W Bly, 14.vi.1984, R. Danielson (UZIL); Lane Co., Andrews EF, Gypsy Camp, 10.viii.2005, J. and A. Skevington (CNC); Lane Co., Siuslaw NF, 10 Mile Ck, 14.vii.1989, BJS (CNC); Lane Co., Skoookum Ck, 3 mi. NE Oakridge, 28.vii.1976, E. Rogers (CAS), 2 mi. N and 11 mi. NE Blue R., H.J. Andrews EF, 17.vii., 15.ix.1982, GWC (WSU); Lane Co., Limberlost Cpgd, 6 km E McKenzie Ridge, Lost Ck, 550 m, 25.vii.1974, PHA (CAS); Marion Co., Silver Falls SP, seep behind falls, 12.vii.1989, BJS (CNC), 23.vi.1974, PHA (CAS); Mt Hood, Hood Rapids, 29.vii.1921, ALM (USNM); Multnomah Co., ck, above falls, 10.vii.1989, BJS (CNC), Multnomah Falls, 26.vi.1974, PHA (CAS); Umatilla Co., 27 km E Ukiah Hwy 244, 24.vii.1974, PHA (CAS). **Pennsylvania:** Erie Co., Four and Six Mile cks, ECM (PSBC); Mineral Spring, 5.ix.1927, ALM (USNM); Pittsburg, 17.vii.1908, H. Kahl (WSU). **South Carolina:** Greenville, 13.vi.1933, H. Townes (USNM). **Tennessee:** Clarksville, 1.vii.1939, D.E. Hardy (SEMC); GSMNP, Chimney Camp, 11.vi.1946, GCS (USNM); Sevier Co., GSMNP, Elkmont Cpgd, 14.vi.1990, BJS (CNC), Le Conte Tr., 16.vi.1990, BJS (CNC), Ramsay Cascade Tr., str., 4.vi.1993, BJS (CNC); Van Buren Co., Fall Ck Falls SP, seep, 19.vi.1990, BJS (CNC). **Utah:** Box Elder Co., Raft R. Mtns, Rock Cyn, 28.iv.1979, Webb (BYU); Cache Co., Blacksmith Fk Cyn, 26.vi.1967, WJH (USUL), Logan Cyn, Turner Ck, 17–18.vii.1985, WJH (USUL), Tony Grave Jct., 8–16.viii.1985, WJH (USUL), W Hodges Cyn, 26.vii.–29.x.1978, WJH, GFK (USUL). **Vermont:** Orleans Co., Jay Peak, 3000–3400′, 20.vii.1968, JRV (CNC). **Virginia:** Augusta Co., Reddish Knob, 29.viii.1953, WWW (USNM); Covington, 25.vi.1933, ALM (USNM); Big Meadows, Shenandoah NP, below Dark Hollow Falls, 7.vii.2004, BJS (ZFMK); Grayson Co., Grayson Highlands SP, nr Big Wilson Ck, 10.vi.1990, BJS (CNC); Luray, 24.vi.1933, ALM (USNM); Patrick Co., Fairy Stone SP, 1000′, 30.v.1962, JRV (CNC). **Washington:** Adna, 10.vii.1917, ALM (USNM); Asotin Co., 6 mi. W Asotin, Asotin Ck, 28.iii.1981, WJT (WSU); Cashmere, 29.vi.1917, ALM (USNM); Clallam Co., Lk. Cresent, 22.vi.1984, R. Danielson (UZIL); Clallam Co., Olympic NP, 650 m, 24.vii.1978, DDW (CAS); Gifford

Pinchot NF, La Wis Wis, 300 m, Ohanapecosh R., 1.x.1939, DDW (CAS); Granite Falls, 5.viii.1917, 25.vii.192?, ALM (USNM); Grant Co., O'Sullivan Dam, 29.v.1955, MTJ (WSU); Greys Harbor Co., Olympic NF, Willaby Cpgd, 21.vii.1978, DDW (CAS); King Co., Seattle, Issaquah, 14.vii.1998, R. Danielson (UZIL); La Center, 8.vii.1917, ALM (USNM); Lewis Co., 23 mi. W Chehalis, Rainbow Falls SP, 300′, 13–14.vii.1977, WJT (WSU), La Wis Wis Cpgd, Grant Percell Falls, 4.vii.1974, PHA (CAS), La Wis Wis Cpgd, Ohanapecosh R., 3.vii.1974, PHA (CAS), Silver Ck, Salem Plant Rd, 1.ix.1981, PHA (CAS); Mt Baker NF, Glacier Ck Rd, Thompson Ck, 400 m, 4.x.1979, DDW (CAS); Mt Constitution, 28.vii.1909, ALM (USNM); Mt Rainier, Longmire Spr. (14.viii.1917), White R. (20.vii.1924), ALM (USNM); Mt Rainier NP, Paradise R. nr Paradise, 20.viii.1977, NEW (WOOC); Pacific Co., Fort Canby SP nr Ilwaco, 19.vii.1978, DDW (CAS), Pluvius, 16.vii.1922, ALM (USNM); Pierce Co., Mt Rainier NP, Comet Falls Tr., 4500′, 12.viii.1977, WJT (WSU), Frog Pond, Paradise, 20.vii.1977, RSZ (WSU), 3.2 mi. NE Sunshine Pt Cpgd, 8.viii.1979, RSZ (WSU); Potlach, Hoods Canal, 28.vii.1917, ALM (USNM); San Juan Co., Moran SP, Cascade Lk., 6.vii.1974, PHA (CAS); Seattle, W.M. Wheeler (AMNH, USNM); Tacoma, 27.v., 29.vii.1917, ALM (USNM); Vancouver, MT, 10.vii.1970 (WSU); Wawawai, 2.vii.1920, ALM (USNM); Whitman Co., Almota, 3.iii.1973, 17.iii.1974, 8.viii.1975, NEW, WJT (WSU), 1 mi. N Almota, 840–900′, 28.iv., 8.vii.1975, NEW, WJT (WOOC), Little Round Top nr Pullman, 10.vi.1974, W.B. Garnett (WSU), 8 mi. SW Pullman, 2120′, Lyle Grove Biological Area, 28.iv.1975, 30.v.1977, WJT (WSU), Steptoe Cyn, 10 mi. SW Colton, 900′, 14.vi.1980, WJT (WOOC), Steptoe Cyn, 10 mi. SW Colton, 28.iii.1980, RSZ (WSU), Lyle Grove, 8 mi. SW Pullman, 8.iii.1975, G.C. Brown (WSU); Yakima Co., Bear Ck, 8 mi. SW Tieton nr Rimrock Lk., 15–17.vi.1982, WJT (WSU). **West Virginia:** Greenbrier Co., Monongahela NF, Summit Lk. Rd, roadcut str., 3.vi.1990, BJS (CNC); Ritchie Co., North Bend SP, 23.vi.1970, GCS (USNM); Tucker Co., 11 mi. S Davis, Hwy 32, 21.v., 2.vii.1986, 7.vii.2004, BJS (DEBU, ZFMK). **Wisconsin:** Dane Co., University W., 2.iv.1953, R.H. Jones (USNM).

Recognition. *Clinocera lineata* is distinguished from all other species of this group by a pair of bluish vittae on the mesonotum and the uniquely wavy veins R_{2+3} and R_{4+5}.

Description (wing length 2.8–3 mm).

Male. Face with shiny blue pruinescence, except upper margin of eye; lower third flat; vertex, gena, and occiput. Arista-like stylus long and slender, subequal to height of eye. Postocular bristles dark.

Lower two-thirds of pleura with blue pruinescence. Scutum with pair of shiny bluish vittae with outer dark border, separated by dark median stripe extending to scutellum. Several short pale acrostichals anterior to first dc; 5 dc; 1 pprn; 1 presut spal; 2 npl, lower weak and short; 1 psut spal; 0 or 1 pal; 2 sctl. Laterotergite with patch of long pale bristles; posterior margin of anepisternum with several short pale setulae.

Wing (Fig. 31*g*) infuscate, lacking dark spots; clear spots at apex and basal corner of cell dm, crossvein r-m, and base of R_4; stigma absent; 1 short basal costal seta. R_{2+3} and R_{4+5} undulating; auxiliary crossvein between R_4 and R_{2+3} absent; stump vein from R_{2+3} to C often present; R_4 bell-shaped; anal vein indistinct or lacking; cell dm produced apically. Halter dark.

Coxae with reddish blue pruinescence; remaining leg segments bright yellowish brown, tarsi increasingly becoming darker. Forecoxa with pale setulae. Forefemur with antero- and postero-ventral row of fine short bristles, longer and more slender near base; anterior preapical comb present; white ventral pile present. Foretibia lacking biserial row of setae beneath. Hind tibia with erect dorsal and ventral bristles on apical third. Fifth tarsomere lacking dorsoapical extension; empodium pale brown laterally.

Abdominal sternites with faint blue pruinescence, tergites brown. Hypandrium broadly triangular. Phallus with slight dip on basal third; distiphallus long, flexible, and whip-like. Clasping cercus finger-like, narrow with rounded tip. Dorsal margin of surstylus with matted microtrichia; tip of surstylus with short nipple-like elongation (Fig. 19*a*).

Female. Similar to male except as follows: ventral setae of forefemur less distinct; legs generally darker; tergite 10 weakly divided medially into two separate sclerites.

Geographic distribution and seasonal occurrence. This species is distributed across North America and is present in most regions except the central arid prairies. It ranges coast to coast and from the arctic region to the southeastern states and northern California (Map 23). Label data indicate that *C. lineata* occurs in a number of habitats, including small streams, cold springs, seepages, and rivers. It appears to have a wide ecological tolerance, which has likely aided its wide distribution.

Clinocera lineata is the only species of this genus recorded from the Cypress Hills, an isolated woodland surrounded by prairie in southeastern Alberta and southwestern Saskatchewan (Newsome and Dix 1968; Campbell and Peck 1990). The hills rise about 650 m from the semi-arid grasslands, dominated by six species of trees and separated from the Rocky Mountain foothills by 320 km to the west (Newsome and Dix 1968; Campbell and Peck 1990). During at least the Wisconsinan glacial period, an ice-free zone existed in the western section. The presence of *C. lineata* in the Cypress Hills could be due to either recent natural dispersal through a boreomontane forest corridor along the front of the retreating ice sheets, or survival from pre-Wisconsinan times. The former scenario is more likely, owing to the cold arctic conditions that would have existed in this ice-free region and given the absence of this species from northern tundra-like regions today. See Campbell and Peck (1990) for further discussion on the biogeographic history of this region.

Landry and Harper (1985) collected *C. lineata* from late June to late

August from emergence traps set in small subarctic (Quebec) streams and rivers. A roadcut seepage in Hamilton (southern Ontario) was monitored by the author (see Sinclair and Marshall 1987) at least once a month for an entire year (1985–1986). Adult specimens of *C. lineata* were collected from this site throughout the year.

Clinocera macdonaldi sp. nov.

Figs. 19*b*, 31*h*; Map 22

Type material. HOLOTYPE, male labelled "USA:AZ: Grand Canyon NP/ CR#7, Elves Canyon/ 10-21.v.1998, J.F. MacDonald"; "HOLOTYPE/ Clinocera/ macdonaldi/ Sinclair [red label]" (CNC).

Paratypes: **USA. Arizona:** Same data as holotype (2 ♀♀, CNC); Grand Canyon NP, CR11, across Matkat, 10–21.v.1998, JFM (1 ♀, PERC); Grand Canyon NP, Vasey's Paradise, 10.v.1998, JFM (3 ♂♂, 1 ♀, PERC). **Utah:** Kane Co., Glen Cyn, Lk. Powell, Escalante R. arm, 15.vii.1997, JFM (3 ♂♂, 1 ♀, USNM, ZFMK); Zion Pk, 29.vi.1940, ALM (1 ♂, USNM). **MEXICO. Baja California:** Sur, Sierra Laguna, La Laguna, 7 air mi. ENE Todos Santos, 6000′, 12–18.xii.1979, C.E. Griswold (1 ♂, ROGC).

Recognition. This species is readily distinguished from all other Nearctic species of *Clinocera* by the reduced length and size of the dorsocentral bristles.

Description (wing length 3.6–3.9 mm).

Male. Face with blue pruinescence on lower two-thirds, lower third flat; gena and occiput with tinge of pruinescence; vertex brown. Arista-like stylus short and stout, less than height of eye; ocellar bristles greatly reduced, slender; postocular bristles dark on upper half of eye.

Scutum lacking vittae; pleura, postpronotal lobe, and notopleural depression with pale-blue pruinescence. Chaetotaxy greatly reduced in size and length; several inconspicuous acrostichal setae anterior to first dc; 5 dc; 1 pprn; 1 presut spal; 2 npl; 1 psut spal; 1 pal; 2 sctl. Laterotergite with patch of long pale bristles. Posterior margin of anepisternum lacking setulae.

Wing (Fig. 31*h*) infuscate, with spots or clouding absent; stigma absent; 1 short basal costal seta. Auxiliary crossvein between R_4 and R_{2+3} absent; stump vein absent; R_4 nearly straight; anal vein indistinct; cell dm produced apically. Base of halter shaft pale, knob dark.

Coxae with pale-blue pruinescence, yellowish brown, especially hind coxa. Femora brown dorsally, pale yellowish brown beneath, remaining leg segments dark. Forefemur with anterior preapical comb reduced to transverse row of 4 or 5 setulae; dense white pile beneath reduced; apical two-thirds with row of 6 stout antero- and postero-ventral bristles. Foretibia with biserial row of very short stout setae beneath. Hind tibia with long erect dorsal and ventral bristles on apical half. Fifth tarsomere lacking dorsoapical extension; empodium brownish laterally.

Abdominal sternites and lateral margin of tergites with blue pruinescence, remaining tergites brown. Hypandrium small, subtriangular. Phallus slightly arched, with short distiphallus (subequal to half length of phallus). Clasping cercus very short, broad, somewhat pointed anteriorly; inner margin with short, spine-like setae. Surstylus short, truncate, extending just beyond epandrium; apex rounded (Fig. 19*b*).

Female. Similar to male except ventral setae of forefemur lacking.

Geographic distribution and seasonal occurrence. This species is known primarily from seepages emerging from the canyon walls of Grand Canyon National Park and Glen Canyon National Recreation Area in Utah and Arizona (J.F. MacDonald, pers. comm., 2006) (Map 22). The flies were aspirated directly from the seepage faces, which were either fully exposed to sunlight or in narrow shaded gorges.

Etymology. The specific name is a patronym in honour of John F. MacDonald, the collector of the type series, who has contributed greatly to the knowledge of Nearctic empidids.

Clinocera madera sp. nov.

Figs. 19*c*, 31*i*; Map 23

Type material. HOLOTYPE, male labelled "USA:AZ: Santa Cruz Co./ Santa Rita Mountains/ Coronado Nat. Forest/ Madera Canyon/ B.J. Sinclair"; "3.VI.1991/ ex. Bog Spring/ 6000 feet"; "HOLOTYPE/ Clinocera/ madera/ Sinclair [red label]" (CNC).

Paratypes: **USA. Arizona:** Cochise Co., Chiricahua Mtns, 5750′, South Fk Cave Ck, 31°51′N, 109°11′W, 18.viii.2007, S.E. Brooks (2 ♂♂, 1 ♀, CNC); Cochise Co., Chiricahua Mtns, 6500′, East Turkey Ck, 1, 2.vi.1991, BJS (25 ♂♂, 8 ♀♀, CNC); Cochise Co., Coronado NF, Chiricahua Mtns, Cave Ck Cyn, Cave Ck, Herb Martyr Dam, 3.vi.1991, BJS (2 ♂♂, CNC); Cochise Co., 12.5 km S Sierra Vista, Ramsey Cyn, MT, oak/alpine/juniper, 10–17.xii.1986, BVB (1 ♂, DEBU); Grand Canyon NP, Saddle Cyn, Fern Cyn, Hance Ck, Elves Cyn, Slimey Tick, Matkat, Mohawk, 10–21.v.1998, JFM (20 ♂♂, 17 ♀♀, PERC); Pima Co., Catalina Mtns, Marshall Gulch nr Summerhaven, 28.v.1986, J. Jenkins (2 ♂♂, 1 ♀, WSU); Santa Cruz Co., Santa Rita Mtns, Madera Cyn, 25.iv.1979, K.N. Barber (4 ♂♂, 3 ♀♀, DEBU); same data as holotype, or Madera Ck (21 ♂♂, 15 ♀♀, CNC).

Additional material examined. USA. Arizona: Mohave Co., Virgin R. + Hwy I15, 515 m, Little Jamaica Springs, SAM (3 ♂♂, 5 ♀♀, DEBU). **California:** Cathedral City, 23.xi.1944, ALM (2 ♂♂, 1 ♀, USNM). **New**

Mexico: Grant Co., Gila NF, 14 mi. NE Silver City, Cherry Ck, 7400′, 5.vi.1990, BJS (1 ♂, CNC). **Utah:** Grand Co., Hwy 128 N Moab, Matrimony Spr., 21.viii.1984, CRN, RWB (1 ♀, BYU); Kane Co., Glen Cyn / Lk. Powell, Bown's Cyn (16.vii.1997), Rana Cyn (17.vii.1997), main lk., Powell body (16.vii.1997), JFM (7 ♂♂, 8 ♀♀, PERC). **MEXICO. Baja California:** Sur Sierra Laguna, La Laguna, 17 air mi. ENE Todos Santos, 6000′, 12–18.xii.1979, C.E. Griswold (1 ♂, 3 ♀♀, ROGC); Norte, Sierra San Pedro, Martir, 17 km E Park entrance, 2900 m, 7.ix.1980, DDW (24 ♂♂, 13 ♀♀, CAS).

Recognition. This species is distinguished by the colour of its face, the form of the clasping cercus, and the distinctive attachment of the whip-like distiphallus.

Description (wing length 2.6–3.5 mm).

Male. Face flat, lower half with pair of lateral blue pruinescent stripes; upper half of face with median blue pruinescent stripe; remaining head dark brown; or lower half of face mostly blue with small central dark patch, upper half often with central stripe to base of antennae. Arista-like stylus long and slender. Postocular bristles dark.

Scutum with pair of pale pruinescent vittae separated by wide dark band; pale vittae expanding medially on prescutellar depression. Lower half of pleura with faint blue pruinescence. Several short acrostichals anterior to first dc; 5 dc; 1 pprn; 1 presut spal; 2 npl, lower short and weak; 1 psut spal; 1 pal; 2 sctl. Laterotergite with patch of long pale bristles. Posterior margin of anepisternum with setulae.

Wing (Fig. 31*i*) lightly infuscate, with clouding or spots absent; stigma absent; 1 long basal costal seta. Veins straight; auxiliary crossvein between R_4 and R_{2+3} absent; stump vein absent; R_4 straight slightly divergent from R_5; anal vein an elongate streak; cell dm greatly produced apically. Halter brown.

Inner margins of coxae and femora yellow–brown, remaining segments yellow to brown. Forecoxa with long pale setulae. Forefemur with anterior preapical comb present; ventral surface with short dense white pile; antero- and posteroventral row of short stout bristles on apical three-quarters and fine, long bristles on basally. Foretibia with biserial row of short bristles beneath. Hind tibia with erect dorsal and ventral bristles on apical half. Fifth tarsomere lacking dorsoapical extension; empodium whitish laterally.

Abdominal sternites with faint blue pruinescence, tergites brown. Hypandrium short, subtriangular. Phallus stocky with expanded tip; distiphallus short, darkly pigmented. Clasping cercus narrow to somewhat broad, apex arched; fine setae on inner tip only. Surstylus shorter than clasping cercus, truncate, with nipple-like tip; dorsal surface with matted microsetae (Fig. 19*c*).

Female. Similar to male except as follows: ventral setae of foretibia reduced; forefemur with setae of basal third reduced.

Geographic distribution and seasonal occurrence. This species is restricted to the southwestern United States and Baja California (Map 23), where it inhabits small streams and seepages in isolated humid valleys and canyons.

Remarks. The clasping cercus appears to be variable in width, ranging from narrow to distinctly broad. In addition, the facial coloration varies from laterally striped to completely blue, and leg coloration varies from dark to yellowish brown.

Etymology. The specific name is derived from the type locality.

Clinocera mexicana sp. nov.

Fig. 19*d*; Map 24

Type material. HOLOTYPE, male labelled "Sierra Juarez/ nr crest/ Oaxaca, Mex[ico]/ May 19, 1963"; "H.E. Robinson/ Collector"; "HOLOTYPE/ Clinocera/ mexicana/ Sinclair [red label]" (USNM).

Paratypes: **MEXICO. Dorrigo:** 10 mi. W El Salto, 9000′, 12.vi.1964, JFMC (2 ♀♀, CNC).

Recognition. This species is distinguished by pale-blue pruinescence along the lateral margin of the face and upper face with median pruinescent stripe. It is distinguished from *C. madera* only on the basis of the male terminalia.

Description (wing length 3.5–3.8 mm).

Male. Face flat, lower half with lateral blue pruinescent stripe; upper half of face with median blue pruinescent stripe; remaining head dark. Arista-like stylus long and slender, subequal to height of eye. Postocular bristles dark.

Thorax dark brown, with pair of dull, pale vittae running beside row of dc; postpronotal lobe somewhat reddish. Lower half of pleura with blue pruinescence. Several acrostichals anterior to first dc; 5 dc; 1 pprn; 1 presut spal; 2 npl, lower short; 1 psut spal; 1 pal; 2 sctl. Laterotergite with patch of pale bristles; posterior margin of anepisternum with pale setulae.

Wing lightly infuscate, with clouding or spots absent; stigma absent; 1 short basal costal seta. Veins straight; auxiliary crossvein between R_{2+3} and R_4 absent; stump vein absent; R_4 straight, gradually divergent from R_5; anal vein pale streak; cell dm produced apically. Halter knob dark, lower shaft pale.

Coxae yellowish brown, thinly covered by blue pruinescence; base and ventral surface of femora yellowish brown, remaining segments dark. Forecoxa with fine pale setulae. Forefemur with antero- and postero-ventral row of dark short bristles; anterior preapical comb present; ventral surface with lacking white pile. Foretibia with single row of stout bristles beneath. Hind tibia with erect dorsal and ventral bristles on apical half. Fifth tarsomere lacking dorsoapical extensions; empodium light brown laterally.

Abdomen dark, brownish. Hypandrium broadly triangular. Phallus straight, apex with broad membranous cap; distiphallus whip-like, flexible, with slender base.

Clasping cercus arrowhead-shaped, strongly tapered; bearing setae on dorsal margin. Surstylus short, projecting free beyond epandrium; dorsal margin with matted microtrichia; apex greatly lengthened and narrowed to beak-like tip (Fig. 19*d*).

Female. Similar to male, including leg chaetotaxy.

Geographic distribution and seasonal occurrence. This species is known only from Mexico (Map 24), collected in May and June. The paratypes were collected in the divide between the east and west sides of the Sierra Madre Occidentale (J.F. McAlpine, unpubl. data). The fauna of this locality was distinguished by many Neotropical elements.

Remarks. The facial coloration of *C. mexicana* is very similar to that of *C. madera*.

Etymology. The specific name is in reference to its apparent restriction to Mexico.

Clinocera nigriscutata sp. nov.

Figs. 20*a*, 31*j*; Map 21

Type material. HOLOTYPE, male labelled "C[osta] R[ica]: Puntarenas/ Monteverde, 1500m/ 1–10.III.1992/ D.M. Wood MT/ beside stream"; "HOLOTYPE/ Clinocera/ nigriscutata/ Sinclair [red label]" (CNC).

Paratypes: **COSTA RICA. Puntarenas Province:** Monteverde, Finca Canada, 1500 m, native for., 1–5.vi.1988, BVB (1 ♂, 1 ♀, CNC); Monteverde, 1700 m, MT, clearcut edge, 28.v.–1.vi.1988, BVB (1 ♂, CNC); same data as holotype (1 ♂, 1 ♀, CNC); Monteverde, 1500 m, waterfall spray zone, 23.ii.1991, BJS (1 ♀, CNC).

Recognition. This species is distinguished from other Costa Rican species by the dark scutum contrasting with the pale yellowish brown postpronotal lobes and pleura and blue pruinescent face.

Description (wing length 2.8–3.1 mm).

Male. Body yellowish brown, scutum and abdomen dark brown; face dark brown, lower two-thirds with blue pruinescence; remainder of head dark brown. Arista-like stylus short and slender, less than height of head. Ocellar bristles weak and short; upper postocular bristles dark.

Scutum dark brown, except yellowish brown postpronotal lobes and notopleuron; black spot at base of wing. Pleura yellowish brown, lacking blue pruinescence. Thoracic setae slender and short; several acrostichals anterior to first dc; 5 dc; 1 pprn reduced to setula; 1 presut spal; 2 npl, lower very short and weak; 1 psut spal; 1 pal; 2 sctl. Laterotergite with several short pale bristles; posterior margin of anepisternum with dark setulae.

Wing (Fig. 31*j*) lightly infuscate, with clouding or spots absent; stigma absent; 1 short basal costal seta, Veins straight; auxiliary crossvein between R_{2+3} and R_4

absent; stump vein absent; R_4 straight, divergent, curved at wing margin; anal vein appearing as small basal smudge; cell dm produced apically. Halter pale brown.

Coxae concolorous with pleura; base and ventral surface of femora yellowish brown, remaining segments dark. Forecoxa with fine pale setulae. Forefemur with antero- and postero-ventral row of dark short bristles; anterior preapical comb present; ventral surface lacking white pile. Foretibia with biserial row of stout setae beneath. Hind tibia with erect dorsal and ventral bristles on apical third. Fifth tarsomere lacking dorsoapical extension; empodium light brown laterally.

Abdomen dark, brownish. Hypandrium broadly triangular. Phallus weakly arched, apical membranous keel ending at distiphallus; distiphallus whip-like, flexible, with slender base, strongly arched toward shaft. Clasping cercus narrow, cylindrical, with rounded tip. Surstylus short, only partially projecting beyond epandrium; dorsal margin with some matted microtrichia; tip tapered to point (Fig. 20*a*).

Female. Similar to male except as follows: most of face dull, pale brown, except dark brown below antennae; ventral bristles of forefemur and foretibia less distinct.

Geographic distribution and seasonal occurrence. This species is recorded only from Costa Rica (Map 21), with the known flight period from February to June.

Etymology. The specific name is from the Latin *nigra* ("dark"), referring to the distinctive dark scutum of this species.

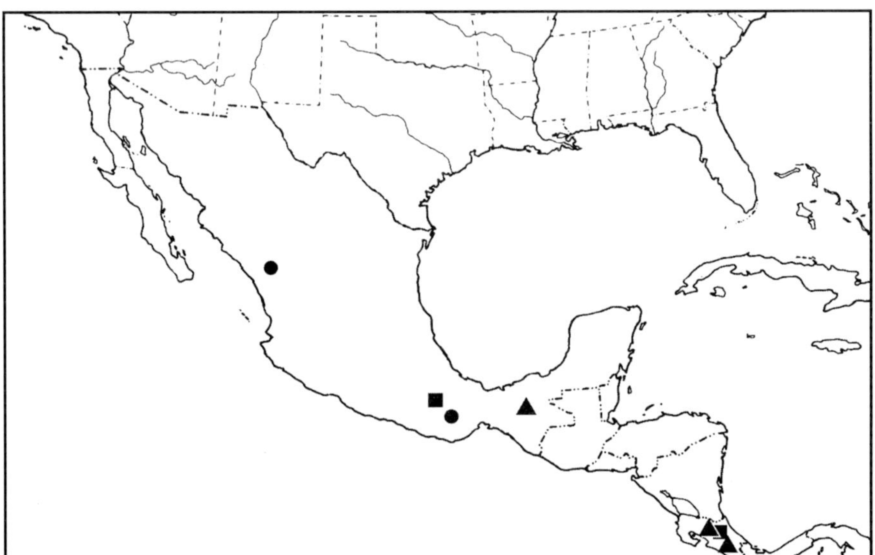

Map 24. Known distribution of *Clinocera mexicana* (●); *C. obunca* (■); *C. woodi* (▲).

Clinocera obunca sp. nov.

Fig. 20*b*; Map 24

Type material. HOLOTYPE, male labelled "C[osta] R[ica]: Puntarenas/ Monteverde, 1700m/ 25.II.1991/ ex. spring/ B.J. Sinclair"; "HOLOTYPE/ Clinocera/ obunca/ Sinclair [red label]" (CNC).

Paratypes: **COSTA RICA. Puntarenas Province:** Same data as holotype (1 ♀, CNC). **San Jose Province:** Zurqui de Moravia, 1600 m, MT, iv.1991, PH (1 ♂, CNC). **MEXICO. Oaxaca:** Valle Nacional, 20.xii.1962, 12–14.v.1963, HER (1 ♂, 3 ♀♀, USNM).

Recognition. This species is distinguished by blue facial pruinescence that extends to the base of the antennae, and a pair of dull, pale vittae on the scutum.

Description (wing length 3–3.2 mm).

Male. Lower half of face with blue pruinescence, upper half with tapering strip of pruinescence continuing to base of antennae; occiput with pruinescence, vertex dark brown. Arista-like stylus long and slender, subequal to height of eye. Upper postocular bristles dark.

Thorax dark brown, with pair of dull, pale vittae running beside row of dc; margin of scutum, including antepronotum from postpronotal lobe to base of wing with strip of pale blue pruinescence. Lower half of pleura with blue pruinescence. Several acrostichals anterior to first dc; 5 dc; 1 pprn; 1 presut spal; 2 npl, lower very short and weak; 1 psut spal; 1 pal; 2 sctl. Laterotergite with patch of short pale bristles; posterior margin of anepisternum with pale setulae.

Wing lightly infuscate, clouding or spots absent; stigma absent; 1 short basal costal seta. Veins straight; auxiliary crossvein between R_{2+3} and R_4 absent; stump vein absent; R_4 straight, divergent apically from R_5; anal vein absent; cell dm produced apically. Halter dark.

Coxae yellowish brown, thinly covered by blue pruinescence; base and ventral surface of femora yellowish brown, remaining segments dark. Forecoxa with fine pale setulae. Forefemur with antero- and postero-ventral row of short dark bristles; anterior preapical comb present; ventral surface with lacking white pile. Foretibia with single row of stout bristles beneath. Hind tibia with erect dorsal and ventral setae on apical half. Fifth tarsomere lacking dorsoapical extensions; empodium light brown laterally.

Abdomen dark, brownish. Hypandrium broadly triangular. Phallus nearly straight, apex with broad membranous cap; distiphallus whip-like, flexible, with slender base, gradually arched toward shaft. Clasping cercus finger-like, narrow, arched subapically; bearing fine setulae on inner dorsal margin. Surstylus short, only partially projecting beyond epandrium; dorsal margin with matted microtrichia; tip gradually tapered (Fig. 20*b*).

Female. Similar to male, including chaetotaxy.

Geographic distribution and seasonal occurrence. This species is recorded from Mexico and Costa Rica (Map 24), with the flight period from February to May. The holotype was collected at the spring source of a cold mountain stream.

Etymology. The specific name is from the Latin *obuncus* ("bent inward", "hooked"), referring to the apically hooked clasping cercus of this species.

Clinocera robinsoni sp. nov.

Fig. 20*c*; Map 21

Type material. HOLOTYPE, male labelled "Sierra Juarez/ 104 km/ Oaxaca, Mex[ico]/ May 19, 1963"; "H.E. Robinson/ Collector"; "HOLOTYPE/ Clinocera/ robinsoni/ Sinclair [red label]" (USNM).

Recognition. This species is very similar to *C. nigriscutata*, but distinguished from the latter on the basis of male terminalia, with the clasping cercus long, slender, and arched medially, with a narrow apex.

Description (wing length 3 mm).

Male. Body yellowish brown, scutum and abdomen dark brown; face dark brown, with narrow stripe of blue pruinescence extending from below antennae and expanding toward base; remainder of head dark brown. Arista-like stylus short and slender, less than height of head. Ocellar bristles weak and short; upper postocular bristles dark.

Scutum dark brown, except yellowish brown postpronotal lobes and notopleuron; black spot at base of wing. Pleura yellowish brown, lacking blue pruinescence. Thoracic setae slender and short; several acrostichals anterior to first dc; 5 dc; 1 pprn reduced to setula; 1 presut spal; 2 npl, lower very short and weak; 1 psut spal; 1 pal; 2 sctl. Laterotergite with several short pale bristles; posterior margin of anepisternum with dark setulae.

Wing lightly infuscate, clouding or spots absent; stigma absent; 1 short basal costal seta, Veins straight; auxiliary crossvein between R_{2+3} and R_4 absent; stump vein absent; R_4 straight, divergent, curved at wing margin; anal vein appearing as small basal smudge; cell dm produced apically. Halter pale brown.

Coxae concolorous with pleura; base and ventral surface of femora yellowish brown, remaining segments dark. Forecoxa with fine pale setulae. Forefemur with antero- and postero-ventral row of dark short bristles on apical two-thirds; anterior preapical comb present; ventral surface lacking white pile. Foretibia with biserial row of stout setae beneath. Hind tibia with erect dorsal and ventral bristles on apical third. Fifth tarsomere lacking dorsoapical extension; empodium light brown laterally.

Abdomen dark, brownish. Hypandrium broadly triangular. Phallus weakly arched, apical membranous keel ending at distiphallus; distiphallus whip-like, flexible, greater than half length of phallus. Clasping cercus slender, tapered apically to narrow truncate tip and arched medially. Surstylus short, only partially projecting beyond epandrium; dorsal margin with some matted microtrichia; tip tapered to beak-like point (Fig. 20*c*).

Female. Unknown.

Geographic distribution and seasonal occurrence. This species is also recorded only from Mexico (Map 21).

Etymology. The specific name is a patronym in honour of Harold Robinson, who collected many rare and unusual *Clinocera* from Central America.

Clinocera sublineata sp. nov.

Figs. 8*c*, 20*d*, 32*a*; Map 22

Type material. HOLOTYPE, male labelled "USA:GA: White Co./ Chattahoochee Nat./ For. 21.VI.1990/ Raven Cliff Falls Tr./ ex. Spring seepage/ BJ Sinclair"; "HOLOTYPE/ Clinocera/ sublineata/ Sinclair [red label]" (CNC).

Paratypes: **USA. Georgia:** Amicalola Falls SP, falls edge, 16.v.1986, BJS (1 ♂, DEBU); Cloudland Canyon SP, cascading str., 15.vi.1986, BJS (1 ♀, DEBU); White Co., Chattahoochee NF, Anna Ruby Falls, streams and seeps, 21.vi.1990, BJS (6 ♂♂, 6 ♀♀, CNC), nr Helen, madicolous zone, 21.vi.1990, BJS (3 ♂♂, 8 ♀♀, CNC), Raven Cliff SP, spring-seep, 21.vi.1990, BJS (7 ♂♂, 13 ♀♀, CNC). **Kentucky:** Whitley Co., Cumberland Falls SP, cascade, 17–18.vi.1990, BJS (3 ♀♀, CNC). **North Carolina:** Highlands and Hwy 64, Bridal Veil Falls, roadcut seep, 17.v.1986, BJS (1 ♂, DEBU), GSMNP, 6 km S Balsam Mtn Cpgd, roadcut seep, 13.vi.1990, BJS (1 ♂, CNC). **Pennsylvania:** O. Sacken (MCZ). **South Carolina:** Oconee Co., Sumter NF, Brasstown Falls, 22.vi.1990, BJS (2 ♂♂, CNC). **Tennessee:** GSMNP, Elkmont Cpgd, roadcut seep, 14.vi.1990, BJS (4 ♂♂, 11 ♀♀, CNC), Roaring Fk Motor Tour, seeps, 17.vi.1990, BJS (1 ♂, CNC), 10 km E Sugarlands Center, seeps, 15.vi.1990, 3–6.vi.1993, BJS (2 ♀♀, CNC). **West Virginia:** Pocahontas Co., Hills Ck Scenic Area, str. over cliff, 3–4.vi.1990, BJS (3 ♂♂, CNC); Tucker Co., 11 mi. S Davis, Hwy 32, 7.vii.2004, BJS (1 ♀, ZFMK).

Recognition: This new species is very similar to *C. lineata*, but the wing veins are straight and not wavy and the body is generally paler, especially the coxae.

Description (wing length 2–2.5 mm).

Male. Body generally brown; face with shiny blue pruinescence. Gena, vertex, and occiput generally dark. Arista-like stylus long and slender, subequal to height of eye. Postocular bristles dark.

Scutum with dark-brown central stripe bordered by pair of shiny brownish stripes from antepronotum to scutellum. Lower half of pleura with faint blue pruinescence. Several acrostichals anterior to first dc; 5 dc; 1 pprn; 1 presut spal; 2

npl, lower short and weak; 0 psut spal; 1 pal; 2 sctl. Laterotergite with patch of short pale bristles; posterior margin of anepisternum with pale setulae.

Wing (Fig. 32*a*) infuscate, clouding or spots absent; apical crossvein of cell dm (separating M_1 and M_2) white, often lacking pigmentation; stigma absent; 1 short basal costal seta. Veins straight; auxiliary crossvein between R_{2+3} and R_4 absent; stump vein absent; R_4 bell-shaped; anal vein absent; cell dm produced apically. Halter yellowish brown.

Coxae and femora bright yellowish brown; tibiae and tarsi dark. Forecoxa with fine pale setulae. Forefemur with anteroventral row of dark short bristles; posteroventral surface with row of short fine pale setulae; anterior preapical comb present; ventral surface with short dense white pile. Foretibia with anteroventral row of stout setae. Hind tibia with erect dorsal and ventral bristles on apical third. Fifth tarsomere slightly projecting (Fig. 8*c*); empodium brown laterally.

Abdomen dark and brownish, except apical segments yellowish. Terminalia very similar to *C. lineata*, but only half the size. Hypandrium broadly triangular. Phallus straight; distiphallus long, whip-like, flexible with distinct curve at base. Clasping cercus narrow, finger-like with rounded tip. Dorsal margin of surstylus with matted microtrichia; tip of surstylus nipple-like (Fig. 20*d*).

Female. Similar to male except ventral bristles of forefemur and foretibia indistinct.

Geographic distribution and seasonal occurrence. *Clinocera sublineata* is confined primarily to the southern Appalachians (Map 22), inhabiting the madicolous zone of seepages, especially roadcuts and small cascading streams. Nearly all known specimens of this species were collected by the author, who has specialized in these microhabitats. A specimen of *C. sublineata* was seen from Erie County, Pennsylvania (Four and Six Mile creeks, ECM), but was destroyed before its identity could be rechecked. Consequently this record is not included in the distribution map of this species.

Etymology. The specific name is derived from the Latin *sub* ("from", "somewhat") and *lineata*, alluding to the structural similarity between this species and *C. lineata*.

Clinocera woodi sp. nov.

Figs. 21*a*, 32*b*; Map 24

Type material. HOLOTYPE, male labelled "C[osta] R[ica]: Puntarenas/ Monteverde, 1500m/ 24.II.1991/ ravine seepage/ B.J. Sinclair"; "HOLOTYPE/ Clinocera/ woodi/ Sinclair [red label]" (CNC).

Paratypes: **COSTA RICA. Cartago–San Jose Province:** Cerro de la Muerte, 2800 m, 20 km S Empalme, MT, x–xii.1989, vii–ix.1990, PH (2 ♂♂, 1 ♀, CNC); 2 km W Empalme, i.1995, 2300 m, PH (1 ♀, CNC); 20 km SE Empalme, 2800 m, ix.1988, PH (2 ♀♀, CNC; 1 ♀, INBIO); San Gerardo de Dota, 2300 m, ck, ypans, iii.1996, LM (2 ♂♂, CNC); San Gerardo de Dota, Savergre Hotel de Montana, 2200 m, 22–23.iii.2005, M. Pollet (2 ♀♀, CNC); 26 km N San Isidiro, 2100 m, ix–x.1992, PH (1 ♂, CNC); Zurqui de Moravia, 1600 m, MT, vii.1989, iv.1991, PH (1 ♂, 2 ♀♀, CNC). **Heredia:** 9 km NE Vara Blanca, Finca Morillo, 1500 m, ypans, rain for., large str., 9–11.iii.2005, M. Pollet (2 ♀♀, CNC). **Puntarenas Province:** Estación Biológica Monteverde, 10.32°N, 84.80°W, 1600 m, 16.iii.2005, M. Pollet (3 ♂♂, CNC); Monteverde, Finca Canada, 1500 m, 5.vi.1988, BVB (3 ♂♂, 3 ♀♀, CNC); same data as holotype (6 ♂♂, 4 ♀♀, CNC); Monteverde, 1500 m, 23–27.ii.1991, ypans on str., MT, BJS (3 ♂♂, 1 ♀, CNC); Monteverde, 1700 m, spr., 25.ii.1991, BJS (2 ♂♂, 2 ♀♀, CNC); Monteverde, 1500 m, MT beside str., 1–10.iii.1992, DMW (7 ♂♂, 8 ♀♀, CNC); Monteverde, 1550 m, ck, ypans, 18.iii.1996, LM (2 ♂♂, 2 ♀♀, CNC). **MEXICO. Chiapas:** nr Puerto Vientes, 24.v.1963, HER (1 ♂, USNM).

Recognition. This species is distinguished by its yellowish brown body coloration and single median stripe on the scutum.

Description (wing length 2.5–3 mm).

Male. Body yellowish brown, abdomen dull brown; face dark brown, lower half with pale pruinescence; remainder of head dark brown. Arista-like stylus long and slender, subequal to height of eye. Upper postocular bristles dark.

Scutum yellowish brown, with brown central stripe bordered by pair of bright yellowish brown stripes, extending onto prescutellar depression; dark stripe beneath dc; small black spot at base of wing. Pleura yellowish brown, lacking blue pruinescence. Several acrostichals anterior to first dc; 5 dc; 1 pprn; 1 presut spal; 2 npl, lower very short and weak; 1 psut spal; 1 pal; 2 sctl. Laterotergite with several short, dark bristles; posterior margin of anepisternum with dark setulae.

Wing (Fig. 32*b*) lightly infuscate, clouding or spots absent; stigma absent; 1 short basal costal seta. Veins straight; auxiliary crossvein between R_{2+3} and R_4 absent; stump vein absent; R_4 divergent, curved at wing margin; anal vein absent; cell dm produced apically. Halter dark.

Coxae concolorous with pleura; base and ventral surface of femora yellowish brown, remaining segments dark. Forecoxa with fine pale setulae. Forefemur with antero- and postero-ventral row of short dark bristles; anterior preapical comb present; ventral surface with short dense white pile. Foretibia with single row of stout setae beneath. Hind tibia with erect ventral bristles on apical third. Fifth tarsomere with dorsoapical extensions; empodium light brown laterally.

Abdomen dark, brownish. Hypandrium broadly triangular. Phallus nearly straight, apical membranous keel extended onto distiphallus; distiphallus whiplike, flexible, with expanded base, before arching toward shaft. Clasping cercus narrow, tapered to rounded tip. Surstylus short, only partially projected beyond epandrium; dorsal margin with matted microtrichia; tip truncate (Fig. 21*a*).

Female. Similar to male except ventral bristles of forefemur and foretibia less distinct.

Geographic distribution and seasonal occurrence. *Clinocera woodi* occurs primarily above 1500 m in Costa Rica (Map 24), likely with a year-round flight period.

Etymology. The specific name is a patronym in honour of D. Monty Wood, a dedicated dipterist whose research station at Monteverde has facilitated the collection of many new species.

Clinocera nigra group

Diagnosis

This species-group is diagnosed by the following characters: wing stigma absent; fifth tarsomere not prolonged; clasping cercus normally strongly attenuated apically (except *C. nigra*), with dense setulae on its inner face surface; distiphallus broadly flattened.

Remarks

The *C. appendiculata* complex of this species-group, which includes *C. aucta* listed below, was revised by Sinclair (1999*b*). The Oriental species are all similar and closely related, characterized by the absence of the preapical comb of the forefemur (also absent in *C. nigra*).

The monophyly of this group remains in doubt, as no unique synapomorphies have been identified.

Distribution

This group is distributed in the Palearctic Region, northern and eastern North America, and the Oriental Region (Nepal).

Included species

This species-group was first recognized by Sinclair (1995) and includes the following species:

PA *Clinocera appendiculata* (Zetterstedt, 1838: 559) (*Wiedemannia*) (holotype examined).

NE *Clinocera appalachia* **sp. nov.**

NE + PA *Clinocera aucta* (Zetterstedt, 1849: 3019) (*Brachystoma*).

OR *Clinocera lunata* Wagner & Leese, 2004: 18 (holotype examined) (see Wagner et al. 2004).

OR *Clinocera lunatoides* Wagner & Leese, 2004: 18 (holotype examined) (see Wagner et al. 2004).

PA *Clinocera megalatlantica* (Vaillant, 1957: 65) (*Atalanta* (*Atalanta*)).

PA *Clinocera nigra* Meigen, 1804: 292 (lectotype examined). syn. *rufipes* Bezzi, 1899: 147 (see Sinclair 2007).

OR *Clinocera nigra orientalis* (Vaillant, 1965: 148) (*Atalanta* (*Atalanta*)).

OR *Clinocera pani* Smith, 1965: 108 (possibly very similar to *C. orientalis*).

PA *Clinocera schumanni* Joost, 1981: 184.

OR *Clinocera setosa* Wagner & Leese, 2004: 18 (holotype examined) (see Wagner et al. 2004).

OR *Clinocera sinclairi* Wagner & Leese, 2004: 19 (holotype examined) (see Wagner et al. 2004).

PA *Clinocera storchi* Mik, 1880: 348 (not described as var. of *appendiculata* as listed by Chvála & Wagner 1989: 330) (lectotype examined).

Key to species of the *Clinocera nigra* group of North America

1 R_{2+3} and R_4 connected by an auxiliary crossvein (Fig. 32c); faint clouding about veins and crossveins; scutum with pair of distinct dark stripes (northern) . . . . **C. aucta** (Zetterstedt)

— R_{2+3} and R_4 not connected by crossvein; wings infuscate, clouding absent; scutum with indistinct vittae (eastern) . . . . **C. appalachia** sp. nov.

Clinocera appalachia sp. nov.

Fig. 21*b*; Map 25

Type material. HOLOTYPE, male labelled "USA:NC: Buncombe Co./ Craggy Gardens Picnic/ Area, Blue Ridge Pkwy/ 12.VI.1990, 1580 m/ roadcut seep, BS9021/ B.J. Sinclair"; "HOLOTYPE/ Clinocera/ appalachia/ Sinclair [red label]" (CNC).

Paratypes: **USA. North Carolina:** Same data as holotype (3 ♂♂, 3 ♀♀, CNC); Swain Co., GSMNP, 6 km S Balsam Mtn Cpgd, roadcut seep, 13.vi.1990, BJS (1 ♂, CNC); Swain Co., GSMNP, Clingmans Dome Rd, roadcut seep, 15.vi.1990, 9.vii.2004, BJS (2 ♂♂, 3 ♀♀, CNC; 1 ♂, 3 ♀♀, ZFMK). **Tennessee:** Sevier Co., GSMNP, Alum Cave Tr., seeps, 16.vi.1990, BJS (5 ♂♂, 2 ♀♀, CNC);

Sevier Co., GSMNP, Mt LeConte Tr., stream, 16.vi.1990, BJS (2 ♀♀, CNC). **West Virginia:** Blackwater Falls SP, seep at Elakala Falls, 2, 3.vii.1986, E. Lippert, BJS (2 ♂♂, 3 ♀♀, DEBU); Blackwater Falls SP, seep at Elakala Falls, 39.11°N, 79.5°W, 6.vii.2004, BJS (6 ♂♂, 3 ♀♀, USNM; 10 ♂♂, 10 ♀♀, ZFMK).

Recognition. A delicate species distinguished from other eastern Nearctic species by the pale coxae and pale ventral face of the femora, contrasting with the dark pleura, strongly curved clasping cercus, and indistinctly vittate scutum.

Description (wing length 3–3.4 mm).

Male. Face flat with blue pruinescence fading dorsally toward base of antennae, remainder of head dark. Arista-like stylus long and slender. Postocular bristles pale throughout.

Scutum dark brown with indistinct vittae; postpronotum and notopleuron light bluish brown; lower two-thirds of pleura with faint blue pruinescence. Several short acrostichals anterior to first dc; 5 dc; 1 pprn; 1 presut spal; 2 npl, lower short and fine; 1 psut spal; 1 pal; 2 sctl. Laterotergite with patch of pale bristles. Posterior margin of anepisternum with pale setulae.

Wing hyaline, clouding or spots absent; stigma absent; 2 long basal costal setae; costal spines present. Veins straight; auxiliary crossvein between R_4 and R_{2+3} absent; stump vein normally absent, occasionally present on one wing; R_4 straight, divergent from R_5; anal vein an elongate streak; cell dm produced apically. Halter pale brown with dark knob.

Coxae and ventral margin of femora light brown, upper face of femora and remaining segments brown. Forecoxa with pale slender setae. Forefemur with anterior preapical comb present; ventrobasal surface with short dense white pile; antero- and postero-ventral row of widely spaced short hairs, increasing in length on posteroventral row. Foretibia with biserial row of short fine bristles beneath. Hind tibia with erect dorsal and ventral bristles on apical half. Fifth tarsomere lacking dorsoapical extension; empodium brown laterally.

Abdomen dark, brownish. Hypandrium broadly triangular. Phallus strongly bent at midlength; distiphallus with expanded apical tip. Clasping cercus geniculate, tapered to narrow apex; inner face densely clothed in fine setulae. Surstylus shorter than clasping cercus, tapered to broad apex; dorsoapical margin with matted microtrichia (Fig. 21*b*).

Female. Similar to male except ventral bristles of forefemur and foretibia reduced.

Geographic distribution and seasonal occurrence. This species is known only from West Virginia, North Carolina, and Tennessee (Map 25) and has the most restricted range among the eastern North American species of *Clinocera*.

Remarks. This species was primarily encountered in seepage habitats, especially roadcuts, wet cliffs surrounding waterfalls, and occasionally in small headwater streams.

Etymology. The specific name refers to the region where this species primarily occurs.

Clinocera aucta (Zetterstedt)

Figs. 21*c*, 32*c*; Map 25

Brachystoma aucta Zetterstedt, 1849: 3019.

Clinocera bivittata Loew, 1864: 258.

Clinocera aucta: Frey, 1913: 61.

Clinocera (*Hydrodromia*) *longifurca* Melander, 1928: 231.

Hydrodromia aucta: Collin, 1935: 370.

Clinocera longifurca: Melander, 1965: 468.

Clinocera aucta: Sinclair, 1999*b*: 227.

Type material examined. *Brachystoma aucta*: HOLOTYPE, male labelled "B. aucta/ Zett., n./p. ♀/ Qvickjock. [Sweden]/ Wahlberg" (UZIL).

Clinocera bivittata: HOLOTYPE, male labelled "Qchotsk/ Shlberg [Russia]"; "10746"; "Clinocera/ bivittata/ m."; "Zool. Mus./ Berlin"; "Holotypus" (ZMB).

Clinocera longifurca: HOLOTYPE, male labelled "MT. WASH'N./[USA]"; "TYPE/ Hydrodromia/ longifurca/ Mel. [red label]" (USNM).

Paratypes: **USA. New Hampshire:** Mt Washington, Alpine Garden, 4.vii.1914, CWJ (2 ♀♀, USNM). **CANADA. Newfoundland and Labrador:** Nain, 18.viii.1908, CWJ (1 ♀, MCZ). A male specimen labelled "paratype" from Mt Washington in MCZ is not part of the original type series.

Nearctic material examined. Some 270 specimens from the following territories, provinces, and states: **CANADA. British Columbia:** Ketchum Lk., 58°22′N, 131°45′W, 26.viii.1960, WWM (CNC); Peterson Ck, Alaska Hwy DC-445, 29.vi.1978, PHA (USNM). **Newfoundland and Labrador:** Hebron, 15–30.vii.1954, JFMC (CNC); Nutak, 24.vii.1954, JFMC (CNC); Tallik Bay, Nagvak Fjord, JFMC (CNC). **Northwest Territories:** Aklavik, 8.vii.1932, OB (CAS); Muskox Lk., 64°45′N, 108°10′W, 4.viii.1953, JGC (CNC); Salmita Mines, 64°05′N, 111°15′W, 17.vi.1953, JGC (CNC); Tununuk, 15.viii.1930, OB (CAS); Yellowknife, 19.v.1953, JGC (CNC). **Nova Scotia:** Cape Breton Highlands NP, nr Pleasant Bay, seep, 14.vii.1983, DMW (CNC). **Nunavut:** Baffin Is., Exeter Sound, 24–25.viii.1962, M.G. Rhodes (BMNH), Lk. Harbour, 9.viii.1935, WJB (CNC); Baker Lk., 64°193′N, 96°02.5′W, 14.vii.2003, P.H. Adler (ZFMK); Bathurst Inlet, Baychimo, 2–26.viii.1966, GES (CNC), Ekalulia Is, 12.viii.1966, GES (CNC), Hiukitak R., 3.viii.1966, GES (CNC); Cape Dorset, 26.viii.1954, H. Huckel (CNC); Frobisher Bay, 10–11.viii.1959, WRMM, WRR (CNC); Wharton Lk., 64°52′N, 99°45′W, 18.viii.1966, JGC (CNC). **Quebec:** SE Akpatok Is., Ungava Bay, 21,

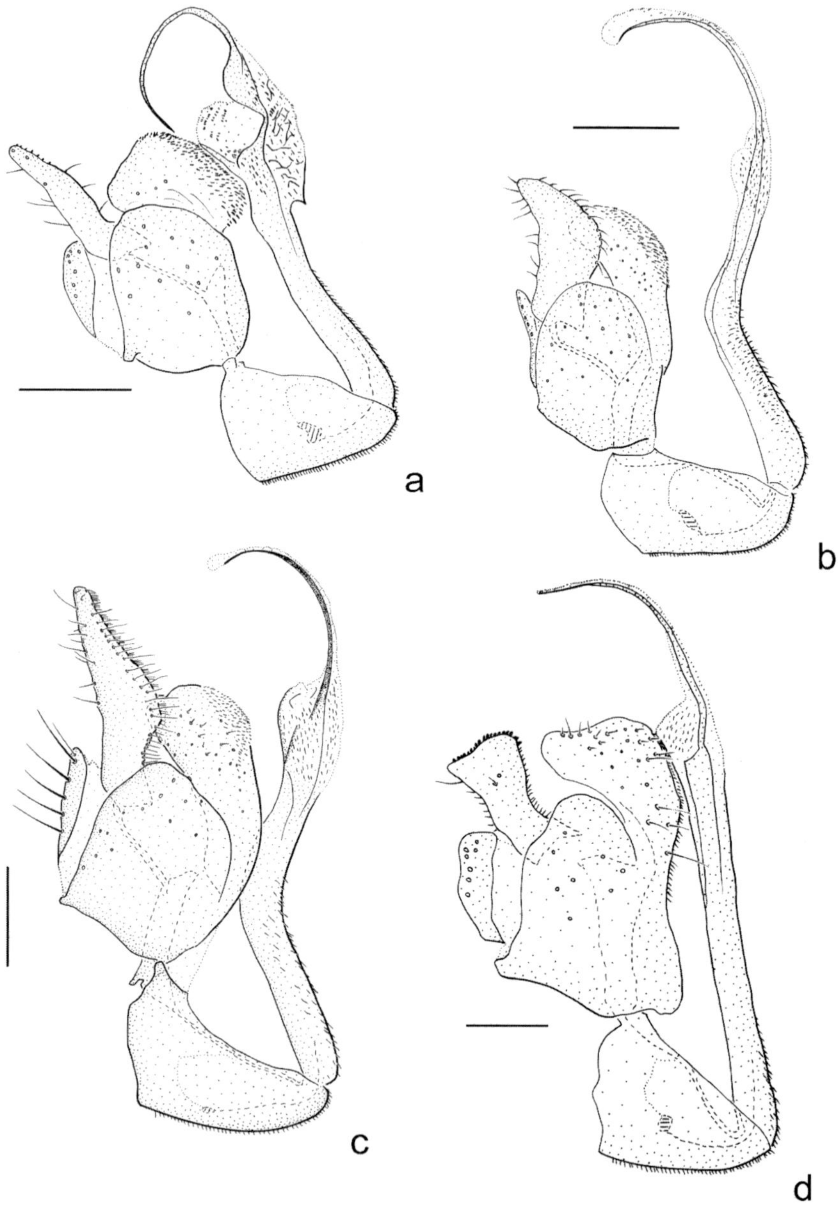

Fig. 21. Male terminalia of *Clinocera*: *a*, *C. woodi*; *b*, *C. appalachia*; *c*, *C. aucta*; *d*, *C. binotata*. Scale bars = 0.1 mm.

25.viii, 2.xi.1931 (HMOU); Indian House Lk., 9–31.vii.1954, WRR (CNC); Payne Bay, 8–19.viii.1958, WRMM (CNC); Port Burwell, 31.viii.1931, WJB (CNC); Sugluk (Saglouc), 8.viii.1954, H. Huckel (CNC); Tasiujaq, 10 km SW, Finger Lk., tundra/pools/str., 17–25.viii., 7–8.ix.1996, BJS (CNC). **Yukon:** Aishihik Lk. Rd, 13 km N Hopkins Lk., 61°28′N, 137°W, 22.viii.1981, C. Guppy (UBC); Dempster Hwy, North Olgilvie Mtns, car netting, taiga, 17.viii.1984, S. and J. Peck (CNC); Dempster Hwy, km 15, 4.vi.2003, AB (CNC); Dempster Jct., 40 km E Dawson, 19.vi.–25.vii.1984, S.B. Peck (CNC); Dempster Hwy, mi. 51, 25–27.vi., 7–12.vii.1973, G. and D.M. Wood (CNC); Dempster Hwy, mi. 87, 4–8.viii.1973, G. and D.M. Wood (CNC); Dempster Hwy, km 95, 19.vii.1982, DMW (CNC);

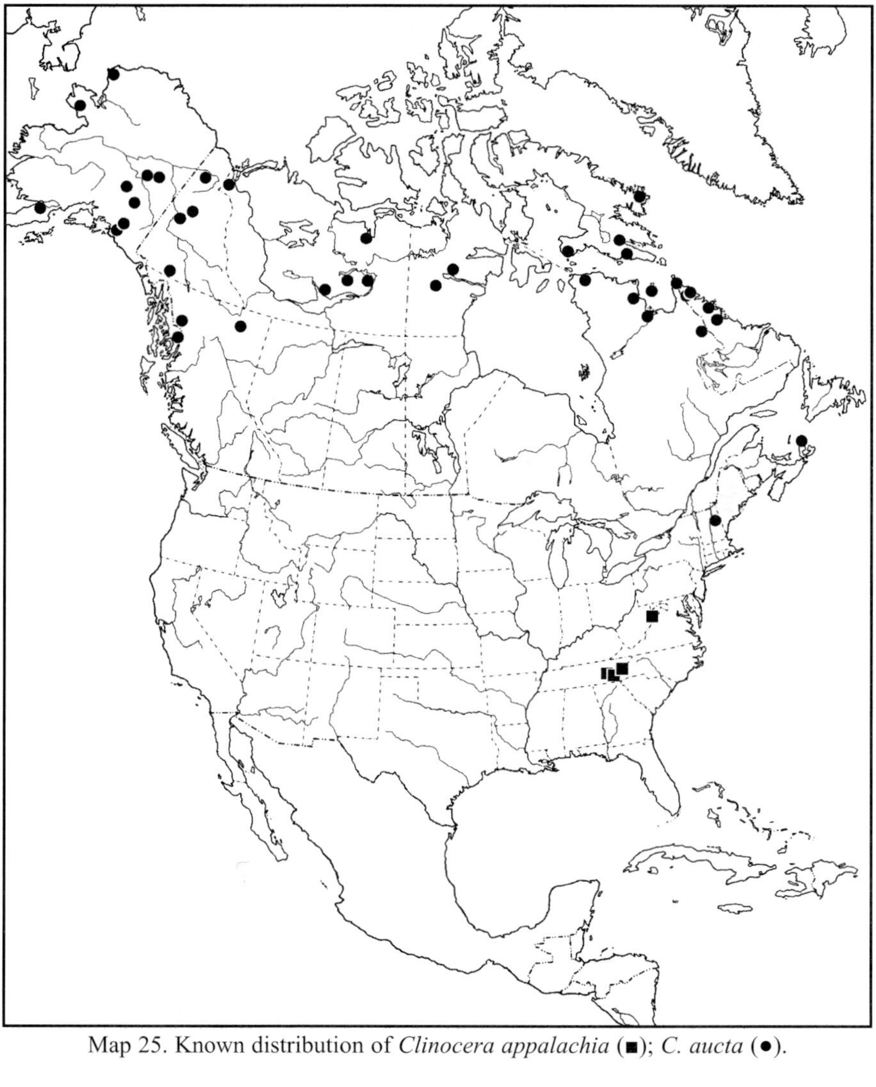

Map 25. Known distribution of *Clinocera appalachia* (■); *C. aucta* (●).

Dempster Hwy, km 152, Windy Pass, 8.vi.2003, AB (CNC); North Fk Crossing, mi. 42, Peel Plt. Rd, 3500′, 24.vi.1962, P.J. Skitsko (CNC); North Fk Pass, Olgilvie Mtns, 4100′, 18.vi.1962, P.J. Skitsko (CNC); Rampart House, 5.vi.1951, C.C. Loan (CNC). **USA. Alaska:** Cape Thompson, 26.vii.1961, R. Madge (CNC); Denali Hwy, mi. 36, Boulder Mtn, 14, 27.vii.1993, TS, OY (BLKU); Eagle Summit, 3800′, 25.vi.1948, R.I. Sailer (USNM); George Pk Hwy, mi. 224, Bison Gulch, 21.vii.1993, TS (BLKU); Gulkana, Panson Lodge, 4000′, 4.viii.1951, WRMM (CNC); Healy, 7.vi.1921, JMA (USNM); Hyder, rd to Rish Ck, 5.viii.1996, PHA (USNM); Mt McKinley NP, Teklanika R. Cpgd, 760 m, 15.vii.1978, PHA (USNM); Naknek R., King Salmon, 10.viii.1952, WRMM (CNC); Paxson, One Mile Ck, 16, 27.vii.1993, TS (BLKU); Richardson Hwy, Miller House, 21.vi.1951, WRMM (CNC), mi. 29, Worthington Glacier, 17.vii.1993, TS (BLKU), mi. 204, Phelan Ck, 15, 27.vii.1993, 8.vi.1994, TS (BLKU), mi. 206, Thompson Pass, 18.vii.1993, TS (BLKU), mi. 206, beaver pond, 20.vi.1987, DMW (CNC); Salmon Lk., 32 mi. N Nome, 4.vii.1993, TS (BLKU); 5 mi. NE Salmon Lk., Nome–Taylor Hwy, 20.vi.1994, TS (BLKU); Steese Hwy, mi. 80, Montana Ck, 11.vii.1993, TS (BLKU), mi. 86, Twelvemile Summit, 29.vii.1993, TS (BLKU), mi. 108, Eagle Summit, 12.vii.1993, TS (BLKU). **New Hampshire:** Coos Co., Mt Washington, Alpine Garden, 18.vi.1982, DSC (UNH); Mt Washington, A.T. Slosson (AMNH); Mt Washington, Lk. of the Clouds, 5000′, 9.viii.1954, WRMM (CNC); White Mtns NF, Mt Washington, 5.viii.1980 (BMNH).

Palearctic material examined. Finland, western Russia, Siberia, Sweden (20 specimens) (see Sinclair 1999*b*).

Recognition. The species is closely related to *C. appendiculata*, but readily distinguished by an auxiliary crossvein between R_{2+3} and R_4, and the face of both males and females with blue pruinescence.

Description (wing length 3.8–4.2 mm).

Male. Face flat, mostly blue, with brown below antennae and along margin of upper half of face. Gena and posterior region of head with blue pruinescence. Arista-like stylus long and slender. Postocular bristles dark.

Scutum with two black vittae separated by faint blue pruinescence. Pleura with blue pruinescence. Several short acrostichals anterior to first dc; 5 dc; 1 pprn; 1 presut spal; 2 npl, lower short and fine; 1 psut spal; 1 pal; 2 sctl. Laterotergite with patch of pale bristles. Posterior margin of anepisternum with pale setulae.

Wing (Fig. 32*c*) infuscate with narrow clouding along veins, especially crossveins; stigma absent; 1 long basal costal seta; costal spines present. R_{2+3} sometimes slightly wavy; auxiliary crossvein between R_4 and R_{2+3} present with cell r_4 often truncate occasionally with short appendix; R_4 straight, divergent from R_5; anal vein faint streak; cell dm produced apically. Halter dark.

Coxae with blue pruinescence; forefemur brown near tip, remainder clothed with blue pruinescence; remaining femora and segments dark. Forecoxa with fine pale setae. Forefemur with anterior preapical comb present; ventrobasal surface with short dense white pile on apical fourth; antero- and postero-ventral row of fine, spine-like bristles, becoming stouter apically. Foretibia with biserial row of short bristles beneath. Hind tibia with short erect dorsal and ventral bristles on apical half. Fifth tarsomere lacking dorsoapical extension; empodium brown laterally.

Abdomen dark, brownish. Hypandrium broadly triangular. Phallus strongly bent at midlength; apex of shaft with oval membranous cap, bearing microtrichia; distiphallus dorsoventrally flattened, apex expanded. Clasping cercus greatly prolonged apically, gradually tapered to narrow apex; inner face with fine setulae; outer dorsolateral margin lined with long fine setae. Surstylus much shorter than clasping cercus; apex expanded, round (Fig. 21*c*).

Female. Similar to male except lacking biserial ventral bristles of forefemur and foretibia.

Geographic distribution and seasonal occurrence. This species is Holarctic in distribution with records from Siberia and eastern Russia throughout arctic North America, and Scandinavia. In North America, several glacial relict populations are known from Cape Breton Highlands National Park and Mount Washington (New Hampshire) (Map 25).

Remarks. Records indicate that adults are present from June to early September, collected in seepages, at edge of tundra pools, and in streams and creeks.

Clinocera stagnalis group

Diagnosis

The *C. stagnalis* group is recognized by the following combination of characters: often distinctive clouding surrounding wing crossveins (usually at base of R_4, apex of cell dm, r-m crossvein, and apex of cell bm); cell dm truncate, auxiliary vein sometimes present between R_4 and R_{2+3}; opaque stigma absent (best viewed in specimens dried from alcohol); legs often reddish with pale apex ("knee"); fifth tarsomere usually with dorsoapical extension; face and thoracic pleura with various amounts of silvery-blue pruinescence; phallus subapically split at base of distiphallus; distiphallus broadly flattened.

Remarks

This species-group is equivalent to the former subgenus *Hydrodromia* Macquart (see under *Clinocera* above). The apices of the femora are pale in many species, the face is often convex, and often the wing is characterized by clouding patterns or spots.

The monophyly of this group is supported on the basis of the following apomorphies: face convex, fifth tarsomere with a dorsoapical extension, and wing with distinct clouding or spots.

Distribution

This species-group is primarily Palearctic and New World in distribution.

Included species

This species-group was first recognized by Sinclair (1995) and includes the following species:

NT	*Clinocera attenuata* **sp. nov.**
NE + NT	*Clinocera binotata* Loew, 1876: 325.
NT	*Clinocera chilensis* **sp. nov.**
NT	*Clinocera collini* **sp. nov.**
NT	*Clinocera colombiana* **sp. nov.**
NT	*Clinocera cubatao* Smith, 1962: 264.
NT	*Clinocera cucama* Smith, 1962: 265.
NT	*Clinocera cummingi* **sp. nov.**
NT	*Clinocera ditaeniata* Bezzi, 1909: 387 (*Clinocera* (*Heleodromia*)).
NT	*Clinocera empodiata* Collin, 1933: 322.
PA	*Clinocera fontinalis* (Haliday, 1833: 160) (*Heleodromia*).
AU	*Clinocera gressitti* Smith, 1964: 327.
NT	*Clinocera insularis* **sp. nov.**
NT	*Clinocera isabellina* **sp. nov.**
NT	*Clinocera lapazensis* **sp. nov.**
NE	*Clinocera maculata* Loew, 1860: 79.
PA	*Clinocera maroccana* (Séguy, 1941: 29) (*Hydrodromia*).
NT	*Clinocera marshalli* **sp. nov.**
NT	*Clinocera meridana* **sp. nov.**
NT	*Clinocera miradorana* **sp. nov.**
NT	*Clinocera naomina* **sp. nov.**
NE + PA	*Clinocera nivalis* (Zetterstedt, 1838: 545) (*Hemerodromia*).
NT	*Clinocera notialis* **sp. nov.**
NT	*Clinocera oriunda* (Collin, 1933: 323) (*Hydrodromia*).
NT	*Clinocera parana* **sp. nov.**
NT	*Clinocera penai* **sp. nov.**
NE + PA	*Clinocera stagnalis* (Haliday, 1833: 159) (*Heleodromia*).
NE	*Clinocera substagnalis* **sp. nov.**
PA	*Clinocera wesmaeli* (Macquart, 1835: 656) (*Paramesia*).

Key to species of the *Clinocera stagnalis* group north of Mexico

1 R_{2+3} and R_4 not connected by an auxiliary crossvein (Figs. 32*e*, *g*), although occasionally present on one wing in some specimens of *C. stagnalis*. 2

— R_{2+3} and R_4 connected by an auxiliary crossvein on both wings (Figs. 32*d*, *f*, *h*) 3

2 Legs dark except for extreme tip of femora; face entirely clothed in blue pruinescence; forefemur lacking basal setae (northern, Rocky Mountains) . . . . *C. stagnalis* (Haliday)

— Femur and tibia bright orange–brown, tarsomeres dark; lower half of face clothed in blue pruinescence; forefemur with stout basal setae (Figs. 9*c*, *d*) (eastern). . . . . *C. maculata* Loew

3 Extreme apex of femora dark, not pale; anterior spiracle encircled by black ring; pleura dark, pruinescence faint (northern). . . . . *C. nivalis* (Zetterstedt)

— Extreme apex of femora reddish brown; anterior spiracle not encircled by black ring; pleura with distinct blue pruinescence 4

4 Forefemur with stout basal setae; clasping cercus expanded apically (Fig. 21*d*) (widespread) . . . . *C. binotata* Loew

— Forefemur lacking stout basal setae; clasping cercus parallel-sided, not expanded apically (Fig. 22*d*) (western) . . . . *C. substagnalis* sp. nov.

Key to species of the *Clinocera stagnalis* group of the Americas south of the United States

1 Auxiliary crossvein between R_{2+3} and R_4 present (at least on one wing) (Fig. 32*i*) 2

— Auxiliary crossvein R_{2+3} and R_4 absent (Fig. 32*j*) 12

2 Scutum with four stripes; outer vitta on supra-alar ridge, distinctly visible when viewed from head to scutellum. 3

— Scutum with two stripes; only central pair of black vittae present 8

3 Scutum light brown laterally, dark brown when viewed from above, pleura lacking blue pruinescence . . . . *C. miradorana* sp. nov.

— Thorax brown, pleura with blue pruinescence 4

4 Forefemur with long setae at base; lower third of face with blue pruinescence; dorsal surface of costa of female wing with series of long erect black setae . . . . *C. isabellina* sp. nov.

— Forefemur lacking long setae at base; at least lower half of face with blue pruinescence; female unknown. 5

5 Lower half only of pleura with blue pruinescence. 6

— Entire pleura clothed in blue pruinescence. 7

6 Lower two-thirds of face with bright-blue pruinescence; clasping cercus very long, finger-like (Fig. 27*a*); phallus straight . . . . ***C. notialis*** sp. nov.
— Lower half of face with pale pruinescence on lateral margin; clasping cercus boot-shaped (Fig. 25d); phallus with strong subbasal bend. . . . . ***C. lapazensis*** sp. nov.

7 Lower face with bright-blue pruinescence; ventral surface of femora yellowish brown . . . . ***C. penai*** sp. nov.
— Lower face with faint, pale pruinescence on lateral margin; all leg segments dark. . . . . ***C. colombiana*** sp. nov.

8 Pleura entirely clothed in blue pruinescence 9
— Lower half of pleura clothed in blue pruinescence 11

9 Apex of femora yellowish brown; forefemur with long basal setae . . . . ***C. binotata*** Loew
— Apex of femora dark or femur entirely pale coloured; long basal setae lacking on forefemur 10

10 Halter yellowish brown; clasping cercus gradually tapered to broadly rounded apex (Fig. 24*d*). . . . . ***C. ditaeniata*** Bezzi
— Halter dark; clasping cercus strongly narrowed and produced on apical half (Fig. 23*a*) . . . . ***C. attenuata*** sp. nov.

11 Mid and hind femora yellowish brown in dorsal view; blue pruinescent spot at base of wing small, not extending to notopleural bristles; posterior margin of clasping cercus evenly tapered, anterior margin straight (Fig. 23*b*) . . . . ***C. chilensis*** sp. nov.
— Mid and hind femora dark in dorsal view; blue pruinescent spot at base of wing large and distinct, extending to notopleural bristles; posterior margin of clasping cercus strongly tapered on apical third, anterior margin curved (Fig. 26*d*) . . . . ***C. naomina*** sp. nov.

12 Scutum with bristles dark and well developed; ocellar bristles longer than basal 3 antennal segments. 13
— Scutum with bristles reduced, shorter, and more slender; postpronotal bristles very fine and pale; length of ocellar bristles subequal to length of basal 3 antennal segments (females difficult to distinguish further). 21

13 Wings with spots or distinct clouding absent about crossveins 14
— Wings with spots or distinct clouding about crossveins (Fig. 33*d*) 16

14 Entire face, pleura, and lateral margin of scutum with bright-blue pruinescence; halter pale . . . . ***C. empodiata*** Collin
— Face with pruinescence confined to lateral margins; pleura with pale pruinescence; lateral margin of scutum brown; halter dark 15

15 Leg segments dark brown, except extreme apex of coxae and trochanters; clasping cercus hammer-shaped and base of distiphallus widely expanded (Fig. 24*b*). . . . . ***C. cucama*** Smith

— Leg segments light brown, including coxae and trochanters; clasping cercus straight, parallel-sided apically (Fig. 24a) . . . . ***C. cubatao*** Smith

16 Femora and tibiae yellowish brown; lateral margins of lower face with bright, silvery pruinescence . . . . ***C. collini*** sp. nov.

— Femora light brown to dark, apex sometimes yellowish brown; tibiae dark; lower face variable 17

17 Entire ventroapical surface of femora or only apical fifth yellowish brown, especially forefemur, contrasting with dark upper surface 18

— Ventroapical surface of femora dark brown, not contrasting with upper surface 19

18 Lower face with bright, silvery pruinescence; surstylus rounded apically (Fig. 27*b*) . . . . ***C. oriunda*** Collin

— Lateral margin of lower face with pale-blue pruinescence; surstylus truncate apically (Fig. 26*b*) . . . . ***C. meridana*** sp. nov.

19 Scutum with two pairs of distinct black vittae and long black spot above base of wing; pleurae with blue pruinescence on lower half, extending anteriorly to postpronotal lobe; clasping cercus short, less than twice as long as wide, nearly half-moon-shaped (Fig. 25*b*) . . . . ***C. insularis*** sp. nov.

— Scutum with one medial pair of black vittae; pleurae entirely clothed in blue pruinescence or mostly absent, postpronotal lobe lacking pruinescence; clasping cercus more than twice as long as wide 20

20 Pleura entirely clothed in blue pruinescence; clasping cercus strongly narrowed and produced on apical half (Fig. 23*a*) . . . . . . ***C. attenuata*** sp. nov.

— Pleura dark brown, faint pruinescence about forecoxae; clasping cercus with arched posterior margin gradually tapered (Fig. 26*a*) . . . . ***C. marshalli*** sp. nov.

21 Clasping cercus very long, narrow, and arched medially; hypandrium not greatly narrowed or lengthened (Fig. 24*c*). . . . . ***C. cummingi*** sp. nov.

— Clasping cercus short, subtriangular, strongly tapered, and directed vertically; hypandrium greatly narrowed and lengthened (Fig. 27*c*) . . . . ***C. parana*** sp. nov.

Nearctic species of the *C. stagnalis* group

Clinocera binotata Loew

Figs. 1*e*, 9*a*, *b*, 21*d*, 32*d*; Map 26

Clinocera binotata Loew, 1876: 325.

Clinocera bicincta (error): Tucker, 1907: 97.

Clinocera (Hydrodromia) binotata: Melander, 1902, 1928, 1965: 468.

Type material examined. HOLOTYPE, female labelled "Type/ 1666 [red label with upper white border]"; "New York[USA]/ O[sten].S[acken]. [handwritten]"; "binotata/ Lw. [handwritten]" (MCZ).

Additional material examined. Some 1600 specimens from the following territories, provinces, and states: **CANADA. British Columbia:** 7.3 km N Armstrong, 12–18.viii.1997, AB (CNC); Mission City, 28.vi.1953, WRMM (CAS, CNC); Pt Grey, Vancouver, 9.ix.1972, JRV (CNC). **Ontario:** Fergus, 25.v., x–xi.1987, SAM (DEBU); Limehouse, Black Ck, 4.iii.1973, 4–16.iii.1975, RER (DEBU); 3 km W Puslinch, Crieff Bog, 29.iv.–4.ix., 5–11.ix., 26.ix.–9.x., 10–30.x.1987, D. Blades (DEBU); Simcoe, 13.vi.1939, GES (CNC); Wellington Co., 1968–1970, PPH (UMQ). **Quebec:** Hull, Fairy Ck, 29.ix.1955, JFMC (CNC); Montreal, 1.ix., CWJ (MCZ); Montreal, 1.ix.1901, WMW (AMNH, USNM). **MEXICO. Baja California Norte:** Arroyo Valladeres, 24 km E Meling Ranch, 8.ix.1980, DDW (CAS). **Chiapas:** San Cristobal, 10 km NE, 7500′, 22.v.1969, HJT (CNC). **Durango:** 10–14 mi. SW El Salto, 8000–9000′, 5.vi.–4.vii.1964, JFMC (CNC). **Michoacan:** Zitacuaro, 13 mi. E, 1.v.1953, E.I. Schlinger (UCB). **Oaxaca:** Oaxaca, 20.v.1963, HER (USNM). **USA. Arizona:** Cochise Co., Coronado NF, Huachuca Mtns, Miller Ck, 3.vi.1991, BJS/JES (CNC, DEBU); Grand Canyon Co., CR 7, Elves Cyn, 10–21.v.1998, JFM (PERC); Greer, 21.vi.1947, J.L. Sperry (USNM); Ramsey Cyn, 6000′, 15 mi. S Sierra Vista, Huachuca Mtns, 13.vii.1967, Sternizky (CNC); Santa Cruz Co., Madera Cyn, 14.iv.1962, P.A. Rude (UCB), 2.5 mi W Pena Blanca, Las Guijas Mtns, 17.x.1978, RER (DEBU); Todds Lodge, Oak Ck, 9.v.1950, J.L. Sperry (USNM); White Mtns, Greer, 9000′, 21.vi.1947, J.L. Sperry (USNM). **California:** Alameda Co., Berkeley Hills, 11.iv.1908 (ANSP), Sunol Regional Wilderness, 10.v.1941, T. Aitken, B. Brookman (CAS), Sunol Regional Wilderness, Indian Joe Ck, 29.iii.1969, DDW (CAS); Alpine, 9.vii.1929, R.H. Beamer (SEMC); Alpine Co., Indian Ck Lk, 15.vi.1979, GEB (USUL), Little Teton Ck, 12.vii.1977, DDW (CAS), West Fk Carson R., 6600′, 19.vi.1971, PHA (USNM); Berkeley, 20–28.v.1915, MCVD (USNM, MCZ, CAS); Berkeley, 17.ix.1958, L.E. Campes (UCD); Berkeley, Strawberry Ck, 20.iv.1948, WWW (USNM); Capitola, 7–12.vi.1940, MTJ, H.B. James (WSU); Contra Costa Co., Briones Regional Pk, Toyon Cyn, 26.v.1979, DDW (CAS), 7 mi. SE Clayton, 1.viii.1982, RH (HSU), Mt Diablo, 1948, WWW (USNM), Mt Diablo SP, Mitchell Cyn, 13.vi.1979, DDW (CAS), Wildcat Cyn, 19.v.1978, C. Griswold

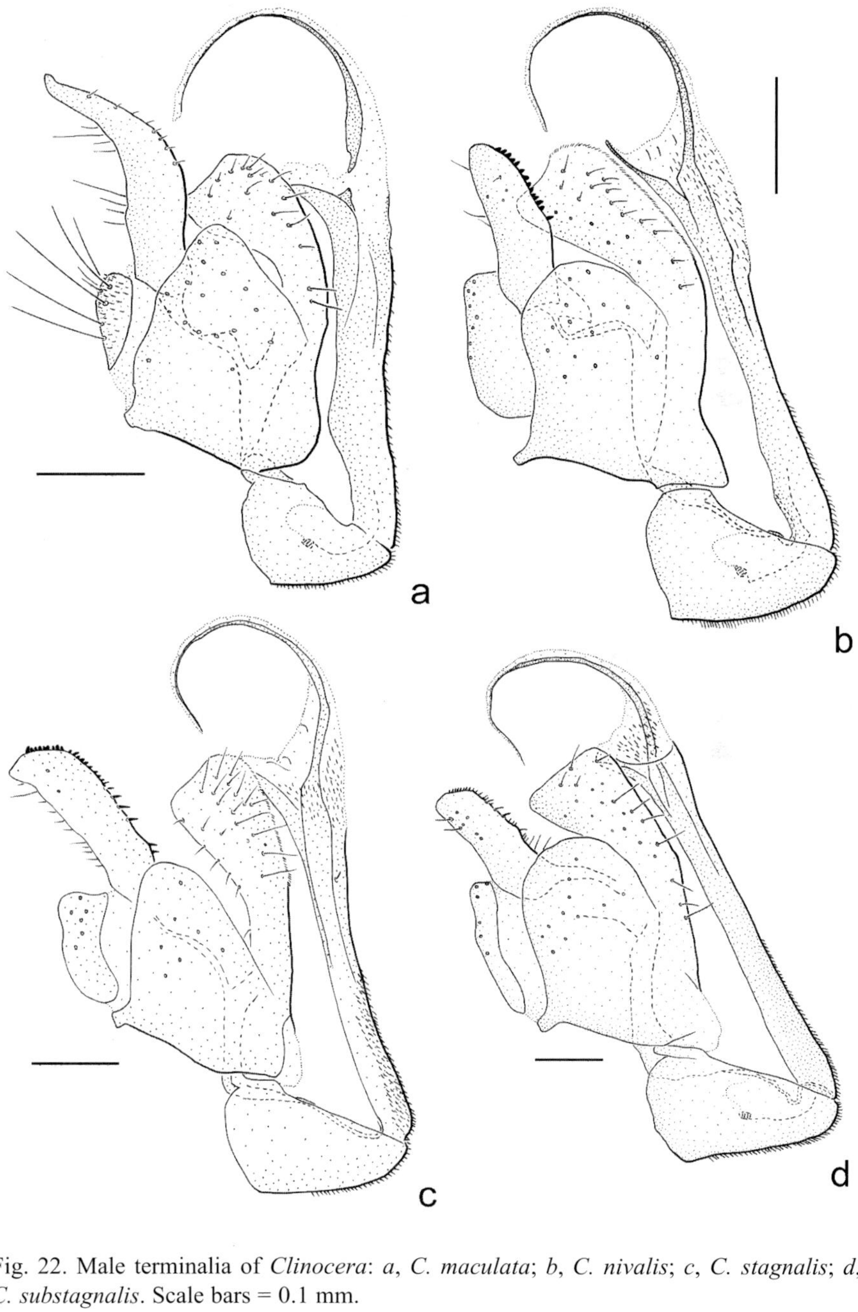

Fig. 22. Male terminalia of *Clinocera*: *a*, *C. maculata*; *b*, *C. nivalis*; *c*, *C. stagnalis*; *d*, *C. substagnalis*. Scale bars = 0.1 mm.

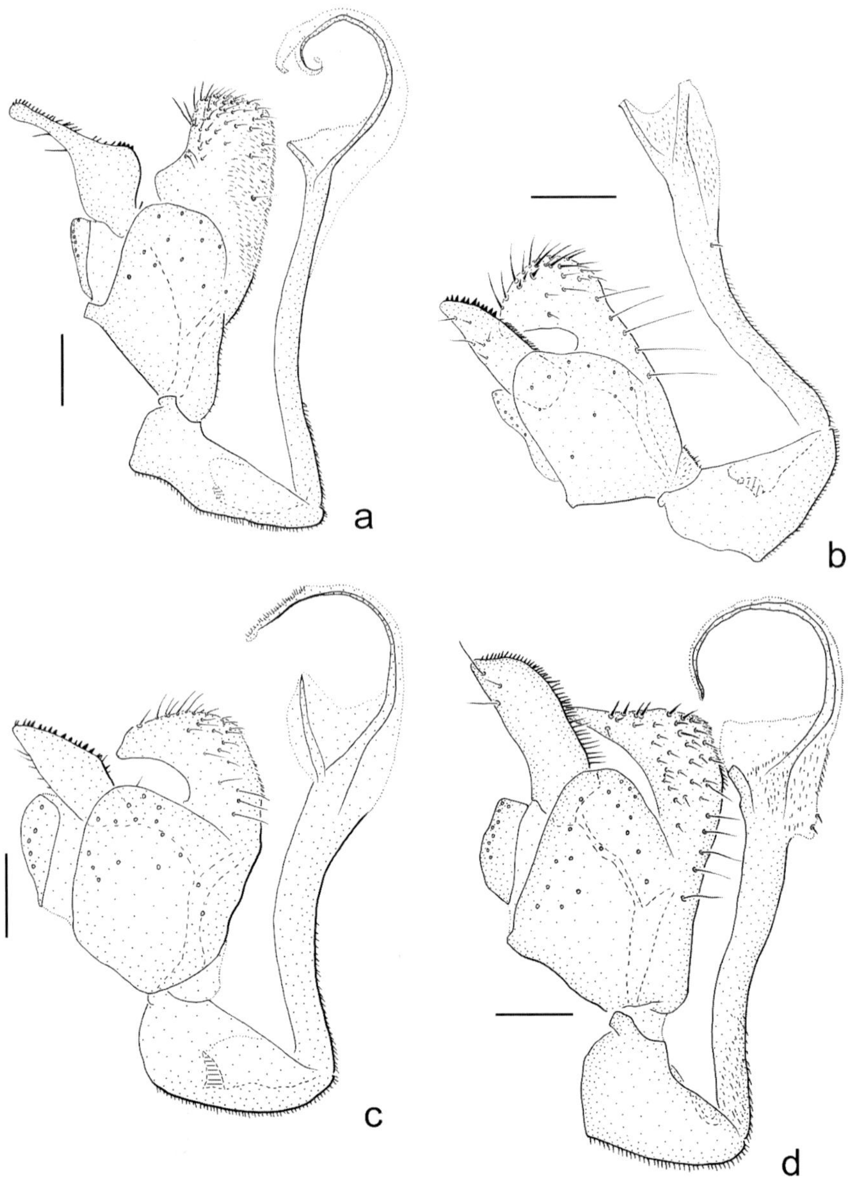

Fig. 23. Male terminalia of *Clinocera*: *a*, *C. attenuata*; *b*, *C. chilensis*; *c*, *C. collini*; *d*, *C. colombiana*. Scale bars = 0.1 mm.

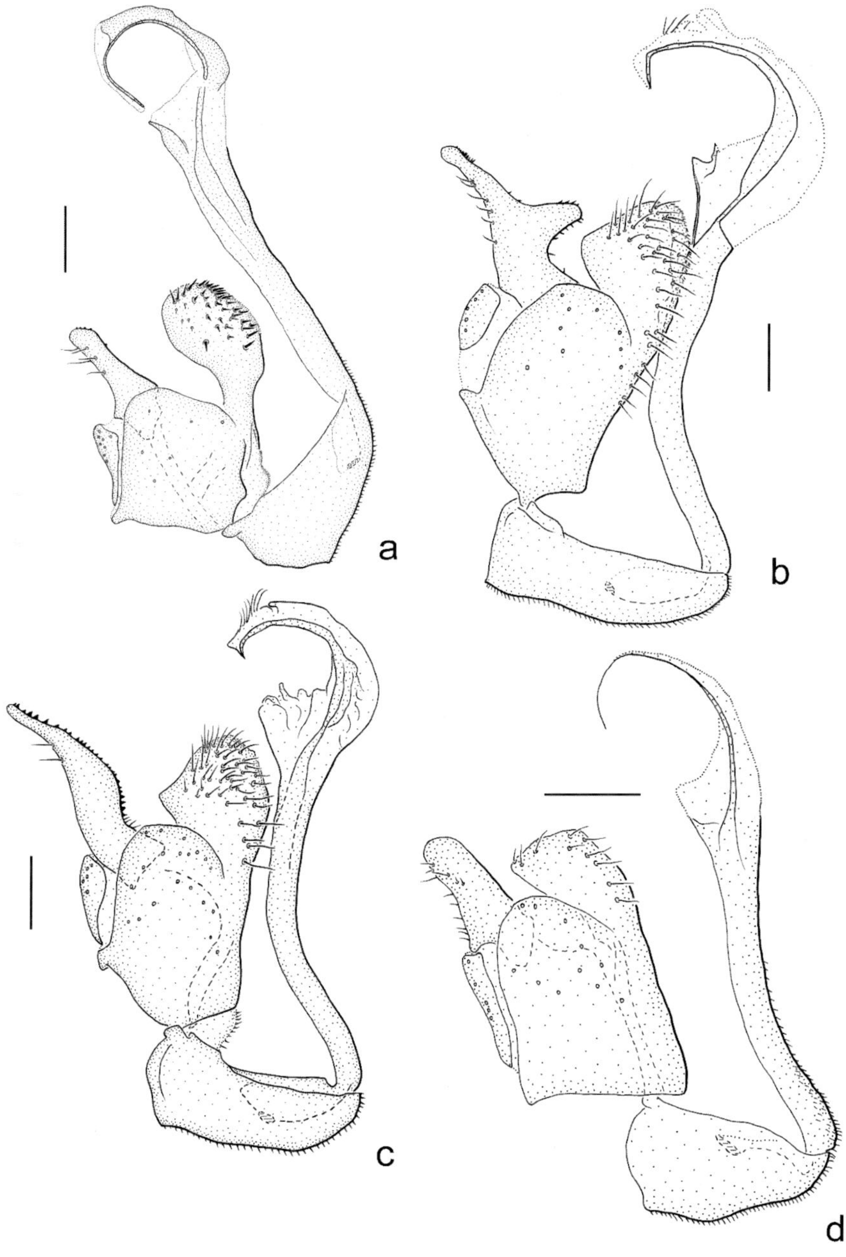

Fig. 24. Male terminalia of *Clinocera*: *a*, *C. cubatao*; *b*, *C. cucama*; *c*, *C. cummingi*; *d*, *C. ditaeniata*. Scale bars = 0.1 mm.

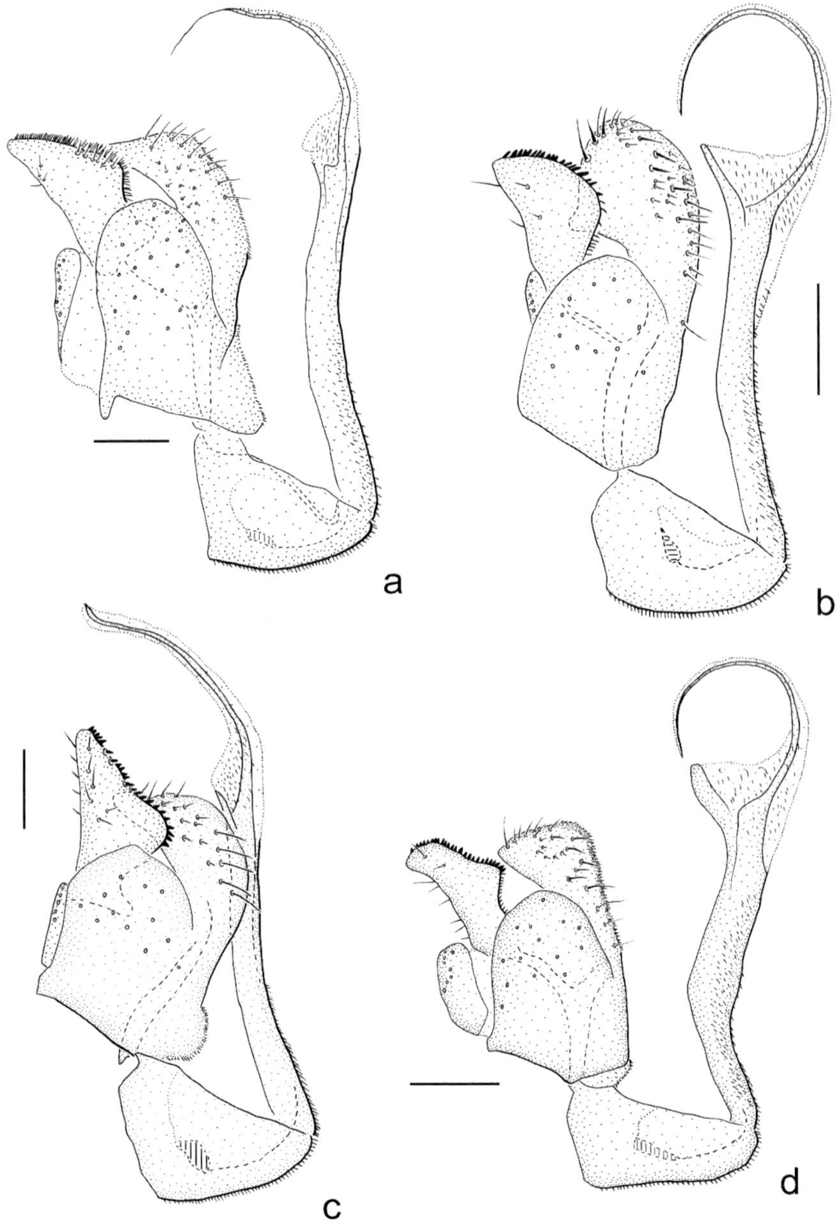

Fig. 25. Male terminalia of *Clinocera*: *a*, *C. empodiata*; *b*, *C. insularis*; *c*, *C. isabellina*; *d*, *C. lapazensis*. Scale bars = 0.1 mm.

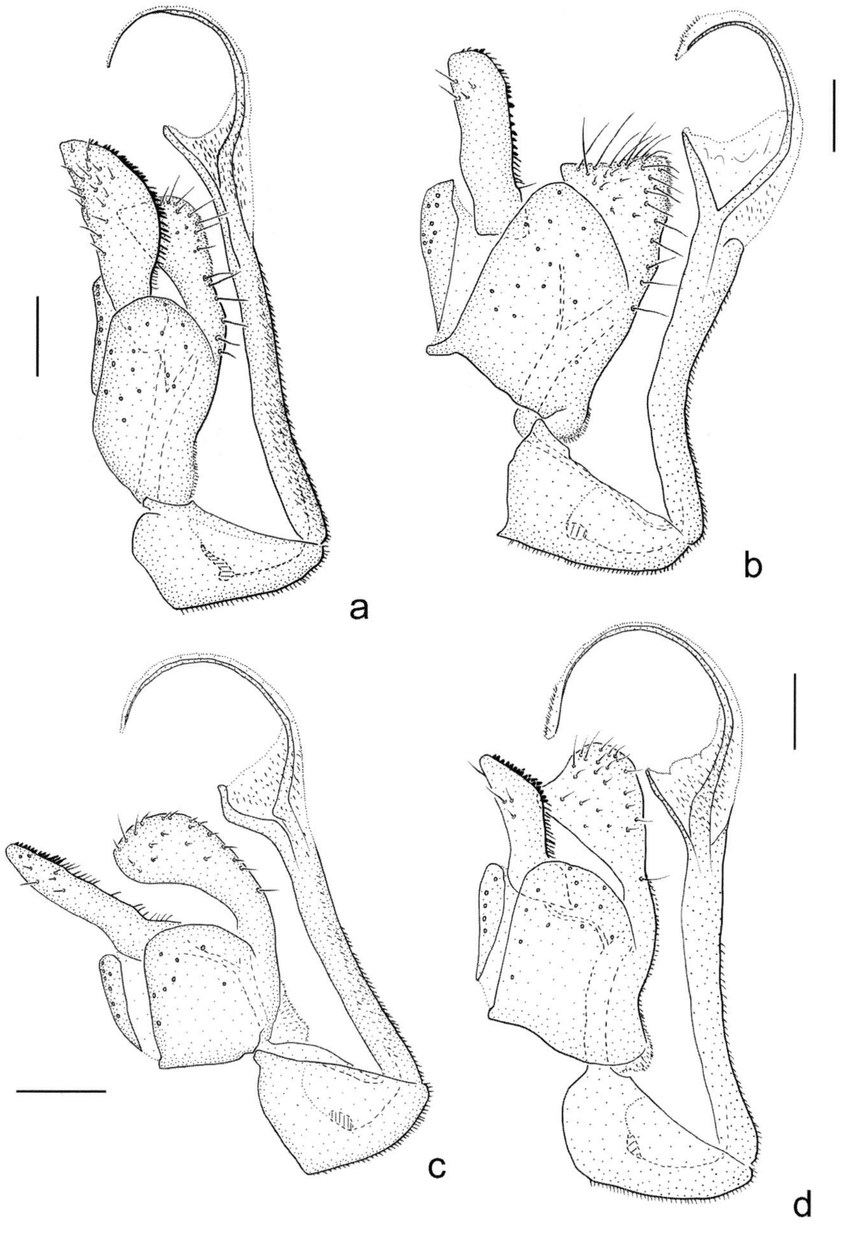

Fig. 26. Male terminalia of *Clinocera*: *a*, *C. marshalli*; *b*, *C. meridana*; *c*, *C. miradorana*; *d*, *C. naomina*. Scale bars = 0.1 mm.

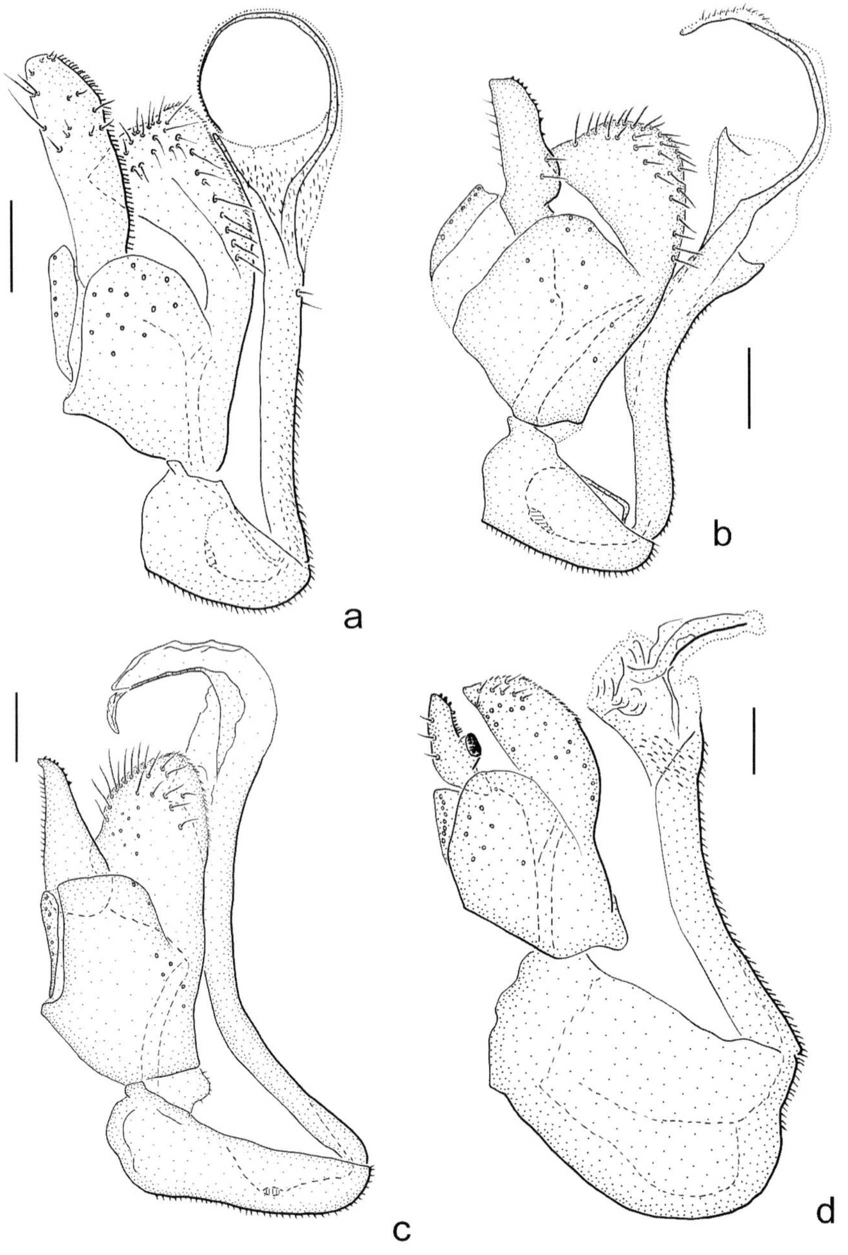

Fig. 27. Male terminalia of *Clinocera*: *a*, *C. notialis*; *b*, *C. oriunda*; *c*, *C. parana*; *d*, *C. penai*. Scale bars = 0.1 mm.

(ROGC); Crestline, 19.vi., 4–13.vii.1942, 13.vii.1944, ALM (USNM); Danville, 23.vi.1951, F.X. Williams (CAS); Del Norte Co., Jedediah Smith, Redwoods SP, Miller Ck, 16.v.1978, DDW (CAS); El Dorado Co., Kyburz, 4080′, 1.v.1978, EJR (ROGC), 2 mi. SE Latrobe, 16.iv.1978, S Kuba (CAS), Pipi Valley, 6 mi. E Omo Ranch, 17.iv.1977, EJR (ROGC), Rocky Bar Rd, 2 mi. NW Omo Ranch, 14.v.1978, EJR (ROGC), Sly Pk, 8 km S Pollock Pines, 1060 m, Hazel Ck, 22.vi.1975, PHA (CAS), S Fk American R., 4.8 km N Placerville, 295 m, 21.vi.1975, PHA (USNM); Fairfax, 11.v.1919, F.P. Van Duzee (CAS); Forest Home, 25.v.1935, ALM (USNM); Fresno Co., Burrough Valley, 18–25.v.1986 (PERC), Dinkey Ck Rd, Summit Ck, 31.vii.1984, JFM (PERC), Dinkey Ck Rd, Laurel Ck, 4.vi.1987, JFM (PERC), Sierra NF, 2200 m, 31.viii.1979, DDW

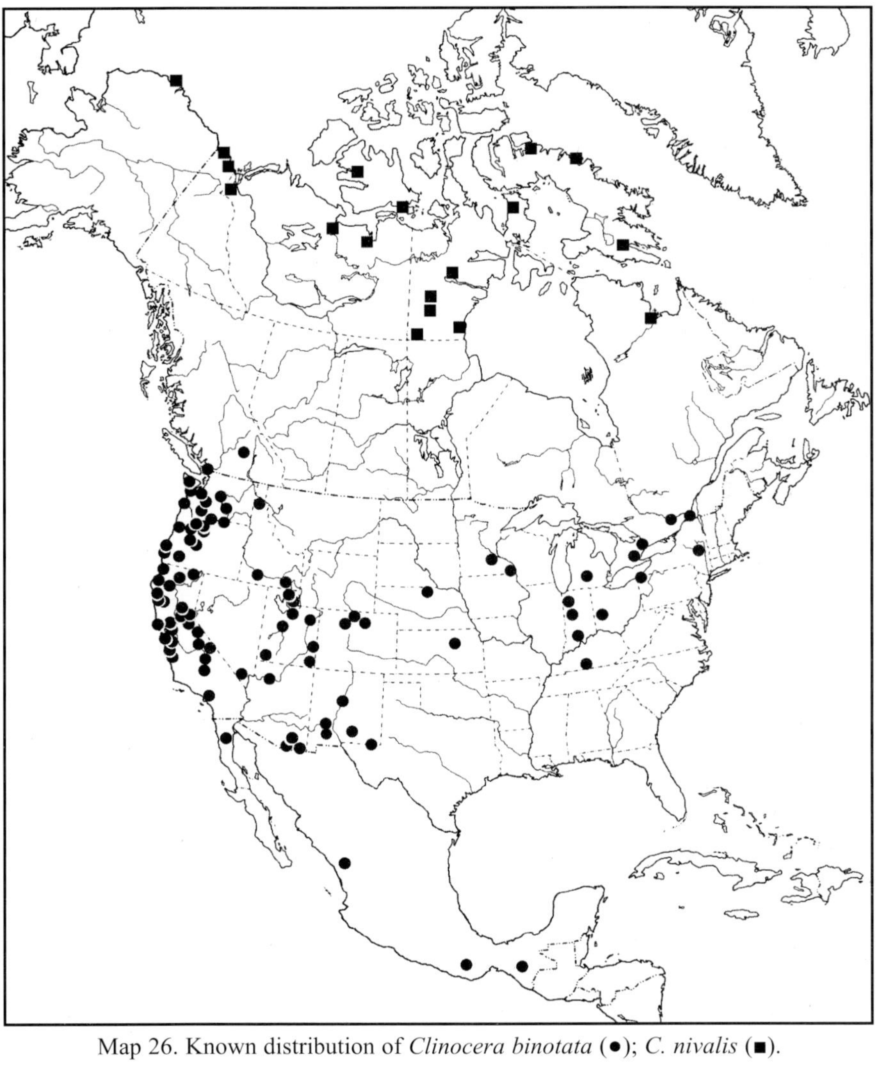

Map 26. Known distribution of *Clinocera binotata* (●); *C. nivalis* (■).

(CAS); Sierra NF, Bear Ck, 2100 m, 1.viii.1979, DDW (CAS), Sierra NF, nr Big Ck, Pitman Ck, 1500 m, 2.viii.1979, DDW (CAS), Sierra NF, Badger Flat Cpgd, Rancheria Ck, 2500 m, 1.viii.1979, DDW (CAS), Sierra NF, Forebay Lk., E and W Fks Camp 61, 2200 m, 31.vii.1979, DDW (CAS), Sierra NF, Oak Flat Cpgd, 22.vii.1984, JFM (PERC); Gen Grant Pk, 1.vi.1935, ALM (USNM); Glenbrook, 21.vi.1955, ALM (USNM); Humboldt Co., Arcata, 9.v.1981, 17.iv.1988, RLH (MSUB), Arcata, 11.vii.1969, C. Slobodchikoff (UCB), Humboldt Redwoods SP, Calf Ck and Bull Ck, 17.v.1978, DDW (CAS), Trinidad, Mill Ck, 13.vi.1984, R.W. Brooks (SEMC); Indio, Mg Spring Cyn, 5.iv.1945, ALM (USNM); Idyllwild, 7.vi.1942, ALM (USNM); Inyo Co., Deep Spring Lk., 5000′, 9.iii.1966, TWF, REO (UCB), Independence, 1.iv., 4.i.1964, J.D. Birchim (CAS); Jalama Beach, 9.vii.1965, J. Powell (UCB); Los Angeles Co., Angeles NF, Windy Spr., 27.vi.1979, DDW (CAS), Hwy 39, 3 mi. S Coldbrook Stn, N Fk Gabriel R., 2.v.1978, DDW (CAS), 2 mi. S N Fk San Gabriel R., 25.iv.1977, DDW (CAS); Madera Co., Bass Lk., 6.vii.1986, R.D. Haines (PERC); Lytle Ck, 10.v.1955, ALM (USNM); Marin Co., below Alpine Dam, Lagunitas Ck, 28.v.1979, DDW (CAS), Bolinas/Fairfax, 10.iv.1978, DDW (CAS), 8 mi. SW Fairfax, 23.vi.1951, A.A. Hubert (SEMC), Hicks Valley, Arroyo Sausal, 18.iv.1981, PHA (USNM), Liberty Gulch, Mt Tamalpais, 27.iv.1978, DDW (CAS), Lily Lk., Mt Tamalpais, 27.iv.1978, DDW (CAS), Little Carson Tr., Mt Tamalpais, 15.vii.1978, DDW (CAS), Mill Valley nr Cascade Res, Mill Ck, 31.v.1978, DDW (CAS), Pt Reyes, 19.iv.1980, SAM (DEBU, ROGC), S.P. Taylor SP, Lagunitas Ck, 20.v.1977, DDW (CAS); Medocino Co., Legget Drive-thru-Tree Rest Area, 17.v.1978, DDW (CAS), Moss Cove Rest Area, S Longville, Hwy 101, 18.v.1978, DDW (CAS); Mendo Co., 2 mi. N Willits, 2.vi.1966, TWF, REO (UCB); Modoc Co., Cedar Pass Cpgd, 1800 m, 11.viii.1967, PHA (CAS); Mono Co., Hot Ck, 17.v.1966, TWF, REO (UCB), Leavitt Meadows, 2470 m, 23.vii.1966, PHA (USNM), 16 mi. W Bridgeport, Twin Lk., Robinson Ck, ypans, 13.v.2002, LM (CNC); Monterey Co., Los Padres NF, Antonio R., 22.v.1977, DDW (CAS), Paraiso Hot Springs, 13 km SW Soledad, 400 m, 20–21.vi.1974, PHA (UNSM); Mount Home, 21.vi.1945, ALM (USNM); Napa Co., 5.5 km NW Moskowite Corner, Capell Ck, 200 m, 25.vii.1977, PHA (USNM); Placer Co., Folsom Lk. S Rec. Area, 26.iv.1978, DDW (CAS), NNE Folsom, 30.iv.1978, EJR (ROGC), nr Westville, N Fk American R., 2.viii.1975, DDW (CAS), Shirttail Ck below Sugar Pine Reservoir, 18.v.1983, RWB, Mauer (BYU); Plumas Co., Grizzly Ck above Hwy 70, 14.ii.1985, RWB, CRN (BYU); Lassen NF, N Fk Feather R., 28–30.v.1977, DDW (CAS), Lassen NF, Domingo Spr., 31.v.1976, DDW (CAS), Massack, Greenhorn Ck, 11.viii.1969, RH (HSU); Redwood City, 7.iv.1906 (USNM); San Benito Co., S Hollister, Limekiln Cyn, 3.iv.1979, DDW (CAS), Pinnacles, 24.iv.1948, WWW (USNM); San Diego Co., Palomar Observatory Cpgd, 1524 m, 26.vi.1968, PHA (USNM); San Luis Obispo Co., Stenner Ck, 55 m, 18.vii.1980, PHA (USNM); San Mateo Co., Woodside, Hubbard County Pk, McGarvey Gulch Ck, 1400′, 8.vi.1986, PHA (USNM); San Mateo Co., Corte de Madera, Portola, 18.vii.1959, 12.v.1974, PHA (CAS), San Bruno, 15.viii.1971,

PHA (USNM); Santa Clara Co., Sandborn Rd, 2.7 km SE Congress Spr. Rd, 440 m, 14.iv.1974, PHA (CAS), Stevens Cyn, 20.vi.1959, EJR (ROGC); Santa Cruz Co., Big Basin SP, Opal Ck, 14.vi.1978, DDW (CAS), Felton, San Lorenzo R., 19.vii.1975, PHA (USNM); SE Fks Sta. Ana, 17.vi.1945, ALM (USNM); Sequoia NP, 6.viii.1940, DEH (SEMC), Big Meadow, 28.vi.1984, JFM (PERC); Shasta Co., Big Spr., 23.v.1941, C.D. Michener (UCB), Hwy 299, Whiskeytown, 15.v.1978, DDW (CAS); Sierra Co., 15 km N Camptonville, Indian Valley Cpgd, N Yuba R., 900 m, 4.vii.1975, PHA (CAS), Salmon Ck Cpgd, 8 km NE Sierra City, 1768 m, 5.vii.1975, PHA (CAS); Siskiyou Co., Elk Ck, 6 km E McCloud, 29.vii.1974, PHA (CAS), Klamath NF, Knownothing Ck, Salmon R., 22.vii.1980, DDW (CAS); Solano Co., Putch Cyn, 11.xi.1965, C.R. Kovacic (UCD); Stanislaus Co., Frank Raines Pk, Del Puerto Ck, 28.iv.1973, PHA (CAS), ditto, 20.ii.1985, RWB, CRN (BYU), Del Peurto Cyn, Frank Raines Pk, 330 m, 1.x.1972, PHA (USNM); Sonoma Co., Kenwood, Sonoma Ck, 105 m, 18.v.1974, PHA (USNM); Sutter Buttes, 5.v.1940, GEB (UCD); Trinity Co., Shasta NF, Trinity R., 2.vii.1978, DDW (CAS); Tulare Co., Johnsondale, 13.viii.1985 (PERC); Tuolumne Co., Lost Claim Cpgd, 975 m, 5, 6.vii.1967, PHA (USNM); Valyermo, 13.v.1944, ALM (USNM); Whitewater Cyn, 11.iv.1965, J. Doyen (UCB); Woodside, 25.iv.1906 (USNM); Yosemite, 10–11.vi.1935, ALM (USNM); Yosemite NP, 1.viii.1940, DEH (SEMC). **Colorado:** Colorado Springs, 5915′, EST (SEMC); Gooding Co., Bliss, 6.vi.1967, E.J. Allen (UIM); Grande Co., S Fk, 8000′, MT, 20.vi.1972, WWW (USNM); Larimer Co., Hewlett Gulch, 24.ix.1995, SF, AF (CSU); Lewiston Co., Lewiston, 26.vi.1921, ALM (USNM), Lewiston Grade, 25.x.1938, H.L. Harris (UIM); Lyons Co., Fort Collins, 24.v., 14.vi.1938, 14.ix.1940, MTJ (CSU). **Idaho:** Latah Co., St. Joe NF, 8 mi. NE Harvard, 11.xi.1981, RSZ (WSU); Mt Moscow, 23.ix.1911, JMA (USNM); Nez Perce Co., 4 mi. S Juliaetta, Pottash Ck, 2.v.1975, D. Veirs (WSU); Nez Perce Co., Juliaetta Falls, 4.ii.1978, RSZ (WSU); Oneida Co., Black Pine, 3.v.1974, GFK (USUL). **Indiana:** Lawrence Co., Mill Spring SP, 19.v.1991, BJS (CNC); Lewiston, 26.vi.1921, ALM (USNM); Tippe Co., MT, viii.1978, JFM (PERC). **Kansas:** Manhattan, 14.iii.1959, N. Marston (USNM). **Kentucky:** Mammoth Cave NP, Echo R., spr., 9.vi.1991, BJS (CNC). **Michigan:** Hillsdale, 21.v.1960, R. and K. Dreisbach (MSU). **Minnesota:** Hennepin Co., Hopkins, 1.iv.1956, H.V. Niemcayk (MSU); Houston, 27.v.1940, W.A. Connell (UMSP); University of Minnesota, Minneapolis, 5.v.1953 (UMSP). **Nebraska:** Keya Paha Co., Meadville, 20.vii.1960, FCH (USNM). **Nevada:** Clarke Co., Red Rock Rec. Area, Spring Mtns, Lost Ck, 21.i.1984, RWB, CRN (BYU). **New Mexico:** Bernallilo Co., E slope Sandia Mtns, 2438 m, 15.vii.1979, T. Davies (CAS); Eddy Co., Black R., Castle Ck, 14.i.1987, RWB, Sargent, Kondratieff (BYU); Grant Co., Iron Ck nr Wright′s Cabin Cpgd, 7600′, 32°54.8′N, 107°46.8′W, 14.viii.2007, JMC (CNC); Lincoln Co., 9.7 km W Alto Lincoln NF, N Fk Rio Ruidoso, 3120 m, MT, 33°23′N, 105°47′W, 21–24.vi.1995, S.D. Gaimari (CNC); Otero Co., 22 mi. SW Cloudcroft, 11 000′, spr., 26.v.1982, W.L. Murphy (USNM); Sierra Co., Hillsboro, 9.vii.1977, WJH, GFK (USUL). **New York:**

Voorheesville, 9.xi.1923, M.D. Leonard (NYSM). **Ohio:** Champaign Co., Dallas Rd, 40°3.2′N, 83°48.0′W, Cedar Ck/Bog, 18.iv.1977, J.T. Ragensburg (USNM). **Oregon:** Curry Co., Humbug Mtn SP, Brush Ck, 17.vii.1989, BJS (CNC); Deschutes Co., Deschutes NF, Cold Spring Cpgd, 22.vi.1978, DDW (CAS); Douglas Co., Canyonville, 20.vi.1977, DDW (CAS); Hood R., 4.vi.1917, F.R. Cole (USNM); Hood R. Co., Mt Hood NF, Hood R. Meadows, 1350 m, 27.vi.1978, DDW (CAS); Hood R. Co., Starvation Ck SP, at falls, 26.vi.1978, DDW (CAS); Lane Co., 11 mi. NE Blue R., H.J. Andrews EF, 800 m, Mack Ck, 14.vi.1982, GWC (OSU); Lane Co., Williamette NF, McKenzie Bridge Cpgd, McKenzie R., 21.vi.1978, DDW (CAS); Lane Co., Siuslaw NF, 10 mi. S Yachats, Rock Ck, 15.vii.1989, BJS (CNC); Marin Co., Silver Falls SP, S Fk Silver Ck, 12.vii.1989, BJS (CNC); 8 km W Sixes, Edson Ck Cpgd, Edson Ck, 16.vii.1989, BJS (CNC); Umatilla Co., Cold Springs, 11.iv.1954, MTJ/J.J. Davis (WSU); Wasco Co., Warm Springs Indian Res., Beaver Ck, 23.vi.1978, DDW (CAS). **Pennsylvania:** Erie Co., Four Mile Ck, ECM (PSBC) (see Table 6). **Utah:** Box Elder Co., 1 mi. N Mantua, 1.viii.1969, WJH (USUL); Cache Co., Franklin Basin, MT, 7–16.vii.1981 (USUL), Green Cyn, MT, 4–7.v.1983 (USUL), Logan, 23.iv., 8.v.1938, 7.xii.1950, DEH (USUL), Logan, 9.iv.1981, CRN (USUL), Logan, Twin Ck, 1–8.vii.1981, WJH (USUL), Logan Cyn, 17.iv.1938, HFK, FCH (USUL), Provo, MT, 18–26.vi.1985 WJH (USUL), Provo, 24.v.1955, L. King (BYU), W Hodges Cyn, 11–15.viii.1978, WJH, GFK (USUL); Moab, 7.v.1941, GFK, FCH (USUL); Monticello, 20.v.1938, GFK, FCH (USUL); Newton, Borippus Spr., 15.vi.1954, GEB (SEMC); Roosevelt, 14.viii.1937, FCH (USUL); Sardine Cyn, 21.v.1938, DEH (USUL); Summit Co., Kamas Fish Hatchery, Beaver Ck, 13.ii.1981, RWB, Stanger (BYU); Toele Co., Mills, 17.iv.1934, GFK, W.L. Thomas (USNM); Washington Co., Sawyer Spr., trib. Ash Ck, 14.i.1976, RWB, Winget (BYU), Toquerville, Hwy 17, Ash Ck, 1.i.1976, RWB, Winget (BYU); Weber Co., Willard Basin, MT, 20.vii.1967, K. Capelle (Utah); Utah Co., Provo, Lions Lk., 19.ii.1993, RWB, CRN (BYU). **Washington:** Adams Co., 3 mi. N Othello nr Lyle Lk., 1.v.1973, D.M. Jackson (WSU); Asotin Co., 6 mi. W Asotin, Asotin Ck, 18.iii., 5.iv.1981, WJT (WSU); Columbia Co., 20 mi. E Dayton, Tucannon R., 3300′, 16.vii.1992, WJY (WSU); Entiat, 26.vii.1919, ALM (USNM); Fields' Spr. SP, 4 mi. S Anatone, 3500–4000′, 12–13.vi.1974, WJT (WSU); Franklin Co., Ringold, 24.ii.1973, NEW (WSU); Garfield Co., 8 mi. S Pomeroy, Pataha Ck and Hwy 128, 24.vi.1978, DDW (CAS); Grant Co., O'Sullivan Dam, 2.iv.–23.v.1954, MTJ (WSU); Grays Harbor Co., Olympic NF, Willaby Cpgd, 20.vii.1978, DDW (CAS); Jefferson Co., Olympic NF, Hoh R., 6.vii.1989, BJS (CNC); Lewis Co., Gifford Pinchot NF, Miller Ck, nr Randle, 290 m, 1.x.1979, DDW (CAS), Mt Rainier NP, Nevada Falls, Paradise R., 4500–5000′, WJT (WSU); Lilliwamp, 23.viii.1917, ALM (USNM); Pacific Co., Fort Canby SP, nr Ilwaco, 19.vii.1978, DDW (CAS); Pulman, ix, ALM (USNM); Tacoma, 27.viii.1911, ALM (USNM); Wawai, 30.v.1921, ALM (USNM); White Swan, 8.x.1941, DEH (SEMC); Whitman Co., Almota, 23.iii.1977, REZ (WSU); Whitman Co., Almota, 3–8.iii.1973, 17.iii.1974, 17–18.iv.1981, NEW, WJT

(WSU), 1 mi. N Almota, 840–900′, 8.iii.1975, WJT (WSU), 10 mi. SW Colton, 28.iii.1980, RSZ (WSU), 8 mi. SW Pullman, Lyle Grove Biological Area, 2120′, 30.v.1977, WJT (WSU), Steptoe Cyn, 10 mi. SW Pullman, 26.iv.1975, 19.ii.1977, 10.iii.1979, WJT (WSU), Steptoe Cyn, 19.ii.1977, S. Woods (WSU).

Recognition. *Clinocera binotata* is readily distinguished by stout ventro-basal setae on the forefemur, auxiliary crossvein present between R_4 and R_{2+3}, and femora with reddish brown "knees".

Description (wing length 3.3–4.3 mm).

Male. Lower half of face flat, with shiny white pruinescence; upper part of face and vertex brown; gena and occiput with blue pruinescence. Arista-like stylus long and slender. Postocular bristles stout and dark on upper half of eye margin.

Pleura with blue pruinescence; scutum with two narrow black vittae. Several pale acrostichals near antepronotum; 5 long dc; 1 pprn; 2 npl, lower shorter; 1 presut spal; 1 psut spal; 1 pal; 2 sctl, 1.5 times longer than dc. Anterior spiracle lacking black ring; laterotergite with patch of long, pale bristles; posterior margin of anepisternum with several short pale setae.

Wing (Fig. 32*d*) hyaline with clouding at tip of cell dm, crossvein between R_4 and R_{2+3}, base of R_4, r-m vein, and tip of cell bm; stigma absent; 1 long basal costal seta; costal spines present. Veins straight; auxiliary crossvein present between R_4 and R_{2+3}; stump vein absent; R_4 straight, diverging from R_5; anal vein long, faint streak; cell dm truncate. Halter dark.

Coxae reddish brown posteroapically, with blue pruinescence anteriorly, concolorous with pleura; femora with pale-blue pruinescence and reddish brown apex. Forecoxa with long pale hairs. Forefemur with 4 or 5 long stout black setae at base; dense white pile beneath; preapical anterior comb present. Foretibia with biserial row of strong, spine-like bristles beneath. Mid and hind femora with row of slender antero- and postero-ventral bristles on apical half. Hind tibia with erect dorsal and ventral bristles on apical third. Fifth tarsomere with small dorsoapical extension (Fig. 9*b*); empodium pale brown laterally.

Abdominal sternites and lateral margin of tergites with blue pruinescence, remaining tergites brown. Hypandrium broadly triangular. Phallus straight; tip of distiphallus flattened, blade-like in dorsal view. Clasping cercus with shallow notch on apical margin; blunt spines on inner apical margin. Surstylus with broad, produced tip; subequal in length to clasping cercus (Figs. 9*a*, 21*d*).

Female. Very similar to male, including leg chaetotaxy.

Geographic distribution and seasonal occurrence. *Clinocera binotata* is a transcontinental species, relatively sparsely distributed in the east compared with the west (Map 26). Its distribution is also unique in that it is rare in southern Canada and ranges southward through Mexico. Apparently the western range of *C. binotata* has not penetrated very far north into previously glaciated regions.

Specimens from emergence traps monitored for a number of years were made available through the generosity of E.C. Masteller (see under *C. fuscipennis*). From the upper site of Fourmile Creek, specimens emerged primarily from May to October, with a single record in March (Table 6).

Remarks. In the eastern range, there are numerous collections from Crieff Bog (Ontario) and spring-fed streams (Kentucky, Indiana), whereas in the west the species is common along warm streams and rivers. There are very few Ontario records, despite the intensity of collections in the southern regions of the province. Most records are from cooler months (including a pair in copula in March) or colder habitats.

Clinocera maculata Loew

Figs. 8*b*, *d*, *e*, 9*c–f*, 22*a*, 32*e*; Map 27

Clinocera maculata Loew, 1860: 79.

Clinocera (*Hydrodromia*) *maculata*: Melander, 1902, 1928, 1965: 468.

Type material examined. HOLOTYPE, female labelled "Type/ 1180 [red label with white upper border]"; "63 [blue label]"; "Loew/ Coll."; "maculata/ m. [handwritten]" (MCZ).

Loew (1860) stated that the holotype of *C. maculata* was collected by O. Sacken from Washington, D.C.

Additional material examined. Some 360 specimens from the following territories, provinces, and states: **CANADA. Ontario:** Arkell Spring, 16.viii.1981, M. Eymann (DEBU); Cambridge, 21.x.1991, R. Rea (DEBU); Dundas, Rock Chapel, 13, 29.iv., 22.vi.1986, BJS (DEBU); Grey Co., Owen Sound, Inglis Falls, spr., 4–6.v., 20.vii.1986, 29.vii.1990, BJS, SAM (CNC, DEBU); Orillia, 13.vii.1923, C.H. Curran (CNC); Peel Co., Forks of the Credit PP, seep, 28.vi.1991, BJS (CNC). **Quebec:** Abbotsford, 2.vii.1936, GES (CNC, CAS); Beechgrove, 18.x.1956, JRV (CNC); Gatineau Pk, 15.iv.1982, Nelson, Roussel (CNC); Knowlton Lodge, 19.vii.1968, JRV (CNC); Lac Philippe, 8.viii.1987, JRV (CNC); Old Chelsea, 5.xi.1958, JGC (CNC); Old Chelsea, 25–29.iii., 15.x.1958, JFMC, 20.ix., 10.x.1961, 26.viii.1964, 27.iv.1987, 7.x.1988, JRV (CNC). **USA. D.C.** (USNM). **Georgia:** Lumpkin Co., Chattahoochee NF, Desoto Falls Scenic Area, Frogtown Ck, 20.vi.1990, BJS (CNC); Rabun Co., Black Mtn SP, Adahi Falls, 3000′, 26.iv.1989, BJS (CNC). **Kentucky:** Mammoth Cave NP, Eco R. spr., 9.vi.1991, BJS (CNC). **Maine:** Mt Desert Is., Hunters Beach Tr., Seal's Harbor, 18.vii.1986, JMC (CNC) **Massachusetts:** Dedham, v.1920, CWJ (USNM): Petersham, vii.1926, ALM (USNM). **Michigan:** Alger Co., Alger Falls, Munising, 9.ix.1992, RH (HSU); Montmorency Co., Hunt Ck, 15.ii.1941, J.W. Leonard (USNM). **New Hampshire:** Mt Washington, Dolly Copp, 4.vii.1956, ALM (USNM). **New York:** Beaverkill, 4.viii.1909, ETC (ANSP, USNM); Colden, 9.vii.1913,

Table 6. Emergence pattern for *Clinocera binotata* in Fourmile Creek, Erie, Pennsylvania, USA (1981–1992).

	Upper site (site 1)		Lower site (site 2)	
	No. of males	No. of females	No. of males	No. of females
January				
February				
March	1			
April				
May	3	6		
June	3	2		
July	9	18	1	2
August	13	7		
September	2	4		
October	3	1	1	
November				
December				
Total	34	38	2	2

Note: Site 2 was downstream from a sewage outflow (see Masteller and Flint 1984).

MCV (USNM); Clinton Co., Chazy Lk., 8.ix.1985, RH (HSU); Cortland Co., Inlet Brook, McLean Res., 13.xi.1924 (CUIC); Wells, 29.vii.1914, D.B. Young (NYSM). **North Carolina:** Buncombe Co., Blue Ridge Pkwy, Craggy Gardens Picnic Area, 1580 m, seepage, 12.vi.1990, BJS (CNC); Clingmans Dome, 18.vi.1939, M.A. Cazier (AMNH); Haywood Co., GSMNP, Cataloochee Ck, seepage, 12.vi.1990, BJS (CNC); Macon Co., Bridal Veil Falls, 5 mi. NW Highlands, 26.iv.1989, JMC (CNC); Swain Co., GSNMP, Balsam Mtn, Flat Ck, 3.vi.1990, BJS (CNC), GSMNP, Newfound Gap, 15.vi.1990, BJS (CNC); Yancey Co., Mt Mitchell SP, Balsam Tr., 2000 m, spr., 11.vi.1990, BJS (CNC). **Pennsylvania:** Erie Co., Four Mile Ck, ECM (see Table 7) (PSBC); Hazleton, 11.viii.1909, ETC (ANSP); Hazleton, 4–27.viii.1910, vii.1911, Dietz (ANSP, USNM). **Tennessee:** Clingmans Dome, Indian Gap, 5200–6600′, 6.viii.1957, JGC (CNC); GSMNP, Alum Cave, 28.vi.1941, ALM (USNM), ditto, 16.vi.1990, BJS (CNC); GSMNP, Chimney Camp, 11.vi.1940, GCS (USNM); GSMNP, Chimney Tops Tr., 5.vi.1993, BJS (CNC); GSMNP, Ramsay Cascade Tr., 4.vi.1993, BJS (CNC). **Virginia:** Augusta Co., Briery Branch, 3.vi.1955, WWW (USNM); Big Meadows, 14.vi.1941, ALM (USNM); Covington, 25.vi.1933, ALM (USNM); Grayson Co., Grayson Highlands SP, Big Wilson Ck, 10.vi.1990, BJS (CNC); Luray, 24.vi.1933, ALM (USNM); Shenandoah, Lewis Falls, 4.vii.1939, ALM (USNM). **West Virginia:** Hardy Co., Lost R. SP, MT, 10.vii.1977, WWW (USNM). **Wisconsin:** Devils Ck, 7.vii.1933 (USNM).

Recognition. This species is distinguished by the spotted wings, bright reddish brown legs, stout ventrobasal bristles on the forefemur, and no auxiliary crossvein between R_4 and R_{2+3}.

Description (wing length 3.3–4 mm).

Male. Lower half of face flat (Fig. 8*b*), with shiny blue pruinescence; upper half of face and vertex dull brown–black; gena and posterior region of head with tinge of blue pruinescence. Arista-like stylus long and stout. Postocular bristles dark and stout on upper half of eye margin.

Scutum faintly bivittate, with median ash-coloured stripe extending to scutellum. Pleura with bright-blue pruinescence; anterior spiracle lacking black ring.

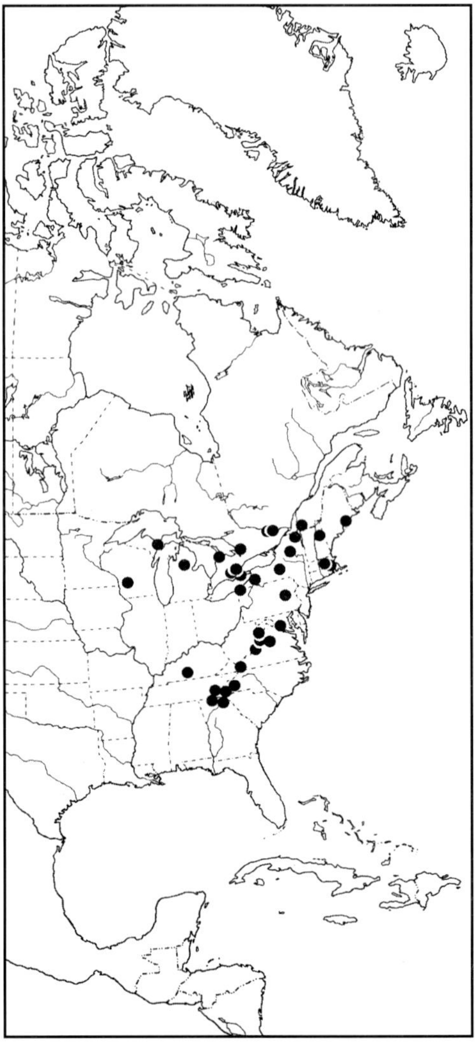

Map 27. Known distribution of *Clinocera maculata*.

Table 7. Emergence pattern for *Clinocera maculata* in Fourmile Creek, Erie, Pennsylvania, USA (1981–1992).

	Upper site (site 1)		Lower site (site 2)	
	No. of males	No. of females	No. of males	No. of females
January	1	1		2
February	2	2		
March	1	1		
April	1	1		
May	1	1		
June	3	2	2	
July	3			
August	1	2		
September	4	4		
October	3	11	4	15
November	6	9	2	2
December	1	6		
Total	21	40	8	19

Note: Site 2 was downstream from a sewage outflow (see Masteller and Flint 1984).

Several short pale acrostichals anterior to first dc; 5 dc; 1 pprn; 1 presut spal; 2 npl, lower shorter; 1 psut spal; 1 pal; 2 sctl subequal to dc. Laterotergite with patch of long pale bristles; posterior margin of anepisternum with several short pale setulae.

Wing (Fig. 32*e*) infuscate, with dark circular spots at base of R_4, base and apex of cell dm, tip of R_{2+3}, r-m, and apex of cell bm; 1 long basal costal seta; costal spines present; stigma absent. All veins straight; auxiliary crossvein between R_4 and R_{2+3} lacking; stump vein absent; R_4 bell-shaped, diverging from R_5; anal vein long and faint; cell dm slightly produced. Halter dark.

Coxae and trochanters with bright-blue pruinescence, concolorous with pleura; femora and tibiae bright reddish brown, tarsi darker, especially apically. Forecoxa with fine pale hairs. Forefemur with anterior preapical comb (Fig. 8*d*); 5 or 6 long stout bristles intermixed with long finer setae at base; dense white pile beneath (Figs. 8*e*, 9*c*, *d*); biserial row of small, fine hairs beneath. Foretibia with single row of short stout black bristles beneath. Mid and hind femora lacking preapical ventral bristles. Hind tibia with erect anteroventral bristles on apical half. Fifth tarsomere with dorsoapical extension; empodium white laterally.

Abdominal sternites and lateral margin of tergites with blue pruinescence, remaining tergites brown. Hypandrium small, subtriangular. Phallus straight, split, and arched anteriorly at its apex; distiphallus relatively long, dorsoventrally flattened. Clasping cercus heavily sclerotized; strongly arched, tapering to medially curved tip. Surstylus arched ventrally with nearly truncate apex; much shorter than clasping cercus (Figs. 9*e*, *f*, 22*a*).

Female. Identical with male, including leg chaetotaxy.

Geographic distribution and seasonal occurrence. This species is confined to eastern North America, ranging from southern Ontario and Quebec to northern Georgia (Map 27). Apparently the range of *C. maculata* has not penetrated very far north into previously ice-covered land. In the northern part of the range, the flight period extends from March to November. Specimens were also collected from a cold spring (2000 m) in Mount Mitchell State Park.

A number of specimens of *C. maculata* were identified from the emergence traps of Ed Masteller (see under *C. fuscipennis*). From the upper site of Fourmile Creek, specimens emerged continuously for nearly the entire year, perhaps with greater frequency in the cooler autumn months (Table 7).

Remarks. The external features (e.g., body coloration, wing pattern, and chaetotaxy of the forefemur) of this species are very similar to those of the European species *C. wesmaeli* and *C. fontinalis*.

This species appears to prefer spring-fed streams and seeps, according to collection data. These habitats provide a low yet steady water temperature, which may explain the long flight period of this species.

Clinocera nivalis (Zetterstedt)

Figs. 22*b*, 32*f*; Map 26

Hemerodromia nivalis Zetterstedt, 1838: 545.

Type material examined. The female holotype is located under *Ardoptera* Macquart (a junior synonym of *Dolichocephala*) in the Zetterstedt collection (UZIL). The identity of this species was kindly verified by I. Shamshev (pers. comm., December 2004).

Nearctic material examined. Some 180 specimens from the following territories, provinces, and states: **CANADA. Northwest Territories:** Aklavik, 13.v.1931, OB (CAS). **Nunavut:** Baker Lk., 64°19.3′N, 96°02.5′W, 14.vii.2003, P.H. Adler (ZFMK); Chesterfield, 4.vi.–2.viii.1950, JGC (CNC); Clyde, Baffin Is., 10.vi.–7.viii.1958, GES, J.E.H. Martin (CNC); Coppermine, 31.v.1951, S.D. Hicks (CNC); Pond Inlet, Baffin Is., 17.ix.1978, 23.vi.1979, G. Johnston (DEBU). **Quebec:** Tasiujaq, 8–27.ix.1996, BJS (CNC). **Yukon:** Herschel Is., 69°34′N, 138°52′W, 21.vii.1983, RAC (UBC); Ogilvie Mtns, 7.viii.1962, P.J. Skitsko (CNC); Shingle Pt, Mackenzie R., 29.viii.1929, OB (CAS). **USA. Alaska:** Pt Barrow, 19.vii.1953, P.D. Hurd (USNM), 1.vii.1969, Bishop (BPBM).

Palearctic material examined. Russia, Scotland, Sweden (four specimens).

Recognition. *Clinocera nivalis* is readily distinguished by the presence of an auxiliary crossvein between R_4 and R_{2+3}, lower half of the face with silvery pruinescence, and a dark ring about the anterior spiracle.

Description (wing length 3–3.5 mm).

Male. Face convex, lower half with silvery pruinescence; remainder of head brown. Arista-like stylus short and stout. Postocular bristles dark and stout on upper half of eye margin.

Scutum brown with black bivittae and median greyish stripe extending onto base of scutellum. Pleura just above coxae with faint blue pruinescence; remaining pleura brown. Anterior spiracle with black outer ring. Several pale acrostichals near antepronotum; 5 dc; 1 pprn, occasionally an additional short, fine seta; 1 presut spal; 2 npl, lower often fine and short; 1 psut spal; 1 pal; 2 sctl subequal to dc. Laterotergite with patch of pale bristles. Posterior margin of anepisternum with pale setulae.

Wing (Fig. 32*f*) infuscate, with clouding at following crossveins: base of R_4, r-m, apex of cells bm and dm; 1 long basal costal seta; costal spines absent; stigma absent. Veins straight; auxiliary crossvein present between R_{2+3} and R_4; stump vein absent; R_4 bell-shaped; anal vein faint streak; cell dm slightly produced apically. Halter dark.

Coxae with dark-blue pruinescence, remaining leg segments dark brown. Forecoxa with pale setae. Forefemur with 4–7 ventrobasal bristles and several long pale bristles; dense white pile beneath; anterior preapical comb present; anteroventral surface with row of short black spine-like setae (hidden in ventral pile). Foretibia with single row of short stout setae beneath. Mid and hind femora lacking preapical ventral setae. Hind tibia with short erect dorsal and ventral bristles on apical third. Fifth tarsomere with dorsoapical extension; empodium white laterally.

Abdominal sternites and lateral margin of tergites with pale-blue pruinescence, remaining tergites brown. Hypandrium broadly triangular. Phallus straight, with relatively long slender distiphallus. Clasping cercus narrow with blunt peg-like setae on inner apical surface. Surstylus strongly curved, shorter than clasping cercus; apex with slight indentation (Fig. 22*b*).

Female. Same as male, including ventral bristles of foreleg.

Geographic distribution and seasonal occurrence. In the Palearctic Region this species is reported from Norway (Jonassen 1998), Finland, Sweden, Scotland, and across the northern areas of both European and Asian Russia. These are the first records of *C. nivalis* from North America, where it is found throughout the arctic region from northern Quebec to Alaska (Map 26).

Remarks. In arctic Quebec, specimens were collected on the tundra from the surface of small pools and by sweeping long grasses bordering these pools.

Nearctic populations are nearly identical with Palearctic populations, based on external characters, although there are minor differences in the male genitalia. The clasping cercus of European species is shorter, with sharply pointed spine-like setae on the inner surface, and the surstylus is not as strongly arched (not extending to the clasping cercus). The phallus appears very similar in these two forms.

Clinocera stagnalis (Haliday)

Figs. 22*c*, 32*g*; Map 28

Heleodromia stagnalis Haliday, 1833: 159.

Brachystoma westermanni Zetterstedt, 1838: 558.

Clinocera genualis Coquillett, 1910: 124. **syn. nov.**

Clinocera (*Hydrodromia*) *stagnalis*: Melander, 1928, 1965: 469.

Clinocera genualis: Landry & Harper, 1985: 1381.

Type material examined. *C. genualis*: HOLOTYPE, male labelled "Banff/ Alb [square label]"; "Clinocera/ genualis/ Coq. [white label with black border in Coquillett's shaky handwriting]" (USNM).

The identity of the holotype of *C. genualis* is tentatively assumed to be a specimen with a C.W. Sabrosky label stating "probably the/ holotype/ CWS 1950". The specimen and locality match the description by Coquillett (1910), although the collector's name (N.B. Sanson) is missing from the label. This species is conspecific with the Nearctic population of *C. stagnalis*. Landry and Harper (1985) listed *C. genualis* as part of the aquatic dance fly fauna of subarctic river systems in Quebec; however, this species is actually *C. stagnalis*.

Material examined. Some 160 specimens from the following countries, territories, provinces, and states: **CANADA. Alberta:** Banff, 5–31.viii.1925, 20.vi.1928, OB (CAS); Banff, 23.viii.1923, CBDG (USNM); Banff NP, 9.vii.1955, GES (CNC). **British Columbia:** Summit Lk., mi. 392 Alaska Hwy, 4200′, 19–21.viii.1959, REL (CNC). **Manitoba:** Churchill, 21.vi.1948, GES (CNC), 24.vii.1950, R. Howe (CNC), 2–9.viii.1938, R.H. Daggy (UMSP); Fort Churchill, 26.v.1952, C.D. Bird, JGC (CNC); Mile 500, Hudson Bay, JGC (CNC); Churchill R., pump house, 15 km S Churchill, Goose Ck Rd, 58°37′43″N, 94°13′42″W, 7.viii.2006 (JBWM). **Northwest Territories:** Aklavik, 12.ix.1929, OB (CAS); Arctic Red R., 24.viii.1929, OB (CAS); Boyd Lk. outlet, 61°39′N, 103°40′W, 3.vii.1966, JGC (CNC); Exmouth Lk., 65°02′N, 115°54′W, 24.viii.1966, GES (CNC); Hay R., 9.ix.1932, OB (CAS); Salmita Mines, 64°05′N, 111°15′W, JGC (CNC) Wholdaia Lk., 24.vii.1966, JGC (CNC); Yellowknife, 21.v.1953, JGC (CNC). **Quebec:** Baie James, R. aux Castor, 4.vi., 5, 8.ix.1975 (UMQ, in alc.); Baie James, Lac Nathalie, Lac Hélène, trib. Nathalie, 17.vi., 9.vii, 6, 14.ix.1974 (UMQ, in alc.); Baie James, Reservoir LG2, 31.viii.1977 (UMQ, in alc.); Great Whale R., 23.vi., 27.viii.1949, JRV (CNC); R. du Tonnerre (50°16′N, 64°47′W), 19.viii.1930, WJB (CNC); Schefferville, 18.vi.1960, B.V. Peterson (CNC); Tasiujaq, riparian vegetation, tundra, 8, 25–27.ix.1996, BJS (CNC). **Yukon:** Dempster Hwy, km 152, Windy Pass, 8.vi.2003, AB (CNC); Lk. Laberge, 60°59′N, 135°11′W, Horse Ck, 8.viii.1981, C. Guppy (UBC); Minto Landing, Klondike Loop, 4.vii.1978, PHA (CAS); Summit Lk., Richardson Mtn, along lk., 3.vii.1987, SAM (DEBU). **GREENLAND.** Angûjartusite, Syaligstravamce, Kids, Diskofjord, 29.iv.1969 (ZMUC); Bjørnesund, Equaluit, 62°54′N, 50°09′W, 28–29.vi.2003, Kissavik

Expedition (ZMUC); Finkenæsfjorden, Amigfik, 63°11′N, 50°25′W, 1–3.vii.2003, Kissavik Expedition (ZMUC); Fgaliko, 31.viii.1889, Sündbeck (ZMUC); Kangersumiq, Pilikerat, 20.viii.1958, J. Böcker (ZMUC); Julianehaab, 23.viii.1951, Chr. Vibe (ZMUC); Kangamiut, 7.vi.1952, Chr. Vibe (ZMUC); Kapisigdlit, 26.vii.–11.viii.1950, Chr. Vibe (ZMUC); Maniitsoq Sukkertoppen, 65°25′N, 52°54′W, 16.vii.2003, Kissavik Expedition (ZMUC); Nasarssuaq (Blue West 1), 30.v.1949, Chr. Vibe (ZMUC); Sonderstrom Air Base, 67°05′N, 50°30′W, 17.vii.1952, WJB (CNC); Tigssaluk, 4–28.vii.1954, T. Andersen (ZMUC). **ICELAND.** 64°23′38″N, 21°32′42″W, 3.viii.1943, T.O. Thatcher (USUL); 25 km

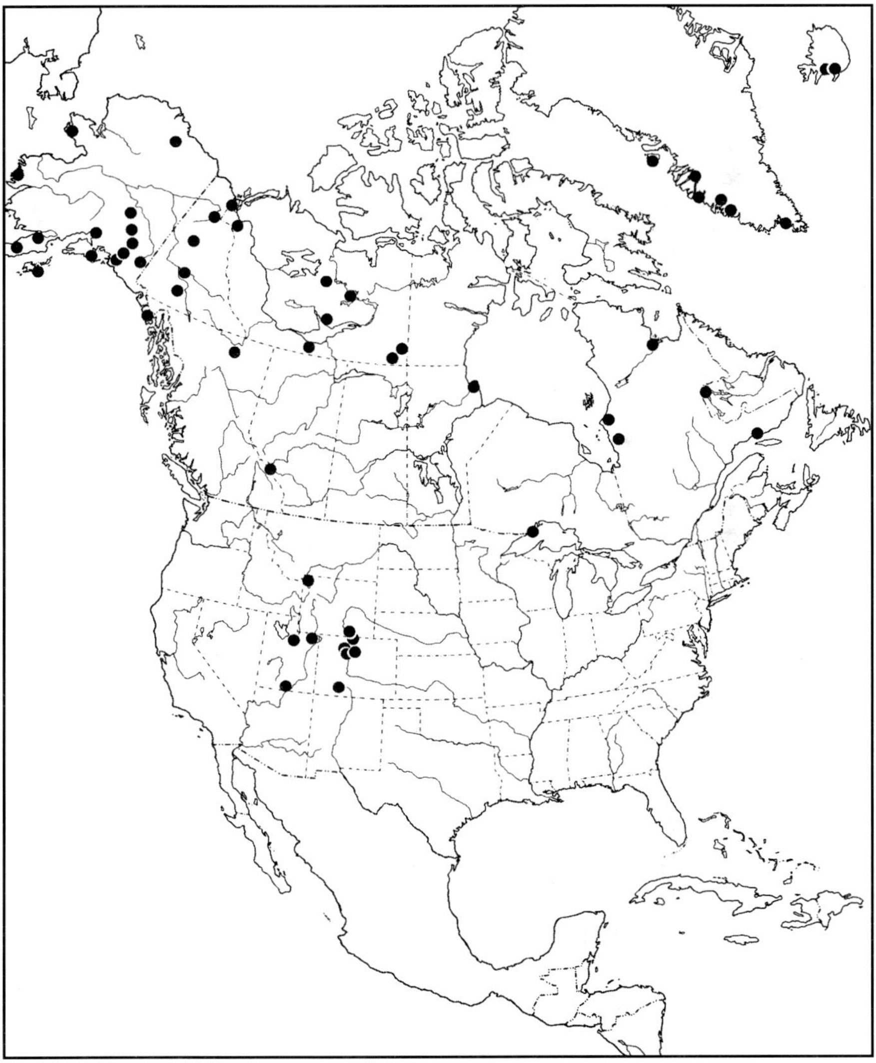

Map 28. Known distribution of *Clinocera stagnalis*.

SW Borgarnes, 1.vii.1962, BVP (CNC); 15 km S Borgarrivki, 2.vii.1962, BVP (CNC); 20 km E Medellandsv, 13.vii.1962, BVP (CNC). **USA. Alaska:** Amphitheater Mtns, mi. 10 Denali Hwy, 8.vi.1994, TS (BLKU); Aniakchak NM, Surprise Lk., 56°56′N, 158°06′W, vii–viii.1988, B. Wisseman (OSU); Boulder Mtn, mi. 335 Denali Hwy, 27.vii.1993, TS (BLKU); Gardiner Ck Camp, Alaska Hwy, DC-1253, 5.viii.1978, PHA (USNM); Healy, 26.vi.1921, JMA (USNM); Kodiak, 21.ix.1919, J.S. Bine (USNM); Isabel Pass, mi. 206 Richardson Hwy, 2900′, 13.vii.1962, REL (CNC); Matanuska Valley, 4.viii.1952, C.O. Berg (USNM); mi. 55 Glenn Hwy, Moose Ck, 6.vi.1993, 4.vi.1994, TS (BLKU); Mendeltna, mi. 153 Glenn Hwy, 6.v.1994, TS (BLKU); Nome, Anvil Valley, 19.vi.1994, TG (BLKU); Phelan Ck, mi. 204 Richardson Hwy, 27.vii.1993, 8.vi.1994, TS (BLKU); Savonoski, Naknek Lk., vii.1919, J.S. Bine (USNM); Sebree Is., 23.vii.1965, D.M. DeLong (USNM); Seward, 26.vii.1921, JMA (USNM); Tiekel R., Richardson Hwy ca. V-47, 1.viii.1978, PHA (USNM); Tununuk, 15.viii.1930, OB (CAS); Umiat, 19.vi.1947, L.A. Jackovski (USNM); Valdez, 10 mi. N, 26.viii.1948, B.L. Morris (USNM). **Colorado:** Boulder Co., Corona Pass, 10 600′, marshy meadow at timberline, 6.vii.1961, JGC (CNC); Gilpin Co., 2 mi. E Rollins Pass, 3048 m, 16.viii.1973, DDW (CAS); Hinsdale Co., Sherman, Cottonwood Ck, 10.ix.1993, SF, T. Eckberg (CSU); Lake Co., Independence Pass, 12 100′, tundra, 31.vii.1961, JGC (CNC); Larimer Co., Cameron Pass, 9.viii.1965, FCH (USUL); Mt Evans, Echo Lk., 10 600′, 11.viii.1961, CHM (CNC); Summit Co., Wheeler Jct., Ten Mile Ck, 25.vi.1995, SF, AF (CSU). **Michigan:** Keweenaw Co., Isle Royale, 9.vii.1938, GCS (USNM). **Utah:** Boulder, 23.viii.1957, FCH (USNM); Heber, 14.viii.1943, GFK, D.R. Maddock (UMSP); Manila, 11.viii.1937, GFK, FCH (USUL). **Wyoming:** Carbon Co., 20 km Centennial, edge of Lk. Marie, 1.viii.1973, PHA (CAS); Yellowstone NP, 19.viii.1918, ALM (USNM).

Palearctic material examined. Czech Republic, England, Finland, Germany, Greece, Ireland, Israel, Nepal, Russia, and Spain (50 specimens).

Recognition. This species has a distinctively convex face clothed entirely with silvery pruinescence, spotted wings, and auxiliary crossvein lacking, and the apices of the femora (or "knees") are pale.

Description (wing length 3.2–3.5 mm).

Male. Face convex, completely clothed in silvery pruinescence; remainder of head brown. Arista-like stylus generally long and slender. Postocular bristles dark and stout on upper half of eye margin.

Scutum with two black vittae extending to suture, separated by faint blue pruinescence extending to scutellum. Pleura and lateral margin of scutum with pale-blue pruinescence. Several scattered acrostichals near antepronotum; 5 dc; 1 pprn; 1 presut spal; 2 npl, lower shorter; 1 psut spal; 1 pal; 2 sctl subequal to dc. Laterotergite with patch of pale bristles. Posterior margin of anepisternum with pale setulae.

Wing (Fig. 32*g*) hyaline with faint clouding at base of R_4, tip of cell dm, and r-m vein; 1 long basal costal seta; costal spines present; stigma absent. Veins straight; auxiliary crossvein between R_4 and R_{2+3} absent, but occasionally present on one wing; stump vein absent; R_4 bell-shaped; anal vein faint streak; cell dm nearly truncate. Halter dark.

Coxae and basal three-quarters of femora with blue pruinescence; apex of femora yellow–brown, remaining leg segments dark. Forecoxa with pale setae. Forefemur with anterior preapical comb; dense white pile beneath; posteroventral row of setae increasing more stout apically; anteroventral face with row of fine short hairs. Foretibia with biserial row of stout bristles beneath. Midfemur with 2 or 3 preapical posteroventral bristles. Hind tibia with erect ventral bristles on apical third. Fifth tarsomere with dorsoapical extension; empodium white laterally.

Abdominal sternites and lateral margin of tergites with blue pruinescence, remaining tergites brown. Hypandrium broadly triangular. Phallus straight, with some stout microtrichae; distiphallus relatively long, dorsoventrally flattened. Clasping cercus slightly arched; stout, peg-like setae on inner face and sharply pointed setae along dorsal margin. Surstylus longer than clasping cercus, projecting free from epandrium; apex truncate (Fig. 22*c*).

Female. Similar to male except as follows: forefemur same as in male, but white pile reduced on ventral surface; foretibia with ventral setae reduced.

Geographic distribution and seasonal occurrence. This species is probably the most widespread clinocerine species, ranging from Great Britain and Europe (including Norway; Jonassen 1987) to Asia, North Africa (Chvála and Wagner 1989), Faeroes (Lyneborg 1968), and Greenland (Lundbeck 1898). In Great Britain it can be encountered almost year-round, peaking at the end of June and early July (Collin 1961; Plant 2003).

This is the first report of this species in North America. Its range is similar to those of other Holarctic species with a primarily arctic distribution, with an extension south through the Rocky Mountains. An isolated population is known from Isle Royale (Lake Superior), which likely represents a glacial relict (Map 28). This distribution is very similar to that of *Wiedemannia simplex* Loew, which also has a relict population in Isle Royale (Sinclair 1998) (see Biogeography section).

Remarks. Nearctic specimens differ from Palearctic populations by the rounded apex of the clasping cercus, which contrasts with the nipple-like tip of specimens from the Palearctic Region (including Iceland).

The author collected specimens in arctic Quebec (Ungava Bay region) on the surface of small pools and by sweeping long grasses bordering these pools. This species was also collected using emergence traps set in the littoral areas of small lakes, tributaries, and rivers of a subarctic region of Quebec (Landry and Harper 1985).

This species is confined to alpine and arctic regions in North America. This is in marked contrast to Europe, where Collin (1961) thought that this species might be found in every county in the United Kingdom, on suitable damp mud or shallow muddy water. *Clinocera stagnalis* was collected in emergence traps nearly throughout the year in a first-order stream in Germany (Wagner and Gathmann 1996).

Clinocera substagnalis sp. nov.

Figs. 22*d*, 32*h*; Map 29

Type material. HOLOTYPE, male labelled "USA: ALASKA, Mile 206/ Richardson Hwy/ Beaver pond margins/ 16.VII.1989/ D.M. Wood"; "HOLOTYPE/ Clinocera/ substagnalis/ Sinclair [red label]" (CNC).

Paratypes: **CANADA. Alberta:** Banff NP, 4.7 mi. S Johnston Cyn, slow str., ROM Survey, 29.vii.1969 (13 ♂♂, 20 ♀♀, ROM); Banff NP, 26 km N Lk. Louise, pools of Mosquito Ck, 28.vi.1989, BJS (8 ♂♂, 8 ♀♀, CNC); Turner Valley, 25.v.1928, G. Bryant (1 ♀, CAS). **British Columbia:** Summit Lk., 16 km E, Hwy 97, 54°22′N, 122°43′W, 30.vi.1977, B. Mckague (1 ♂, CNC). **Yukon:** Klondike Hwy, km 562, Moose Ck Cpgd, 63°30′N, 137°01′W, 18–20.vi.1979, ROM field party (1 ♀, DEBU); Whitehorse, Fisk Lk., 8.viii.1981, C.S. Guppy (1 ♂, 1 ♀, UBC). **USA. Alaska:** Cold Bay, tundra, 25.vii.1952, WRMM (2 ♂♂, CNC); Dutch Harbor, 13.viii.1907, E.C. Van Duzee (1 ♀, CAS); Isabel Pass, Richardson Hwy, mi. 206, 2900′, 13.vii.1962, REL (1 ♀, CNC); Woods Ck, mi. 153, Glenn Hwy, 1.vii.1993, TS (1 ♂, BLKU). **Colorado:** Cameron′s Pass, 20.viii.1940, R.H. Beamer (1 ♀, SEMC); Jackson Co., N Platte R., Verner, 19.viii.1993, SF (1 ♂, CSU); Larimer Co., Chambers Lk., 16.viii.1966, GFK (1 ♀, USUL); Summit Co., Wheeler Jct. nr Ten Mile Ck, ponds, 27.vi.1995, SF and AF (9 ♂♂, 7 ♀♀, CSU). **Utah:** Cache Co., 1 mi. W Logan, 5.v.1978, WJH (3 ♂♂, 6 ♀♀, USUL); Cache Co., Franklin Basin, 7–16.vii.1981 (1 ♂, USUL); Logan, 25.x.1964, G. Weller (1 ♀, Utah); Provo, 25.v.1955, L. King (1 ♂, BYU). **Wyoming:** Yellowstone NP, Kepler Cascades, Firehole R., 20.vii.1989, BJS (2 ♀♀, CNC).

Additional material examined. USA. Alaska: Mt McKinley NP, Horseshoe Lk., 470 m, 14.vii.1978, PHA (1 ♀, USNM); N Fk Chena R., Chena Hot Springs Rd, 11.vii.1978, PHA (1 ♂, USNM); Richardson Hwy, mi. 206, beaver pond, 20.vi.1987, DMW (1 ♀, CNC). **Utah:** Garfield Co., Panguitch, Sevier R., 1.x.1981, RWB and Clark (1 ♀, BYU, in alc.).

Recognition. This species is similar to *C. stagnalis* and *C. binotata*, but differs by the antero- and postero-ventral row of short setae and lack of ventrobasal setae on the forefemur, lower half of face with silvery pruinescence (sometimes divided medially), and presence of an auxiliary crossvein between R_{2+3} and R_4.

Description (wing length 3.8–4.5 mm).

Male. Face convex; lower half of face with silvery pruinescence, sometimes divided medially, remainder of head dark brown. Arista-like stylus short and stout. Postocular bristles dark and stout on upper half of eye margin.

Scutum bivittate, separated by faint metallic-blue pruinescence. Pleura and lateral margin of scutum with dull blue pruinescence. Several acrostichals anterior to first dc; 5 dc; 1 pprn; 1 presut spal; 2 npl, subequal to dc; 1 psut spal; 1 pal; 2 sctl. Laterotergite with patch of pale setae. Posterior margin of anepisternum with setulae.

Wing (Fig. 32*h*) infuscate with dark veins; clouding around major crossveins: base of R_4, auxiliary vein, and apex of cell dm; 1 long basal costal bristle; costal spines present; stigma absent. Veins straight; auxiliary crossvein present between R_{2+3} and R_4; occasionally additional crossvein between R_{2+3} and C; stump vein absent; R_4 gently arched, diverging from R_5; anal vein a long streak; cell dm truncate. Halter dark.

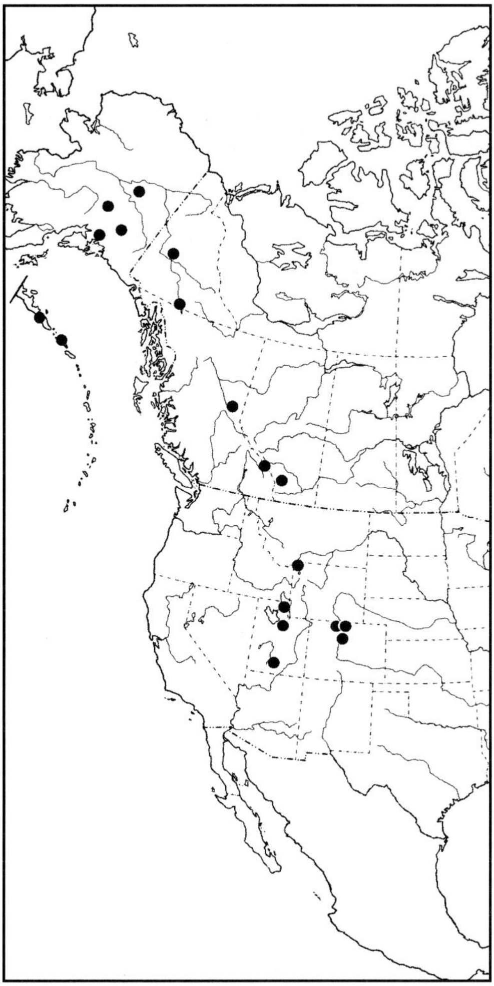

Map 29. Known distribution of *Clinocera substagnalis*.

Coxae and basal three-quarters of femora with blue pruinescence; apex of femora yellow–brown, all remaining leg segments dark. Forecoxa with pale setae. Forefemur with subapical anterior comb present; dense white pile beneath; antero- and postero-ventral row of short setae, basal half comprised of narrow white hairs. Foretibia with biserial row of short bristles beneath. Mid and hind femora with antero- and postero-ventral row of bristles on apical half. Hind tibia lacking erect bristles. Fifth tarsomere with dorsoapical extension; empodium white laterally.

Abdominal sternites and lateral margin of tergites with blue pruinescence, remaining tergites brown. Hypandrium broadly triangular. Phallus split apically; distiphallus dorsoventrally flattened. Clasping cercus in shape of short, finger-like lobe; dorsal margin bearing short stout spine-like setae. Surstylus shorter than clasping cercus, with truncate apical margin (Fig. 22*d*).

Female. Similar to male except ventral bristles on all femora shorter.

Geographic distribution and seasonal occurrence. This species is restricted to the western Nearctic Region, ranging from the arctic region south through higher altitudes of the Rocky Mountains (Map 29). In the southern part of its range, the flight period is from spring through the summer months and into October.

Remarks. This species was frequently encountered flying low over small pools in slow streams and along the edge of ponds.

Etymology. The specific name is derived from the Latin *sub* ("from", "somewhat") and *stagnalis*, alluding to the structural similarity between this species and *C. stagnalis*, which is sympatric over the range of this new species.

Neotropical species of the *C. stagnalis* group

Clinocera attenuata sp. nov.

Fig. 23*a*; Map 30

Type material. HOLOTYPE, male labelled "Cruzeiro Brazil/ XII.1962/ F. Plauman"; "HOLOTYPE/ Clinocera/ attenuata/ Sinclair [red label]" (CNC).

Recognition. This species is distinguished by blue pruinescence on the lower half of the face, which is divided medially, pleura entirely clothed in blue pruinescence, spotted wings, and unique male terminalia.

Description (wing length 3 mm).

Male. Face flat, lower half of face with broad lateral bands of blue pruinescence narrowly divided medially, upper half brown; gena, postgena, and occiput with blue pruinescence; remainder of head dark brown. Arista-like stylus long and slender. Upper postocular setae dark.

Thorax brown, with pair of black vittae extending to prescutellar depression.

Pleura with blue pruinescence. Several acrostichals anterior to first dc; 5 dc; 1 slender pprn; 1 slender presut spal; 2 npl, lower very short and weak; 1 slender psut spal; 1 slender pal; 2 sctl. Laterotergite with patch of pale setae; posterior margin of anepisternum with pale setulae.

Wing infuscate, with faint clouding at base of R_4, r-m vein, and tip of cells dm and bm; 1 basal costal seta; costal spines present; stigma absent. Veins straight; auxiliary crossvein between R_4 and R_{2+3} present or absent; stump vein absent; R_4 straight, curved just prior to wing margin; anal vein distinct streak extending to wing margin; cell dm nearly truncate apically. Halter dark.

Coxae yellowish brown, clothed in blue pruinescence; ventral surface of femora yellowish brown, remaining segments dark. Forecoxa with fine pale setae. Forefemur with antero- and postero-ventral row of pale short slender setae; anterior preapical comb present; ventral surface with dense white pile. Foretibia with single row of stout setae beneath. Midfemur lacking preapical setae. Hind legs missing. Fifth tarsomere lacking dorsoapical extensions; empodium light brown laterally.

Abdominal sternites and lateral margin of tergites with blue pruinescence, remaining tergites brown. Hypandrium long and slender. Phallus slightly arched, apex split; distiphallus whip-like, flexible, broad laterally; membranous sheath broad extending onto shaft. Clasping cercus strongly tapered to long, slender, parallel-sided extension from broad oval base; inner face densely clothed in peg-like setae. Surstylus with greatly extended apical margin projecting far forward, apex truncate, clothed in long setae; dorsal margin with matted microtrichia; base with nose-like projection (Fig. 23*a*).

Female. Unknown.

Geographic distribution and seasonal occurrence. This species is known only from the type locality (Map 30), situated in the biogeographic region known as the Brazilian Atlantic Forest (see Morrone 2005).

Remarks. This species is based only on the male holotype, and because the auxiliary crossvein is present on one wing and absent on the other, *C. attenuata* comes out in two places in the key to species.

Etymology. The specific name is derived from the Latin *attenuatus* ("drawn out", "tapered"), referring to the long, narrow prolongation of the clasping cercus.

Clinocera chilensis sp. nov.

Fig. 23*b*; Map 32

Type material. HOLOTYPE, male labelled "Puerto Ibanez/ Lag. Buenos Aires/ Aysen, CHILE/ 12-15.I.'[19]61. L. Pena"; "HOLOTYPE/ Clinocera/ chilensis/ Sinclair [red label]" [damaged: face collapsed, terminalia fractured, missing distiphallus] (CNC).

Map 30. Known distribution of *Clinocera attenuata* (★); *C. cummingi*, *C. parana* (♦), *C. meridana* (▲); *C. naomina* (■); *C. oriunda* (●).

Recognition. This species is distinguished by the bright-blue pruinescence on the face, lower half of pleura with blue pruinescence, infuscate wing with spots about crossveins, auxiliary crossvein present, and yellowish brown legs.

Description (wing length 3.6 mm).

Male. Face convex, lower face with bright -lue pruinescence; gena, postgena, and occiput with blue pruinescence; remainder of head dark brown. Antenna long and slender. Postocular setae dark.

Thorax brown; scutum with pair of narrow black vittae between dc, extending to prescutellar depression. Lower half of pleura and small region near base of wing with blue pruinescence. Several pale acrostichals anterior to first dc; 5 long dc; 1 pprn; 1 presut spal; 2 npl, lower slender and short; 1 psut spal; 1 pal; 2 sctl. Laterotergite with patch of long pale setae. Posterior margin of anepisternum with setulae.

Wing infuscate with spots about tip of cells bm and dm, base of $R_{4,}$ and margin of r-m vein; costal spines present; stigma absent; 1 basal costal seta. R_{2+3} somewhat wavy; auxiliary crossvein between R_4 and R_{2+3} present; stump vein absent; R_4 straight, parallel to R_5, divergent at wing margin; anal vein short faint streak; cell dm slightly produced apically. Halter dark.

Coxae yellowish brown, clothed in blue pruinescence, concolorous with pleura; femora and tibiae yellowish brown; remaining segments brown. Forecoxa with short pale setae. Forefemur with antero- and postero-ventral row of short setae; anteroventral setae very short, most conspicuous at base; posteroventral setae longer than width of femur, decreasing in length apically; dense pile beneath; preapical anterior comb present. Foretibia with biserial row of stout setae beneath. Hind tibia with erect ventral setae on apical half. Fifth tarsomere with dorsoapical extension; empodium pale laterally.

Abdominal sternites and lateral margin of tergites with blue pruinescence, remaining tergites brown. Hypandrium subtriangular. Phallus long, arched, or distinctly bent at midlength, apex split; distiphallus damaged. Clasping cercus straight, finger-like; tapered on one side apically, spines on inner and inner dorsal margins. Surstylus projecting slightly beyond epandrium; strongly curved anteriorly to rounded apex; long setae on apical margin and arising from lateral ridge (Fig. 23*b*).

Female. Unknown.

Geographic distribution and seasonal occurrence. This species is known only from the type locality (Map 32).

Etymology. The specific name is in reference to the type locality of this species.

Clinocera collini sp. nov.

Fig. 23*c*; Map 32

Type material. HOLOTYPE, male labelled "Type/ male [red bordered circle]"; "Hydrodromia/ oriunda Collin/ Type male [handwritten]"; "Bariloche./ 28.xi.–1.xii.1926"; "Argentina/ Terr.Rio Negro./ F.& M. Edwards./ B.M. 1927-63"; "PARALECTOTYPE/ of Hydrodromia/ oriunda Collin/ des. B.J. Sinclair 1994"; "HOLOTYPE/ Clinocera/ collini/ Sinclair [red label]" (BMNH).

The holotype was designated from a paralectotype of *C. oriunda* because of the good condition of the specimen.

Paratypes: **ARGENTINA.** Bariloche, Rio Negro, ix.1926, RES (1 ♂, USNM). **CHILE.** Concepción, A.M. Thaxter (1 ♀, USNM); Nuble: 22.7 km ESE Recinto, 1330 m, *Nothofagus* for., 10.xii.1982–3.i.1983, A.F. Newton, M.K. Thayer (1 ♀, USNM); Osorno: Pucatrihue, 27–30.i.1978, W.N. Mathis (1 ♂, USNM).

Additional material examined. ARGENTINA. Terra Rio Negro, Lk. Gutierrez, 3–14.xi.1926, FME (1 ♂, BMNH); Bariloche, Rio Negro, ix.1926, RES (1 ♀, USNM). **CHILE.** Ancud, Chiloe Is., 17–19.xii.1926, FME (1 ♀, BMNH); Marga-Marga, ix.1929, F. Jaffuel (1 ♀, BMNH) [these four specimens are paralectotypes of *C. oriunda*]; Malleco: Cordillera Nahuelbuta, 1300 m, 6–21.i.1982, LEP (1 ♀, KMNH). **ECUADOR.** Cerro Tinajillas, Azuay, 3200 m, 15.iii.1965, LEP (1 ♀, CNC). [This specimen appears to be conspecific, but without an associated male I am reluctant to include it as a paratype.]

Recognition. This species has the bright pruinescence of the lower half of the face divided medially by a brown stripe, spotty wings, yellowish brown legs, and auxiliary crossvein absent.

Description (wing length 3.8–4.1 mm).

Male. Face convex, lower half with shiny white pruinescence, divided medially by brown stripe; upper part of face brown; gena, postgena, and occiput with pale-blue pruinescence; vertex brown. Arista-like stylus long and slender. Upper postocular setae dark.

Thorax brown; scutum with two narrow black vittae between dc. Lower half of pleura with pale-blue pruinescence, distinctly separated from upper half. Several pale acrostichals anterior to first dc; 5 long dc; 1 pprn; 1 presut spal; 2 npl, lower slender and short; 1 psut spal; 1 pal; 2 sctl. Laterotergite with patch of pale setae. Posterior margin of anepisternum with long setulae.

Wing hyaline with clouding about following crossveins: tip of cells bm and dm, base of R_4, and margin of r-m vein; short costal spines present; stigma absent; 1 long basal costal seta. Auxiliary crossvein between R_4 and R_{2+3} absent; stump vein absent; R_4 bell-shaped; anal vein elongate fold; cell dm truncate apically. Halter dark.

Coxae yellowish brown, with pale-blue pruinescence, concolorous with pleura; legs dark yellowish brown, apical tarsomeres dark. Forecoxa with short pale setae. Forefemur with antero- and postero-ventral row of slender setae, becoming longer and paler toward base; dense short pile beneath; preapical anterior comb present. Foretibia with biserial row of stout setae beneath. Hind tibia with erect ventral setae on apical third. Fifth tarsomere with short dorsoapical extension; empodium pale laterally.

Lateral and ventral surfaces of abdomen with blue pruinescence, dorsum dark. Hypandrium slender, subtriangular. Phallus strongly arched; distiphallus relatively short, flattened, blade-like; apex with fine microtrichia. Clasping cercus elongate, strongly tapered, pointed apically; blunt spines on inner margin. Surstylus strongly arched, nearly C-shaped, outer margin with long setae (Fig. 23*c*).

Female. Similar to male except as follows: ventral setae of forefemur stouter and darker; ventral setae of foretibia reduced.

Geographic distribution and seasonal occurrence. This species is known primarily from the extreme southwestern region of South America, with a possible record from Ecuador (Map 32). Flight records indicate that this species is present at least from September to January.

Etymology. The specific name is a patronym in honour of J.E. Collin, who provided the framework for all taxonomic studies of Neotropical Empidoidea.

Clinocera colombiana sp. nov.

Fig. 23*d*; Map 32

Type material. HOLOTYPE, male labelled "COLOMBIA. Caldes:/ Termales de Ruiz/ (5.3km W), 3000m/ 28Feb1984 WNMathis"; "HOLOTYPE/ Clinocera/ colombiana/ Sinclair [red label]" [antennae missing] (USNM).

Recognition. This species is distinguished by a face that is mostly brown, except for a narrow faint pruinescence along the lower margin of the eye, pleura completely clothed in blue pruinescence, infuscate wing with clouding about crossveins, and auxiliary crossvein present.

Description (wing length 4.3 mm).

Male. Face flat, lower third of face with pale pruinescence on lateral margin; gena, postgena, and occiput with blue pruinescence; remainder of head dark brown. Antenna damaged, postpedicel and arista-like stylus lost. Postocellar setae dark.

Thorax brown; scutum with pair of narrow black vittae between dc, extending to prescutellar depression; supra-alar with black stripe from presut spal to scutellum. Pleura completely with blue pruinescence. Several pale acrostichals anterior to first dc; 5 long dc; 1 very short, fine pprn; 1 short presut spal; 2 npl, lower slender and short; 1 psut spal; 1 pal; 2 sctl. Laterotergite with patch of long pale setae. Posterior margin of anepisternum with setulae.

Wing darkly infuscate with clouding about tip of cells bm and dm, base of R_4, and margin of r-m vein; costal spines present; stigma absent; 1 long, basal costal seta. Veins straight; auxiliary crossvein between R_4 and R_{2+3} present; stump vein absent; R_4 straight, parallel to R_5; anal vein long faint streak; cell dm truncate apically. Halter dark.

Coxae with blue pruinescence, concolorous with pleura; remaining segments dark. Forecoxa with short pale setae. Forefemur with posteroventral row of short setae, basal setae pale and slender, apical setae dark, short, and stout; very short dense pile beneath; preapical anterior comb present. Foretibia with row of stout setae beneath. Hind tibia with erect dorsal and ventral setae on apical half. Fifth tarsomere lacking dorsoapical extension; empodium brown laterally.

Abdominal sternites and lateral margin of tergites with blue pruinescence, remaining tergites brown. Hypandrium subtriangular. Phallus relatively short, slightly bent near base, apex split; apical membranous cap with posterior, setose horn; distiphallus relatively short, dorsoventrally flattened, narrow. Clasping cercus straight, elongate, tapered on one side apically, spines on inner and inner dorsal margins. Surstylus projecting beyond epandrium; strongly curved anteriorly, with long, slender, beak-like extension; outer margin with stout setae (Fig. 23*d*).

Female. Unknown.

Geographic distribution and seasonal occurrence. This species is known only from the type locality (Map 32).

Etymology. The specific name is in reference to the type locality.

Clinocera cubatao Smith

Fig. 24*a*; Map 31

Clinocera (*Hydrodromia*) *cubatao* Smith, 1962: 264.

Clinocera cubatao: Smith, 1967: 44.

Type material examined. HOLOTYPE, male labelled “SÃO PAULO/ Cubatão/ Pereira. Marinez/ Werner. d’Andretta/ 15-XII-[19]55 [written on side of label]”; “Paratype/ Clinocera/cubatao/ n.sp. K.G.V.Smith”; “HOLOTYPE [red label]” [dissected, mounted on slide below] (MZUSP). The holotype of this species was not identified as such in the collection, despite the clear statement by Smith (1962: 265) that a holotype was designated. Consequently a red holotype label has been attached to the originally dissected male specimen, which was illustrated in Smith (1962).

Paratypes: **BRAZIL.** Same data as holotype (3 ♀♀, BMNH; 1 ♂, 6 ♀♀

Map 31. Known distribution of *Clinocera cubatao* (♦); *C. ditaeniata* (▲); *C. empodiata* (■); *C. insularis*, *C. marshalli*, *C. miradorana*, *C. notialis* (□); *C. isabellina* (★); *C. penai* (●).

(including allotype) MZUSP); same locality as holotype, 15.ii.55 (2 ♀♀, MZUSP).

An additional two males (collected in February, perhaps a labelling error?) and four males (collected in December) are also in MZUSP identified under this species, but were likely not seen by Smith. Since Smith (1962) clearly stated that there were only 1 male and 12 female paratypes, these additional 6 male specimens are not considered part of the original type series and have no type status.

Recognition. This species is distinguished by the blue pruinescence on the lateral margins of the lower half of the face, absence of the auxiliary crossvein, indistinct clouding about wing crossveins and at base of R_4, and brightly pale brown legs.

Description (wing length 3–3.2 mm).

Male. Lateral margins of lower half of face with blue pruinescence; upper part of face brown. Arista-like stylus long and slender. Upper postocular setae dark.

Thorax brown; scutum with pair of dark vittae between dc, extending to prescutellar depression. Pleura with bright-blue pruinescence. Acrostichals absent; 5 dc; 1 weak pprn; 1 presut spal; 2 npl, lower slender and short; 1 psut spal; 1 pal; 2 sctl. Laterotergite with patch of pale setae. Posterior margin of anepisternum with setulae.

Wing infuscate, indistinct clouding about crossveins: tip of cells bm and dm, base of R_4 and margin of r-m vein; stigma absent; 1 basal costal seta; costal spines present. Auxiliary crossvein between R_4 and R_{2+3} absent; stump vein absent; R_4 bell-shaped; anal vein an elongate streak; cell dm produced slightly apically. Halter knob dark, base of stalk pale brown.

Coxae yellowish brown; femora pale brown; remaining segments dark brown. Forecoxa with short pale setae. Forefemur with anteroventral row and more evident posteroventral row of slender setae, becoming longer and paler toward base, more than half width of femur; dense short pile beneath nearly inconspicuous; preapical anterior comb present. Foretibia with biserial row of stout setae beneath. Hind tibia with erect ventral setae on apical half. Fifth tarsomere with slight dorsoapical extension; empodium brown laterally.

Abdominal sternites and lateral margin of tergites with blue pruinescence, remaining tergites brown. Hypandrium subrectangular, tapered toward apex. Phallus very long, weakly arched apically; distiphallus short, dorsoventrally flattened, somewhat expanded subapically. Epandrium approximately square. Clasping cercus finger-like, straight, with broad base; apex broadly rounded. Surstylus broadly rounded, apex margin clothed with stout setae (Fig. 24*a*).

Female. Similar to male.

Geographic distribution and seasonal occurrence. This species is known only from the type locality (Map 31).

Remarks. No additional specimens of this species were available, other than those described above.

Clinocera cucama Smith

Fig. 24*b*; Map 32

Clinocera (*Hydrodromia*) *cucama* Smith, 1962: 265.

Clinocera cucama: Smith, 1967: 44.

Type material examined. HOLOTYPE, male labelled "Holo-/ type [red bordered circle]"; "Brasilien/ Nova Teutonia/ 27°11´B 52°23´L/ Fritz Plaumann/ 28.10.1938"; "HOLOTYPE/ Clinocera/ cucama male symbol/ det.1961 sp.n./ K.G.V. Smith" [genitalia dissected and mounted] (BMNH).

Additional material examined. ARGENTINA. Balneario San Juan, 10 km ENE Oberá, Misiones, 13.xi.2004, TS (1 ♂, 1 ♀, BLKU).

Recognition. This species is distinguished by blue pruinescence on the lateral margins of the lower half of the face, auxiliary crossvein absent, wings unspotted, thoracic chaetotaxy reduced, and clasping cercus hammer-shaped.

Description (wing length 3.1 mm).

Male. Lateral margins of lower half of face with blue pruinescence; upper part of face brown. Arista-like stylus long and slender. Ocellar bristles short and weak. Upper postocular setae dark.

Thorax brown; scutum with pair of dark vittae between dc, extending to prescutellar depression. Pleura with blue pruinescence. Chaetotaxy weak and reduced: acrostichals absent; 5 dc (anteriormost less than half length of second); 1 pprn, reduced and slender; 1 presut spal; 2 npl, lower slender and short; 1 psut spal; 1 pal; 2 sctl. Laterotergite with patch of pale setae. Posterior margin of anepisternum with setulae.

Wing hyaline, lacking clouding or spots; stigma absent; 1 basal costal seta; costal spines present. Auxiliary crossvein between R_4 and R_{2+3} absent; stump vein absent; R_4 straight until near wing margin; anal vein an elongate streak; cell dm nearly truncate apically. Halter dark.

Coxae with pale-blue pruinescence, concolorous with pleura; remaining segments light brown. Apex of coxae and trochanters yellowish brown. Forecoxa with short pale setae. Forefemur with antero- and postero-ventral margins with row of slender setae, becoming longer and paler toward base, more than half width of femur; dense short pile beneath nearly inconspicuous; preapical anterior comb present. Foretibia with biserial row of stout setae beneath; apical half with stout posterodorsal setae. Hind tibia with erect dorsal

Map 32. Known distribution of *Clinocera chilensis* (■); *C. collini* (●); *C. colombiana* (★); *C. cucama* (♦); *C. lapazensis* (▲).

and ventral setae on apical half. Fifth tarsomere lacking dorsoapical extension; empodium pale laterally.

Abdominal sternites and lateral margin of tergites with blue pruinescence, remaining tergites brown. Hypandrium long and slender, subrectangular, tapered toward apex. Phallus strongly arched, apex dilated with wide membrane; distiphallus short, dorsoventrally flattened, blade-like. Clasping cercus with wide base, apex hammer-shaped; anterior-directed apex greatly narrowed and curved medially; stout setae on inner margin. Surstylus broadly rounded, apex angular; clothed with long setae (Fig. 24*b*).

Female. Similar to male, except ventral setae of forefemur and foretibia not modified.

Geographic distribution and seasonal occurrence. This species is confined to the Paraná region of Brazil (Map 32).

Remarks. This species is very similar to *C. attenuata* from Brazil, which also has a greatly attenuated clasping cercus and modified phallus.

Clinocera cummingi sp. nov.

Fig. 24*c*; Map 30

Type material. HOLOTYPE, male labelled "BRAZIL: Parana/ Serra-do-Mar near/ Morretes, 29.viii.2000/ 25°21.80′S 48°52.55′W/ J.M. Cumming/ ex. Mae Cotira river"; "HOLOTYPE/ Clinocera/ cummingi/ Sinclair [red label]" (MZUSP).

Paratypes: **BRAZIL. Paraná:** Same data as holotype (1 ♂, 1 ♀, CNC).

Recognition. This species is distinguished by blue pruinescence on the lateral margins of the lower half of the face, lacking auxiliary crossvein, wings unspotted, reduced thoracic chaetotaxy, and clasping cercus greatly attenuated and arched medially.

Description (wing length 3.3 mm).

Male. Lower third of face with blue pruinescence; upper part of face brown. Arista-like stylus short and slender. Ocellar bristles short and weak. Upper postocular setae dark.

Thorax brown; scutum with pair of dark vittae between dc, extending to prescutellar depression. Pleura with blue pruinescence. Chaetotaxy weak and reduced: acrostichals absent; 5 dc, short and slender; 1 pprn, reduced and slender; 1 presut spal; 2 npl, lower slender and short; 1 psut spal; 1 pal; 2 sctl. Laterotergite with patch of pale setae. Posterior margin of anepisternum with setulae.

Wing narrow and infuscate, clouding or spots absent; stigma absent; 1 slender basal costal seta; costal spines present. Auxiliary crossvein between R_4 and R_{2+3} absent; stump vein absent; R_4 straight to wing margin; anal vein distinct basally; cell dm produced apically. Halter knob dark, base of stalk paler than knob.

Coxae with pale-blue pruinescence, concolorous with pleura; femora pale brown beneath, remaining segments light brown; apex of coxae yellowish brown. Forecoxa with short pale setae. Forefemur with ventral setae greatly; dense short pile beneath nearly inconspicuous; preapical anterior comb present. Foretibia with biserial row of stout setae beneath. Hind tibia with erect dorsal and ventral setae on apical half. Fifth tarsomere lacking dorsoapical extension; empodium brown laterally.

Abdominal sternites and lateral margin of tergites with blue pruinescence, remaining tergites brown. Hypandrium long and slender, subrectangular, tapered toward apex. Phallus strongly arched near midlength, apex dilated with wide membrane; distiphallus short, dorsoventrally flattened, arched with ragged dorso-apical margin. Clasping cercus greatly attenuated on apical half, arched medially; stout setae on inner margin. Surstylus broadly rounded, projected anteriorly; clothed with long setae (Fig. 24*c*).

Female. Similar to male, except ventral setae of forefemur and foretibia not modified.

Geographic distribution and seasonal occurrence. This species also is recorded only from the state of Paraná (Map 30) in the Brazilian Atlantic Forest biogeographic province (see Morrone 2005). The species was swept from large boulders in a sunny, exposed river (Fig. 3*b*).

Remarks. The shape of the surstylus (rounded and projected) is similar to that in *C. attenuata* and *C. insularis*.

Etymology. The specific name is a patronym in honour of Jeff Cumming, a friend and fellow empidoid specialist, who collected the type series.

Clinocera ditaeniata Bezzi

Fig. 24*d*; Map 31

Clinocera (*Heleodromia*) ditaeniata Bezzi, 1909: 387.

Clinocera ditaeniata: Smith, 1967: 44.

Clinocera ditaeniata: Rafael, 1996: 437.

Type material examined. LECTOTYPE, male labelled "Bolivia/ 7.XII.[19]02./ La Paz [green label]"; "Heleodromia/ ditaeniat./ n.sp. [handwritten]"; "Coll. W. Schnuse/ 1911 - 3 [green label]"; "SYNTYPUS/ des. Kallweit [red label]"; "Clinocera/ ditaeniata/ BEZZI"; "LECTOTYPE/ des. J.A. Rafael male [blue label]" [dissected] (MTD).

Paralectotype: **BOLIVIA.** Same data as lectotype [right wing mounted on slide pinned beneath specimen; abdomen dissected] (1 ♀, MTD).

Recognition. This species is distinguished by the presence of the auxiliary crossvein, wing with dark clouding at major crossveins, and features listed in the key to species.

Description (wing length 3.6 mm; from Rafael 1996).

Male. Lateral margin of lower face with narrow white pruinescence; upper part of face brown.

Pleura, including postnotum, with pale-blue pruinescence; scutum with two narrow black vittae (inside of dc). Several pale acrostichals anterior to first dc; 5 dc; 1 pprn; 1 presut spal; 2 npl, lower slender; 1 psut spal; 1 pal; 2 sctl. Laterotergite with patch of pale setae.

Wing (Rafael 1996, fig. 18) infuscate with clouding about crossveins: tip of cell dm, base of R_4, apex of cell bm, margin of vein r-m. Stigma absent; single basal costal seta. Auxiliary crossvein present between R_4 and R_{2+3}; stump vein absent; costal spines present; R_4 straight; anal vein faint; cell dm nearly truncate apically. Halter orange–brown.

Legs orange–brown to dark, especially coxae; coxae with pale pruinescence. Forefemur with posteroventral row of short setae on apical half; dense short pile beneath; preapical anterior comb present. Hind tibia with dorsal and ventral setae on apical third. Fifth tarsomere lacking dorsoapical extension.

Lateral and ventral surfaces of abdomen with blue pruinescence, dorsum dark. Hypandrium subtriangular, tapering to apex. Phallus slightly arched near base; distiphallus relatively short, flattened, blade-like; apex with fine microtrichia. Clasping cercus finger-like; blunt spines on inner margin along base and distal edge. Surstylus broadly triangular, outer margin with stout setae (Fig. 24*d*).

Female. Similar to male, except ventral setae of forefemur less pronounced.

Geographic distribution and seasonal occurrence. Known only from the type series in La Paz, Bolivia (Map 31).

Remarks. Rafael (1996) redescribed this species and designated a lectotype. No additional specimens of this species were available.

Clinocera empodiata Collin

Fig. 25*a*; Map 31

Clinocera empodiata Collin, 1933: 322.

Clinocera empodiata: Smith, 1967: 44.

Type material examined. HOLOTYPE, female labelled “Type/ female [red bordered circle]”; “Clinocera/ empodiata, Collin/ Type female [handwritten]”; “Potrerillos./ 6.i.1927.”; “Argentina:/ Mendoza Prov./ F. & M. Edwards./ B.M. 1927—63” (BMNH).

Additional material examined. ARGENTINA. Jujuy, La Quiaca, 3500 m,

23.x.1968, LEP (1 ♂, CNC); Jujuy, 2 km SW Cienaga, rd to Pirquitas, 3900 m, 3.xi.1968, LEP (1 ♂, CNC).

Recognition. This species is readily distinguished by the clear or hyaline wings, convex face with shiny, silvery pruinescence extending along the eye margin above the antennae, and lacking the preapical comb of the forefemur.

Description (wing length 4.2 mm).

Male. Face convex, entire face with shiny, silvery pruinescence extending above antenna along margin of eye; gena, postgena, and occiput with blue pruinescence; vertex brown. Arista-like stylus short and stout. Upper postocular setae dark.

Thorax brown; scutum with pair of dark vittae between dc, extending to prescutellar depression. Pleura and lateral margin of scutum, including postpronotal lobe and notopleuron, with pale-blue pruinescence. Acrostichals absent; 5 dc; 1 pprn; 1 presut spal; 2 npl, lower shorter; 1 psut spal; 1 pal; 2 sctl. Laterotergite with patch of long pale setae. Posterior margin of anepisternum with long setulae.

Wing hyaline with clouding or spots absent; stigma absent; 2 basal costal setae; costal spines short. Auxiliary crossvein between R_4 and R_{2+3} absent; stump vein absent; R_4 bell-shaped; anal vein an elongate fold; cell dm truncate apically. Halter pale.

Coxae and all leg segments with pale-blue pruinescence, except for apical two tarsomeres; inner apex of forefemur pale. Forecoxa with short pale setae. Forefemur with antero- and postero-ventral row of short slender setae, shorter than half width of femur; dense short pile beneath; preapical anterior comb absent. Foretibia with biserial row of stout setae beneath. Hind tibia lacking erect setae. Fifth tarsomere with short dorsoapical extension; empodium very broad, white laterally.

Lateral and ventral surfaces of abdomen with blue pruinescence, narrow strip of dorsum dark. Hypandrium subtriangular, tapered toward apex. Phallus slightly arched; distiphallus short, undulated, dorsoventrally flattened. Clasping cercus subtriangular; slender spine-like setae on inner apical margin. Surstylus narrowly produced apically, outer margin with slender setae (Fig. 25*a*).

Female. Similar to male, including leg chaetotaxy.

Geographic distribution and seasonal occurrence. This species is known only from Mendoza and Jujuy provinces of Argentina (Map 31).

Clinocera insularis sp. nov.

Fig. 25*b*; Map 31

Type material. HOLOTYPE, male labelled "CHILE: JuanFernandez Islands/ Robinson Crusoe Island/ 1.i.1993; seep; S. side/ Mirador de Selkirk/ S.A. Marshall"; "HOLOTYPE/ Clinocera/ insularis/ Sinclair [red label]" (MNHNS).

Paratypes: **CHILE. Juan Fernandez Islands:** Same data as holotype (12 ♂♂, 4 ♀♀, CNC, DEBU, MNHNS); Robinson Crusoe Is., N side Mirador, pans in mossy seep, 1–10.i.1993, SAM (1 ♀, MNHNS); Robinson Crusoe Is., open for. above Plazoleta, MT, 1–9.i.1993, SAM (3 ♂♂, 1 ♀, DEBU); Robinson Crusoe Is., Villigra, seep, 4.i.1993, SAM (3 ♂♂, 2 ♀♀, DEBU); Más a Tierra N side, Yunque, 1–9.ii.1973, LEP (1 ♀, CNC).

Recognition. This species is distinguished by pale pruinescence along the eye on the lower half of the face, distinctly spotted wings, auxiliary crossvein absent, scutum with two pairs of vittae, and lower half of the pleura with faint blue pruinescence extending anteriorly to the postpronotal lobe.

Description (wing length 3–3.2 mm).

Male. Face flat, lower third of face with pale pruinescence along margin of eye, widely separated by brown median stripe; upper face brown; gena, postgena, and occiput with faint blue pruinescence; vertex brown. Arista-like stylus long and slender. Upper postocular setae dark.

Thorax brown; scutum with pair of narrow black vittae between dc; supra-alar with black stripe from presut spal to scutellum; dark spot below postsutural supra-alar; margin of scutellum dark. Lower half of pleura extending anteriorly to postpronotal lobe with blue pruinescence. Thoracic setae very long; several pale acrostichals anterior to first dc; 5 long dc; 1 pprn; 1 presut spal; 2 npl, lower slender and short; 1 psut spal; 1 pal; 2 sctl. Laterotergite with patch of setae. Posterior margin of anepisternum with setulae.

Wing darkly infuscate with distinct spots about tip of cells bm and dm, base of R_4, and margin of r-m vein; costal spines present; stigma absent; 1 long, basal costal seta. Veins straight; auxiliary crossvein between R_4 and R_{2+3} absent; stump vein absent; R_4 bell-shaped; anal vein dark streak; cell dm truncate apically. Halter dark.

Coxae dark brown with faint blue pruinescence, concolorous with pleura; remaining segments dark brown. Forecoxa with short dark setae. Forefemur with posteroventral rows of slender setae on basal two-thirds, decreasing in length apically; anteroventral margin with short row of setae on basal fifth; short dense pile beneath; preapical anterior comb present. Foretibia with biserial row of stout setae beneath. Hind tibia with erect ventral setae on apical half. Fifth tarsomere with dorsoapical extension; empodium light brown laterally.

Abdomen dark, except for faint blue pruinescence on ventral surface. Hypandrium tapered, subtriangular. Phallus slightly curved, apex split; distiphallus relatively long, dorsoventrally flattened. Clasping cercus short, broad, strongly tapered to pointed apex; short spines on inner margin. Surstylus projecting free from epandrium; apex broadly rounded, with short ventral projection; outer margin with long stout setae (Fig. 25*b*).

Female. Similar to male except as follows: posteroventral setae of forefemur longer, nearly half width of femur; ventral setae of foretibia reduced.

Geographic distribution and seasonal occurrence. This species is recorded only from the Juan Fernandez Islands (Map 31).

Etymology. The specific name is from the Latin *insularis* ("of an island"), referring to the restriction of this species to the Juan Fernandez Islands.

Clinocera isabellina sp. nov.

Figs. 25*c*, 32*i*; Maps 31, 33

Type material. HOLOTYPE, male labelled "COSTA RICA: San Jose Prov./ San Gerardo de Dota/ 7-10.viii.1995, 2200m/ J.M. Cumming/ ex. trib. R. Savegre"; "HOLOTYPE/ Clinocera/ isabellina/ Sinclair [red label]" (CNC).

Paratypes: **COSTA RICA. Cartago:** Cerro de la Muerte, La Cangreja, MT, 1950 m, vii.1991, PH (1 ♀, CNC); Reserva Forestal Rio Macho, La Esperanza del Guarco, 2750 m, 1.vi.1998, SAM LN-185000, 550700 (1 ♂, DEBU). **Heredia:** Vara Blanca, Finca Georgina, 2100 m, ix.1989, PH (2 ♂♂, CNC); Volcan Barva, nr Sacramento, 6600′, 15.iii.1991, Grimaldi and Stark (1 ♂, AMNH). **Puntarenas Province:** Estación Biológica Monteverde, emerg. tp, 8.vi.1998, SAM (1 ♀, CNC). **San Jose:** same data as holotype (3 ♂♂, 4 ♀♀, CNC); same data as holotype, 2200–2600 m, MT, 7–10.viii.1995 (9 ♂♂, 13 ♀♀, CNC); same data as holotype, sweeping disturbed area, 9.viii.1995, SAM (1 ♂, DEBU); same data as holotype, 2200–2300 m, ck, ypans, iii.1996, LM (1 ♂, CNC); same data as holotype, 2500–2600 m, 9.viii.1995, D.C. Caloren (1 ♀, DEBU); Cerro de la Muerte, 2600 m, 19 km S, 3 km W Empalme, iv–v.1993, PH (1 ♀, CNC); 6 km N Gerado, 2800 m, MT, v.1991, PH (1 ♀, CNC); San Geraldo de Dota, domain Savegre Hotel de Montana, 2200 m, 22–23.iii.2005, M. Pollet (2 ♂♂, 3 ♀♀, CNC); 2 km W Empalme, 2300 m, i.1995, PH (3 ♀♀, CNC). **VENEZUELA. Merida:** La Aguada, 3000 m, 13.v.1981, LM (1 ♂, 1 ♀, CNC); S. Bolivar NP, nr La Aguada, 3000 m, 13.v.1981, LM (1 ♂, CNC).

Recognition. This species is distinguished by the lower half of the face with bright silver-blue pruinescence, scutum with four pairs of black stripes; auxiliary crossvein present, facial pruinescence in females confined to narrow strips along the lower eye and 3–6 long, slender basal setae on the forefemur.

Description (wing length 3.2–3.5 mm).

Male. Face flat, lower third with silver-blue pruinescence; dorsal margin of pruinescence with median notch or deep cleft; occiput, postgena, and gena with pale-blue pruinescence, remainder of head brown. Arista-like stylus generally long and slender. Postocular setae dark on upper half.

Scutum with two black vittae between dc extending to suture, separated by faint blue pruinescence on anterior margin; supra-alar with black stripe extending from presut spal to scutellum. Lower half of pleura with light-blue pruinescence.

Several scattered acrostichals near antepronotum; 5 dc; 1 pprn; 1 presut spal; 2 npl, lower shorter; 1 short psut spal; 1 pal; 2 long sctl. Laterotergite with patch of pale setae. Posterior margin of anepisternum with pale setulae.

Wing (Fig. 32*i*) infuscate with faint clouding at base of R_4, r-m vein, tip of cells dm and bm; 1 long basal costal seta. Two-thirds length of basal costal seta; costal spines present; stigma absent. Veins straight; auxiliary crossvein between R_4 and R_{2+3}; stump vein absent; R_4 straight; anal vein distinct streak; cell dm nearly truncate. Halter dark, except for yellowish brown base of shaft.

Forecoxa and ventral surface of forefemur and foretibia yellowish brown; mid and hind coxae with faint blue pruinescence; remaining leg segments brown. Forecoxa with pale setae. Forefemur with anterior preapical comb present; dense white pile beneath; base with 3–6 long stout setae, several longer than width of femur. Foretibia with row of stout setae beneath. Hind tibia with erect dorsal and ventral setae on apical half. Fifth tarsomere with dorsoapical extension; empodium brown laterally.

Abdominal sternites and lateral margin of tergites with faint blue pruinescence, remaining tergites brown. Hypandrium broadly triangular. Phallus with slight arch; distiphallus not recurved, apex curved dorsally; apical half of distiphallus dorsoventrally flattened. Clasping cercus triangular; stout, peg-like setae on inner face and dorsal margin. Surstylus shorter than clasping cercus, projecting free from epandrium; apex tapered to round tip (Fig. 25*c*).

Female. Similar to male except as follows: lower face with narrow

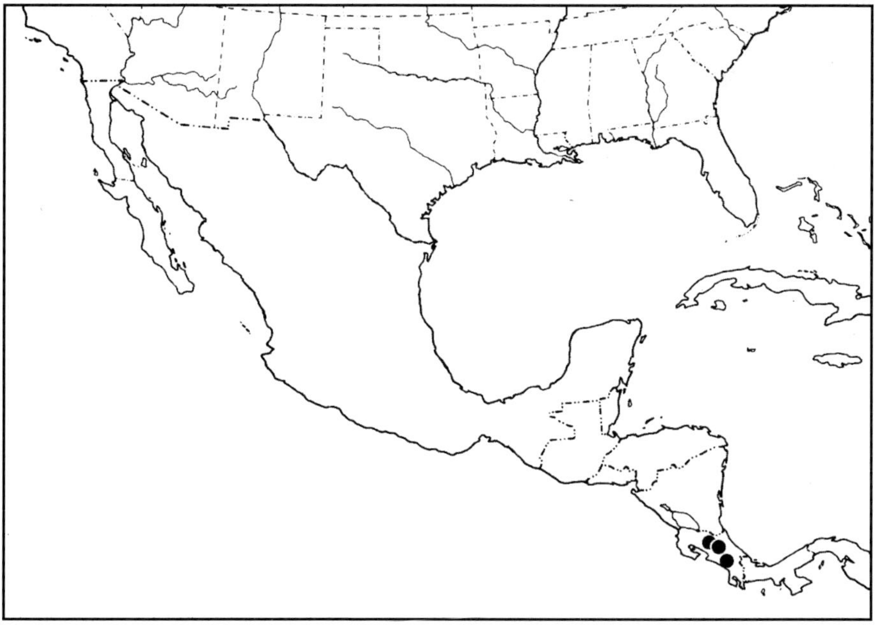

Map 33. Known distribution of *Clinocera isabellina in* Central America.

pruinescence along eye margin only; forefemur same as in male; 4–6 long, slender setae projecting perpendicular from costa between humeral crossvein and junction of Sc and C; erect dorsal setae of hind tibia reduced.

Geographic distribution and seasonal occurrence. This species is known only from high-altitude regions of Costa Rica (Map 33) and Venezuela (Map 31), probably present year-round.

Etymology. The specific name *isabellina* is proposed in appreciation of the author's eldest daughter, for her assistance with museum curation.

Clinocera lapazensis sp. nov.

Fig. 25*d*; Map 32

Type material. HOLOTYPE, male labelled "BOLIV[ia]. Unduavi/ 3500-4000m./ La Paz, Yungas/ 29.X.[19]68. [L. E.]Pena"; "HOLOTYPE/ Clinocera/ lapazensis/ Sinclair [red label]" [missing arista-like stylus, hind legs, left wing broken; stored in microvial] (CNC).

Recognition. This species is distinguished by the mostly brown face, except for narrow faint pruinescence along lower margin of eye, infuscate wing with clouding about crossveins, and auxiliary crossvein present.

Description (wing length 3.2 mm).

Male. Face convex, lower third of face with pale pruinescence on lateral margin; gena, postgena, and occiput with blue pruinescence; remainder of head dark brown. Antenna damaged, arista-like stylus lost. Upper postocular setae dark.

Thorax brown; scutum with pair of narrow black vittae between dc, extending to prescutellar depression; supra-alar with faint black stripe from presutural supra-alar to scutellum. Lower half of pleura with pale-blue pruinescence. Thoracic setae very long; several pale acrostichals anterior to first dc; 5 long dc; 1 pprn; 1 presut spal; 2 npl, lower slender and short; 1 psut spal; 1 pal; 2 sctl. Laterotergite with patch of long pale setae. Posterior margin of anepisternum with setulae.

Wing darkly infuscate with clouding about tip of cells bm and dm, base of $R_{4,}$ and margin of r-m vein; costal spines present; stigma absent; 2 long basal costal setae. Veins straight; auxiliary crossvein between R_4 and R_{2+3} present; stump vein absent; R_4 straight, divergent at wing margin; anal vein dark streak; cell dm truncate apically. Halter dark.

Coxae with blue pruinescence, concolorous with pleura; remaining segments dark. Forecoxa with short pale setae. Forefemur with posteroventral row of short setae; short dense pile beneath; preapical anterior comb present. Foretibia with row of stout setae beneath. Fifth tarsomere with dorsoapical extension; empodium brown laterally.

Abdominal sternites and lateral margin of tergites with blue pruinescence, remaining tergites brown. Hypandrium slender, tapered. Phallus with strong bend near base, followed by slight lateral expansion; apex split; distiphallus relatively long, dorsoventrally flattened, narrow. Clasping cercus boot-shaped; short blunt spine-like setae on inner and inner dorsal margins. Surstylus projecting beyond epandrium; straight, apex truncate; outer margin with stout setae (Fig. 25*d*).

Female. Unknown.

Geographic distribution and seasonal occurrence. This species is known only from the type locality (Map 32).

Etymology. The specific name is in reference to the type locality.

Clinocera marshalli sp. nov.

Figs. 26*a*, 32*j*; Map 31

Type material. HOLOTYPE, male labelled "CHILE: JuanFernandez Islands/ Robinson Crusoe Island/ 4.i.1993/ Villigra, seep/ S.A. Marshall"; "HOLOTYPE/ Clinocera/ marshalli/ Sinclair [red label]" (MNHNS).

Paratypes: **CHILE. Juan Fernandez Islands:** Same data as holotype (3 ♂♂, 1 ♀, CNC, DEBU); Robinson Crusoe Is., open for. above Plazoleta, MT, 1–9.i.1993, SAM (1 ♂, DEBU); Robinson Crusoe Is., Bahia Cumberland, MT, 1–11.i.1993, SAM and Gonzalez (1 ♀, MNHNS).

Recognition. This species is distinguished by pale pruinescence along the eye on the lower half of the face, infuscate wing with often very faint spots, auxiliary crossvein absent, one distinct pair of scutal vittae present, and the absence of pruinescence on the postpronotal lobe.

Description (wing length 3.3–3.7 mm).

Male. Face flat, lower third of face with pale pruinescence along margin of eye, widely separated by brown median stripe; remaining head dark brown. Arista-like stylus short and stout. Postocellar setae dark.

Thorax brown; scutum with pair of narrow black vittae between dc; supra-alar with faint black stripe from presutural supra-alar to scutellum; postpronotal lobe reddish brown. Pleura mostly brown with faint pruinescence near forecoxa; upper katepisternum pale brown. Thoracic setae very long; several pale acro-stichals anterior to first dc; 5 long dc; 1 pprn; 1 presut spal; 2 npl, lower slender and short; 1 psut spal; 1 pal; 2 sctl. Laterotergite with patch of short, dark setae. Posterior margin of anepisternum with setulae.

Wing (Fig. 32*j*) darkly infuscate, often with very faint clouding about tip of cells bm and dm, base of R_4, and margin of r-m vein; costal spines present; stigma absent; 1 long, basal costal seta. Veins straight; auxiliary crossvein between R_4 and R_{2+3} absent; stump vein absent; R_4 bell-shaped; anal vein dark streak; cell dm truncate apically. Halter dark.

Coxae dark brown, concolorous with pleura; remaining segments dark brown. Forecoxa with short dark setae. Forefemur with posteroventral row of slender setae on basal two-thirds, decreasing in length apically; anteroventral margin with short row of setae on basal fifth; length of basal setae nearly equal to basal width of femur; short dense pile beneath; preapical anterior comb present. Foretibia with biserial row of stout setae beneath. Hind tibia with erect ventral setae on apical half. Fifth tarsomere with dorsoapical extension; empodium light brown laterally.

Abdomen dark, except for faint blue pruinescence on ventral surface. Hypandrium strongly tapered, subtriangular. Phallus long and straight, apex split; distiphallus short, dorsoventrally flattened. Clasping cercus long, broad, gradually tapered to broad apex; short spines on inner margin. Surstylus projecting free from epandrium; long and slender; apex truncate, with pointed ventral apex; outer margin with long setae (Fig. 26*a*).

Female. Similar to male except as follows: posteroventral setae of forefemur longer than width of femur; ventral setae of foretibia reduced.

Geographic distribution and seasonal occurrence. This species is recorded only from the Juan Fernandez Islands (Map 31).

Etymology. This species is named in honour of the author's colleague and friend Steve Marshall, who collected nearly all the *Clinocera* material known from these islands.

Clinocera meridana sp. nov.

Figs. 26*b*, 33*a*; Map 30

Type material. HOLOTYPE, male labelled "Venez[uela]. Merida/ S. Bolivar N.P./ nr. La Aguada/ 3000m, 13.v.[19]81/ L. Masner"; "HOLOTYPE/ Clinocera/ meridana/ Sinclair [red label]" (CNC).

Paratypes: **VENEZUELA.** Same data as holotype (6 ♀♀, CNC); La Aguada, 3000 m, 13.v.1981, LM (6 ♀♀, CNC).

Recognition. This species is distinguished by blue pruinescence on the margins of the lower half of the face, spotted wings, auxiliary crossvein lacking, and the entire pleura with blue pruinescence.

Description (wing length 4–4.5 mm).

Male. Face flat, lower half with blue pruinescence on lateral margin, divided medially by brown stripe; upper part of face brown; gena, postgena, and occiput with pale-blue pruinescence; vertex brown. Arista-like stylus short and slender. Upper postocular setae dark.

Thorax brown; scutum with two narrow black vittae between dc. Pleura with blue pruinescence; postnotum brown. Several pale acrostichals anterior to first dc; 5 long dc; 1 pprn; 1 presut spal; 2 npl, lower slender and short; 1 psut spal; 1

pal; 2 sctl. Laterotergite with patch of long pale setae. Posterior margin of anepisternum with setulae.

Wing (Fig. 33*a*) infuscate with clouding about the following crossveins: tip of cells bm and dm, base of R_4, and margin of r-m vein; short costal spines present; stigma absent; 1 short and 1 long basal costal setae. Auxiliary crossvein between R_4 and R_{2+3} absent; stump vein absent; R_4 bell-shaped; anal vein dark streak; cell dm slightly produced apically. Halter dark.

Coxae with pale-blue pruinescence, concolorous with pleura; remaining segments dark brown; femora with yellowish brown apicoventral tips. Forecoxa with short pale setae. Forefemur with posteroventral row of slender setae, becoming darker apically; dense short pile beneath; preapical anterior comb present. Foretibia with biserial row of stout setae beneath. Hind tibia with erect ventral setae on apical half. Fifth tarsomere lacking dorsoapical extension; empodium brown laterally.

Sternites and thin lateral margin of abdomen with blue pruinescence, dorsum dark. Hypandrium tapered, subtriangular. Phallus arched near base, apex split; distiphallus relatively short, flattened, blade-like; apex with fine microtrichia. Clasping cercus finger-like, straight, parallel-sided, apex bluntly rounded; spines on inner margin. Surstylus triangular, apex truncate, hardly extending beyond epandrium; outer margin with long setae (Fig. 26*b*).

Female. Similar to male except lacking ventral setae of forelegs.

Geographic distribution and seasonal occurrence. This species is recorded only from Mérida Province in Venezuela (Map 30).

Etymology. The specific name is derived from the type locality of this species.

Clinocera miradorana sp. nov.

Figs. 26*c*, 33*b*; Map 31

Type material. HOLOTYPE, male labelled "CHILE: JuanFernandez Is./ Robinson Crusoe Is./ 1-4.i.1993; pans in wet/ moss N side Mirador/ S.A. Marshall"; "HOLOTYPE/ Clinocera/ miradorana/ Sinclair [red label]" (MNHNS).

Paratypes: **Chile. Juan Fernandez Islands:** Same data as holotype, 1–10.i.1993 (1 ♂, 2 ♀♀, DEBU); Robinson Crusoe Is., quebrada S side Mirador, fern for., 1–10.i.1993, SAM (1 ♂, CNC); Más a Tierra N side, Yunque, 1–9.ii.1973, LEP (1 ♀, CNC).

Recognition. This species is distinguished by two pairs of black vittae and a short median stripe of blue pruinescence on the scutum, infuscate wing with clouding about crossveins, and auxiliary crossvein present.

Description (wing length 2.7–3.2 mm).

Male. Face flat, lower third of face with pale-blue pruinescence; remainder of head dark brown. Arista-like stylus long and slender. Postocellar setae dark.

Thorax brown; scutum with pair of narrow black vittae between dc, extending to prescutellar depression; supra-alar with black stripe from presutural supra-alar to dc 5; short blue pruinescent strip between vittae from anterior margin to just beyond dc 1. Pleura mostly brown to pale brown, lacking pruinescence; upper katepisternum pale brown. Thoracic setae very long; several pale acrostichals anterior to first dc; 5 long dc; 1 pprn; 1 presut spal; 2 npl, lower slender and short; 1 psut spal; 1 pal; 2 sctl. Laterotergite with patch of short dark setae. Posterior margin of anepisternum with setulae

Wing (Fig. 33*b*) darkly infuscate with clouding about tip of cells bm and dm, base of R_4, and margin of r-m vein; costal spines present; stigma absent; 1 long basal costal seta. Veins straight; auxiliary crossvein between R_4 and R_{2+3} present; stump vein absent; R_4 bell-shaped; anal vein dark streak; cell dm slightly produced apically. Halter dark.

Coxae and femora light brown, concolorous with pleura; remaining segments dark brown. Forecoxa with short dark setae. Forefemur with posteroventral row of slender setae on basal two-thirds, decreasing in length apically; anteroventral margin with short row of setae indistinct; length of posteroventral basal setae more than half width of femur; short dense pile beneath; preapical anterior comb present. Foretibia with biserial row of stout setae beneath. Hind tibia with erect ventral setae on apical half. Fifth tarsomere lacking dorsoapical extension; empodium pale laterally.

Abdomen dark brown. Hypandrium subtriangular. Phallus straight, apex split; distiphallus relatively long, dorsoventrally flattened, narrow. Clasping cercus long, straight, finger-like; tapered to rounded apex; short spines on inner dorsal margin and inner apex. Surstylus projecting well free from epandrium; arched, with broadly rounded apex; outer margin with stout setae (Fig. 26*c*).

Female. Similar to male except as follows: blue pruinescence on lower face divided medially by broad brown stripe; posteroventral setae of forefemur shorter; ventral setae of foretibia reduced.

Geographic distribution and seasonal occurrence. This species is recorded only from the Juan Fernandez Islands (Map 31).

Etymology. The specific name is in reference to the type locality of this species.

Clinocera naomina sp. nov.

Figs. 26*d*, 33*c*; Map 30

Type material. HOLOTYPE, male labelled “ECU[ador]: 2,Imbabura (IM)/

Cayambe, Olmedo, Quellen/ Planada San Pietro, 3450m/ 30.x.1992, leg. GER"; "HOLOTYPE/ Clinocera/ naomina/ Sinclair [red label]" (ZSM).

Paratypes: **COLOMBIA.** Caldes, Mt Ruiz, 3000 m, 21.viii.1969, D.H. Messer-Smith (2 ♂♂, 4 ♀♀, USNM). **EQUADOR.** Same data as holotype (1 ♂, 2 ♀♀, ZSM).

Recognition. This species is distinguished by blue pruinescence on the lower two-thirds of the face, wings with distinct clouding, auxiliary crossvein present, and lower half of pleura and region near the base of the wing with blue pruinescence.

Description (wing length 2.8–3.2 mm).

Male. Face convex, lower two-thirds with blue pruinescence; upper face brown; gena, postgena, and occiput with pale-blue pruinescence; vertex brown. Arista-like stylus short and slender. Postocellar setae dark.

Thorax brown; scutum with two narrow black vittae between dc. Lower half of pleura and region near base of wing extending to notopleuron with blue pruinescence. Several pale acrostichals anterior to first dc; 5 long dc; 1 pprn; 1 presut spal; 2 npl, lower slender and short; 1 psut spal; 1 pal; 2 sctl. Laterotergite with patch of setae. Posterior margin of anepisternum with setulae.

Wing (Fig. 33*c*) infuscate with distinct clouding about tip of cells bm and dm, base of R_4 and margin of r-m vein, and basal third of M_1; costal spines present; stigma absent; 1 basal costal seta. R_{2+3} slightly wavy; auxiliary crossvein between R_4 and R_{2+3} present; stump vein absent; R_4 bell-shaped; anal vein dark streak; cell dm produced apically. Halter dark.

Coxae with pale-blue pruinescence, concolorous with pleura; forefemur with blue pruinescence, except for reddish brown apex; mid and hind femora reddish brown; remaining segments dark. Forecoxa with short dark setae. Forefemur with posteroventral row of slender setae; dense pile beneath lacking; preapical anterior comb reduced to 4 setulae. Foretibia with biserial row of stout setae beneath. Hind tibia with erect ventral setae on apical half. Fifth tarsomere with dorsoapical extension; empodium light brown laterally.

Ventral margin of abdomen with blue pruinescence, dorsum dark. Hypandrium tapered, subtriangular. Phallus nearly straight, apex split; distiphallus relatively long, dorsoventrally flattened, blade-like; apex with fine microtrichia. Clasping cercus slender, curved apically and strongly tapered; apex bluntly rounded; spines on inner margin. Surstylus round apicodorsally, projected free from epandrium; apex rounded; outer margin with long setae (Fig. 26*d*).

Female. Similar to male except ventral setae of forelegs reduced, especially on tibia.

Geographic distribution and seasonal occurrence. *Clinocera naomina* is currently known only from above 3000 m in Ecuador and Colombia (Map 30).

Etymology. The specific name *naomina* is proposed in appreciation of the author's youngest daughter and for her cheerful companionship on "bug hikes".

Clinocera notialis sp. nov.

Fig. 27*a*; Map 31

Type material. HOLOTYPE, male labelled "CHILE: JuanFernandez Islands/ Robinson Crusoe Island/ pans in mossy seep; N side/ Mirador; 1-10.i.1993/ S.A. Marshall"; "HOLOTYPE/ Clinocera/ notialis/ Sinclair [red label]" (MNHNS).

Possible additional material. ARGENTINA. Tierra del Fuego, beside Lago Fagnano, 13.ii.1992, SAM (1 ♀, DEBU). There is some doubt whether this species is conspecific, especially given the ocean gaps separating the localities. Consequently, without an associated male, I am reluctant to include it as a paratype.

Recognition. This species is distinguished by the bright-blue pruinescence on the lower two-thirds of the face, lower half of pleura with blue pruinescence, infuscate wing with spots about crossveins, and auxiliary crossvein present.

Description (wing length 4.3 mm).

Male. Face convex, lower two-thirds of face with bright-blue pruinescence; gena, postgena, and occiput with blue pruinescence; remaining head dark brown. Antenna long and slender. Upper postocular setae dark. Thorax brown; scutum with pair of narrow black vittae between dc, extending to prescutellar depression; supra-alar with black stripe from presutural supra-alar to scutellum. Lower half of pleura with blue pruinescence. Several pale acrostichals anterior to first dc; 5 long dc; 1 very short, fine pprn; 1 short presut spal; 2 npl, lower slender and short; 1 psut spal; 1 pal; 2 sctl. Laterotergite with patch of long pale setae. Posterior margin of anepisternum with setulae.

Wing darkly infuscate with spots about tip of cells bm and dm, base of R_4, and margin of r-m vein; costal spines present; stigma absent; 1 basal costal seta. Veins straight; auxiliary crossvein between R_4 and R_{2+3} present; stump vein absent; R_4 straight, parallel to R_5, divergent at wing margin; anal vein short dark streak; cell dm slightly produced apically. Halter dark, base of stalk yellowish brown.

Coxae with blue pruinescence, concolorous with pleura; apex of femora and ventrobasal portion of tibiae yellowish brown; remaining segments dark brown. Forecoxa with short pale setae. Forefemur with antero- and postero-ventral row of short setae; anteroventral setae very short, most conspicuous at base; posteroventral setae longer than width of femur, decreasing in length apically; short dense pile beneath; preapical anterior comb present. Foretibia with biserial row of stout setae beneath. Hind tibia with erect ventral setae on apical half. Fifth tarsomere lacking dorsoapical extension; empodium pale laterally.

Abdominal sternites and lateral margin of tergites with faint blue pruinescence, remaining tergites brown. Hypandrium subtriangular. Phallus short, straight, apex split; distiphallus relatively long, dorsoventrally flattened, narrow. Clasping cercus very long, straight, finger-like; tapered on one side apically, spines on inner and inner dorsal margins. Surstylus projecting beyond epandrium;

strongly curved anteriorly, with truncate apex; long setae on apical margin and arising from lateral ridge (Fig. 27*a*).

Female. Unknown.

Geographic distribution and seasonal occurrence. This species is recorded with confidence only from the Juan Fernandez Islands (Map 31).

Clinocera oriunda (Collin)

Figs. 27*b*, 33*d*; Map 30

Hydrodromia oriunda Collin, 1933: 323.

Type material examined. LECTOTYPE (here designated) male labelled "Bariloche./ 6-7.xi.1926"; "Argentina/ Terr.Rio Negro./ F.& M. Edwards./ B.M. 1927-63 [terminalia dissected and stored in glycerin beneath specimen]" (BMNH). My lectotype label "LECTOTYPE/ of Hydrodromia/ oriunda Collin/ des. B.J. Sinclair 1994 [red label]" has been attached to this specimen.

The lectotype is designated in order to clarify the application of the name to this taxon, since two different species are found in the original syntype series.

Paralectotypes: **ARGENTINA.** Same data as lectotype (3 ♀♀, BMNH) [conspecific with lectotype]; Bariloche, 25–28.x.1926, FME (1 ♀, BMNH) [conspecific with lectotype]; Bariloche, 28.xi.–1.xii.1926, FME (1 ♂, BMNH) [= *C. collini* sp. nov.]; Bariloche, Rio Negro, xi. 1926, RES (1 ♀, USNM) [conspecific with lectotype]; Bariloche, Rio Negro, ix.1926, RES (1 ♀, USNM) [= *C. collini* sp. nov.]; Correntoso, Rio Negro, ix.1926, RES (1 ♀, USNM) [conspecific with lectotype]; Lake Nahuel Huapi, Puerto Blest, 2–3.xii.1926, FME (1 ♀, BMNH) [conspecific with lectotype]; Lake Gutierrez, 3–14.xi.1926, FME (1 ♀, BMNH) [= *C. collini* sp. nov.]. **CHILE.** Ancud, Chiloe Is., 17–19.xii.1926, FME (1 ♀, BMNH) [= *C. collini* sp. nov.]; Llanquihue, Casa Pangue, xii.1926, RES (1 ♂, USNM) [conspecific with lectotype]; Marga-Marga, ix.1929, F. Jaffuel (1 ♀, BMNH) [= *C. collini* sp. nov.]. My paralectotype labels have additionally been attached to these specimens.

Additional material examined. ARGENTINA. Arroyo Casa de Piedra, 15 km W Bariloche, Rio Negro, 17.xi.2004, TS (1 ♀, BLKU); Arroyo Lopéz, 20 km W Bariloche, Rio Negro, 18.xi.2004, M. Sugimoto (2 ♂♂, BLKU); Arroyo Torrontegui, Gutiérrez, Bariloche, Rio Negro, 18.xi.2004, TS (3 ♂♂, 2 ♀♀, BLKU); Isla de los Estados, Puerto San Juan, 12–15.v.1971, OSF, G.F. Hevel (1 ♀, USNM); Neuquen: Pucara, 8, 15.i.1972, J.P. Duret (3 ♂♂, 2 ♀♀, MNHN). **BOLIVIA.** Alto de La Paz, 4500 m, La Paz, 29.x.1968, LEP (2 ♀♀, CNC). **CHILE.** Aysen: Puerto Cisnes, 77°40′W, 44°45′S, 16–28.ii.1961, LEP (1 ♂, CNC). Cautin: Volcan Villarrica, 1120 m, 15–29.xii.1982, native for., A.F. Newton, M.K. Thayer (2 ♂♂, 1 ♀, USNM). Llanquihue: Rio Los Cuarteles, 15 m, nr Ralún, at light, 41°23′S, 72°17′W, 30.i.1998, NEW (1 ♂, USNM). Punta Arenas, 2.ii.1908, coll. Schnuse (MTD). Tierra del Fuego, Ruephen, 30 km SE

Cameron, 20–30.xi.1960, LEP (1 ♂, CNC). **EQUADOR.** Pichincha, 20.5 km E Pifo, 3700 m, 26–28.ix.1990, OSF (1 ♂, USNM).

Recognition. This species is distinguished by the bright pruinescence on the lower half of the face, usually not divided medially by a brown stripe, spotted wings, and lack of an auxiliary crossvein.

Description (wing length 3.5–4.7 mm).

Male. Face convex, lower half with shiny white pruinescence; upper part of face brown; gena, postgena, and occiput with blue pruinescence. Arista-like stylus long and slender. Upper postocular setae dark.

Thorax brown; scutum with pair of dark vittae between dc, extending to prescutellar depression. Lower half of pleura with pale-blue pruinescence, distinctly separated from dull upper half. Several pale acrostichals anterior to first dc; 5 long dc; 1 pprn; 1 presut spal; 2 npl, lower slender and short; 1 psut spal; 1 pal; 2 sctl. Laterotergite with patch of pale setae. Posterior margin of anepisternum with long setulae.

Wing (Fig. 33*d*) hyaline with clouding about the following crossveins; tips of cells bm and dm, base of $R_{4,}$ and margin of r-m vein. Stigma not visible absent; 1 long basal costal seta; costal spines present. Auxiliary crossvein between R_4 and R_{2+3} absent; stump vein absent; R_4 bell-shaped; anal vein an elongate streak; cell dm nearly truncate apically. Halter dark.

Coxae with pale-blue pruinescence, concolorous with pleura; femora with reddish brown apex and ventral surface; remaining segments dark. Forecoxa with short pale setae. Forefemur with posterodorsal margin with pale-blue pruinescence; antero- and postero-ventral margins with row of slender setae, becoming longer and paler toward base, length more than half width of femur; dense short pile beneath; preapical anterior comb present. Foretibia with biserial row of stout setae beneath. Hind tibia with erect ventral setae on apical third. Fifth tarsomere with short dorsoapical extension; empodium white laterally.

Lateral and ventral surfaces of abdomen with blue pruinescence, dorsum dark. Hypandrium subtriangular. Phallus strongly arched; distiphallus relatively long, dorsoventrally flattened, blade-like; apex with fine microtrichia. Clasping cercus finger-like; blunt spines on inner margin. Surstylus broadly triangular, outer margin with stout setae (Fig. 27*b*).

Female. Similar to male, including chaetotaxy.

Geographic distribution and seasonal occurrence. *Clinocera oriunda* is the most common and widespread species in South America, ranging from the southern tip of the continent to Ecuador (Map 30). Collection dates indicate that this species may be present at most times of the year.

Remarks. Unfortunately the syntype series comprises two species. The circular type labels with a red border were not placed on the specimens by Collin (see Pont 1995) and consequently do not represent holotype designation. I have chosen to attach my lectotype label to the species distinguished from *C. stagnalis* in Collin's diagnosis, which also corresponds to the type locality as selected by Collin (1933) and identified in his text with an asterisk (*). The species is distinguished by having the lower half of the face with the white/blue pruinescence undivided and a yellowish brown apex of the forefemur. The second species in the syntype series (see *C. collini* sp. nov.) is distinguished by the bluish pruinescence of the face divided medially by a brown stripe and uniformly dark yellowish brown legs.

Clinocera parana sp. nov.

Fig. 27*c*; Map 30

Type material. HOLOTYPE, male labelled "BRAZIL: Parana/ Serra-do-Mar near/ Morretes, 29.viii.2000/ 25°21.80′S 48°52.55′W/ J.M. Cumming/ ex. Mae Cotira river"; "HOLOTYPE/ Clinocera/ parana/ Sinclair [red label]" (MZUSP).

Recognition. This species is distinguished by blue pruinescence on the lateral margins of the lower half of the face, lack of an auxiliary crossvein, unspotted wings, reduced thoracic chaetotaxy, and subtriangular clasping cercus.

Description (wing length 3 mm).

Male. Lower third of face with blue pruinescence; upper part of face brown. Arista-like stylus short and slender. Ocellar bristles short and weak. Upper postocular setae dark.

Thorax brown; scutum with pair of dark vittae between dc, extending to prescutellar depression. Pleura with blue pruinescence. Chaetotaxy weak and reduced: acrostichals absent; 5 dc, short and slender; 1 pprn, reduced and slender; 1 presut spal; 2 npl, lower slender and short; 1 psut spal; 1 pal; 2 sctl. Laterotergite with patch of pale setae. Posterior margin of anepisternum with setulae.

Wing narrow and infuscate, with clouding or spots absent; stigma absent; 1 slender basal costal seta; costal spines present. Auxiliary crossvein between R_4 and R_{2+3} absent; stump vein absent; R_4 straight to wing margin; anal vein distinct basally; cell dm produced apically. Halter knob dark, base of stalk paler than knob.

Coxae with pale-blue pruinescence, concolorous with pleura; femora pale brown beneath, remaining segments light brown; apex of coxae yellowish brown. Forecoxa with short pale setae. Forefemur with ventral setae greatly; dense short pile beneath nearly inconspicuous; preapical anterior comb present. Foretibia with biserial row of stout setae beneath. Hind tibia with erect dorsal and ventral setae on apical half. Fifth tarsomere lacking dorsoapical extension; empodium brown laterally.

Abdominal sternites and lateral margin of tergites with blue pruinescence, remaining tergites brown. Hypandrium very long and narrow, subrectangular, tapered toward apex. Phallus long, strongly arched near midlength, apex dilated with wide membrane; distiphallus short, not whip-like, dorsoventrally flattened, arched with wavy dorsoapical margin. Clasping cercus subtriangular, finger-like, tapered to narrow apex; stout setae on inner margin. Surstylus broadly rounded, projected anteriorly; clothed with long setae (Fig. 27*c*).

Female. Similar to male, except ventral setae of forefemur and foretibia not modified.

Geographic distribution and seasonal occurrence. This species is also recorded only from the state of Paraná (Map 30) in the Brazilian Atlantic Forest biogeographic province (see Morrone 2005). The species was swept from large boulders in a sunny, exposed river (Fig. 3*b*).

Remarks. This species is distinguished from *C. cummingi* only on the basis of the male terminalia (see the key to Neotropical species).

Etymology. The specific name is in reference to the type locality of this species.

Clinocera penai sp. nov.

Fig. 27*d*; Map 31

Type material. HOLOTYPE, male labelled "CHILE: Chiloe Is./ Ahoni Alto/ April 1988/ L.E. Pena, MT/ primary forest"; "HOLOTYPE/ Clinocera/ penai/ Sinclair [red label]" (CNC).

Recognition. This species possesses very slender wings, an auxiliary crossvein, pleura and lower half of the face with blue pruinescence, and highly modified male terminalia.

Description (wing length 3.2 mm).

Male. Condition of holotype poor, all setae of head and scutum lost. Face flat, lower half with blue pruinescence; upper face and vertex brown; occiput and postgena with faint blue pruinescence. Arista-like stylus short and slender.

Scutum with median pair of black vittae extending to prescutellar depression; faint black supra-alar stripes extending from presutural supra-alar to scutellum; anterior margin with dusting of blue pruinescence. Pleura and postnotum with blue pruinescence. Setae lost. Laterotergite with patch of long pale setae. Posterior margin of anepisternum with setulae.

Wing greatly narrowed, infuscate; distinct clouding at base of R_4, r-m vein, tip of cells dm and bm; stigma absent; basal costal seta lost; costal spines present. Veins straight; auxiliary crossvein between R_4 and R_{2+3}; stump vein absent; R_4 divergent from R_5; anal vein an elongate streak; cell dm truncate apically. Halter brown.

Coxae with blue pruinescence; ventral surface of femora yellowish brown, remaining segments brown. Forecoxa with short pale setae. Forefemur with anterior preapical comb present; ventral surface with short white pile; posteroventral row of short fine setae, with 4 or 5 stout setae on apical half. Foretibia with biserial row of setae beneath. Hind tibia with erect dorsal and ventral setae on apical half. Fifth tarsomere lacking dorsoapical extension; empodium white laterally.

Abdominal sternites and lateral margin of tergites with blue pruinescence, remaining tergites brown. Hypandrium broad, rectangular, and much larger than epandrium. Phallus slightly arched with spit tip; distiphallus as short as expanded apex of phallus, reduced, not whip-like. Clasping cercus narrow, tapered; stout setae on inner dorsal margin; basal tubercle with crown of short blunt setae. Surstylus longer than clasping cercus, with short tip; apex with several short setae (Fig. 27*d*).

Female. Unknown.

Geographic distribution and seasonal occurrence. This species is known only from its type locality (Map 31) and its occurrence in April.

Remarks. The male terminalia are very unusual and unlike those of any known New World species.

Etymology. The specific name is a patronym in honour of L.E. Peña, who collected the holotype and a great many fine and valuable species of Chilean Empididae.

Phylogenetic analysis

A total of 14 morphological characters have been compiled for the analysis of the species-groups of *Clinocera* (Table 2). The matrix (Table 3) was analysed as described in the Materials and methods (branch-and-bound search), which resulted in one phylogenetic tree (Fig. 28) with length 25 steps (CI = 0.76; RI = 0.60; RC = 0.60). Two exemplars each were scored for the *fuscipennis*, *conjuncta*, *lineata*, *nigra*, and *stagnalis* groups (Fig. 28).

Results must be viewed as tentative because of the low number of characters found to be informative and the limited data set. The *C. tripunctata* group was placed as sister to the remaining *Clinocera* species-groups, but this is questionable, based on the available information on the distribution and age of the genus (see Biogeography section below). The phylogenetic position of the *C. tripunctata* group is due to their apparently derived male genitalia, especially the absence of the distiphallus. The remaining species-groups of *Clinocera* form a monophyletic clade defined by the presence of a whip-like distiphallus. The *C. fuscipennis* group (*fuscipennis* + *gubernans* in Fig. 28) is assigned as sister to the remaining *Clinocera* (exclusive of the *C. tripunctata* group). The monophyly of the *C. nigra* group is not supported and further analysis should be undertaken. The next phase of this study should focus on analyses of the species relationships within each species-group on a world basis. Resolution of these relationships will likely aid an analysis of the complex biogeographic history of the species in the New World, where there have presumably been multiple invasions eastward and westward through Beringia.

Biogeography

Geographical patterns of Nearctic species of Clinocera

Presented below are descriptions of a number of patterns or distributional elements exhibited by Nearctic species of *Clinocera*. These general patterns are similar to those observed and discussed in the family-level treatments in the monograph *Insects of the Yukon* (Danks and Downes 1997) and in Kavanaugh's (1988) treatment of insect biogeography in the Pacific Northwest. When the distributions plotted in this work are compared with the distributions described in

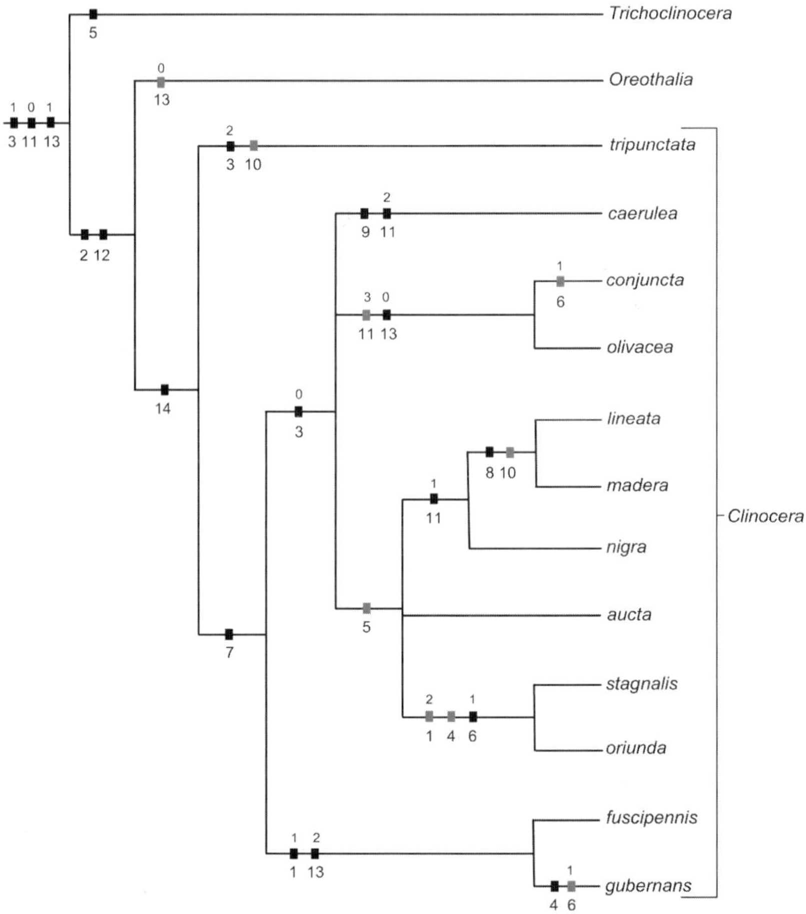

Fig. 28. Cladogram of the relationships of the *Clinocera* [*tripunctata*, *caerulea*, *conjuncta* (*conjuncta*, *olivacea*), *fuscipennis* (*fuscipennis*, *gubernans*), *lineata* (*lineata*, *madera*), *nigra* (*nigra*, *aucta*), *stagnalis* (*stagnalis*, *oriunda*)] species-groups. Numbers below the line refer to character numbers listed and defined in Table 2. Numbers above the line refer to states of multistate characters. Solid rectangles are non-homoplastic apomorphies and shaded rectangles are homoplastic characters.

those two works, they coincide very closely. Not all species could be assigned confidently to a particular pattern because of collecting gaps; nevertheless the distributional categories for the species are listed in Table 8 and discussed below.

1. Holarctic elements

This category includes species that are widespread in both the Nearctic and the Palearctic regions. The species here have a circumpolar or circumboreal distribution (e.g., *C. aucta, C. nivalis*) and some are distributed much farther south in the Palearctic Region (e.g., *C. stagnalis*). The Nearctic range of *C. aucta* includes Alaska and northern Canada, with southern glacial relict disjuncts (Map 25) (see below); the Palearctic range includes Finland, Sweden, western Russia, and Siberia. *Clinocera stagnalis* is confined to boreal, alpine, and arctic habitats in North America (Map 28), but is very widespread in the Palearctic Region, especially Europe, where it is an opportunistic species adapted to a range of water qualities and temperatures.

2. Nearctic elements

This category includes several species that are widely distributed or transcontinental in North America. This category includes *C. binotata* (Map 26) and *C. lineata* (Map 23). Excluded from this element are Boreal Holarctic and Nearctic disjunct species (see below).

3. Nearctic disjunct elements

Most Nearctic elements are widely disjunct, and divided into two broad populations primarily confined to the eastern states and provinces and the western Cordillera (Maps 8, 16). This is in marked contrast to the clinocerine genus *Trichoclinocera*: only *T. dasycoxa* Sinclair is widely disjunct (all other species of *Trichoclinocera* are confined to either the eastern or the western region of North America (Sinclair 1994)). Disjunct ranges are known in other dipteran families, e.g., Culicidae, Muscidae, and Simuliidae (Currie 1997; Danks 1981), and in many other insects. Ricker (1964) considered similar disjunct ranges in the Plecoptera to be examples of postglacial or interglacial dispersal across the plains followed by extinction in the middle regions as these dried up and habitats disappeared.

4. Appalachian (= Carolinian or Alleghanian) elements

Only three species of *Clinocera* are endemic to eastern North America, and all are confined to the Appalachian region or the eastern deciduous-forest areas (Maps 22, 25). Other species in this region are more widespread, either Nearctic or Nearctic disjunct elements.

Table 8. List of Nearctic species of *Clinocera* and their distributional patterns.

	Distributional pattern*
Clinocera amarilla	C
C. appalachia	A
C. arizonica	SW
C. arnaudi	C (Rocky Mountains only)
C. aucta	H (Boreal)
C. binotata	N
C. brevicauda	SW
C. borkenti	C
C. brunnea	ND
C. bulbosa	C (Rocky Mountains only)
C. caerulea	C
C. coloradensis	C (Rocky Mountains only)
C. conjuncta	ND
C. cordillerana	C (Rocky Mountains only)
C. daniellae	P
C. disjuncta	ND (Boreal)
C. femorata	P
C. fuscipennis	ND
C. gubernans	C
C. guttipennis	P
C. lecta	C
C. lineata	N
C. longicauda	P
C. maculata	A
C. macdonaldi	SW
C. madera	SW
C. melanderi	P
C. mexicana	SW
C. nivalis	H (Boreal)
C. olivacea	C
C. pacifica	C
C. prasinata	C
C. rainiericola	P
C. robinsoni	SW
C. sabroskyi	C (Rocky Mountains only)
C. schlinkerti	C
C. setitibia	C
C. stagnalis	H (Boreal)
C. stigma	C
C. sublineata	A
C. substagnalis	C
C. subtrunca	C
C. trunca	P
C. wallowa	P

*A, Appalachian; C, Cordilleran; H, Holarctic; N, Nearctic; ND, Nearctic disjunct; P, Pacific coast-centred; SW, southwestern Mexico.

5. Cordilleran elements

Most Nearctic species are broadly distributed in the western Cordillera, including the Rocky Mountain and Coastal regions, with 18 endemic species. As discussed by Kavanaugh (1988), some species extend widely along the coast, while others are restricted to northern or southern parts. Several particular patterns are observed. Firstly, the range extends along the Coastal and Rocky Mountain regions, and is connected across southern British Columbia south to central Oregon (e.g., Map 10, *C. olivacea*). Secondly, the range extends much farther south along the Rocky Mountains than in the coastal area (e.g., Map 6, *C. borkenti*). In the third pattern, the species reach the coast only in the north in Alaska (e.g., Map 29, *C. substagnalis*).

Often there is a gap in central British Columbia between southern and northern populations (e.g., Maps 17, 18, 20, 29, *C. gubernans*, *C. lecta*, *C. stigma*, *C. substagnalis*). This gap could represent a collecting artefact but is more likely a vicariant pattern between Beringian and "southern" populations. There is no morphological divergence between populations, which suggests a Late Wisconsinan vicariant event (Courtney 1994). Among stream-dwelling aquatic insects a similar pattern is known in the mountain midges (Deuterophlebiidae) (Courtney 1990, 1994), Plecoptera (Ricker 1964, figs. 8–10), and Trichoptera (Ross 1965, fig. 5). Each of these patterns has been considered a collecting artefact, but in the case of mountain midges, no further records have been discovered during additional surveys into this region (G.W. Courtney, pers. comm., 2006). The distribution of several of the species of *Clinocera* listed above may also be due to the same vicariant event. This distribution has been termed a bicentric pattern, with populations of these species surviving the last glacial episode in a northern Beringian refugium and southern coastal and interior refugia (see Ball and Currie 1997; Currie 1997; Kavanaugh 1988). The Beringian population has subsequently expanded eastward and the southern population has subsequently dispersed northward, following the retreat of the Laurentide and Cordilleran ice sheets, but they have not yet completely rejoined. Further collections of clinocerines in the gap region are encouraged in order to investigate this biogeographic pattern. Several additional examples of northern and southern refugia are listed in Tangelder (1988), where specimens from these two refugia differ in colour pattern.

6. Pacific coast-centred elements

Pacific coast-centred elements are confined to the coastal mountains and also extend eastward into the continental interior via the Columbia River drainage and Fraser River valley (Kavanaugh 1988). Eight species exhibit this pattern (Table 8), including *C. daniellae*, *C. femorata*, *C. trunca*, and perhaps *C. longicauda* (Maps 1, 3, 13, 15). *Clinocera trunca* (Map 13, ●) is classified as a Pacific coast-centred element and its sister species, *C. subtrunca* (Map 13, ■), is a Cordilleran element and their ranges overlap in Washington State.

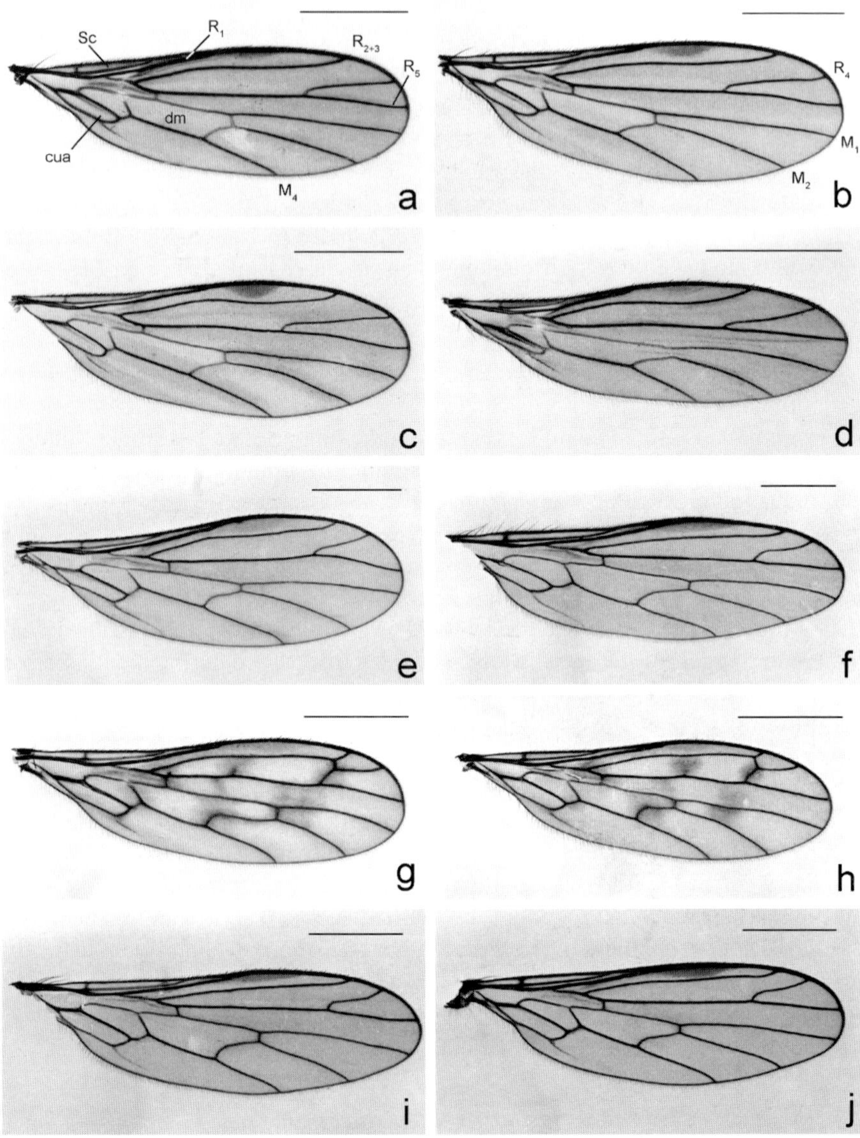

Fig. 29. Wings of *Clinocera*: *a*, *C. brunnea*; *b*, *C. caerulea*; *c*, *C. daniellae*; *d*, *C. longicauda*; *e*, *C. amarilla*; *f*, *C. borkenti*; *g*, *C. conjuncta* (Alaska); *h*, *C. conjuncta* (North Carolina); *i*, *C. cordillerana*; *j*, *C. olivacea*. Scale bars = 1 mm.

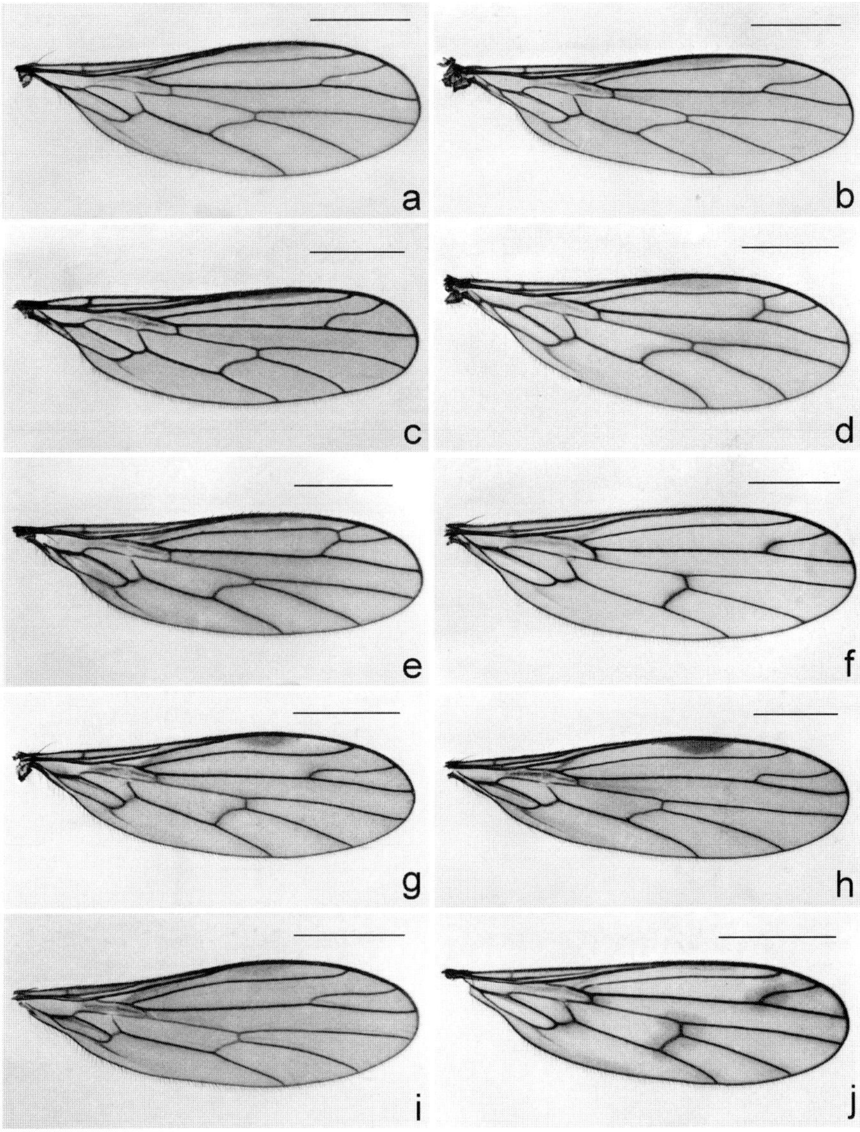

Fig. 30. Wings of *Clinocera*: *a*, *C. pacifica*; *b*, *C. prasinata*; *c*, *C. setitibia*; *d*, *C. subtrunca*; *e*, *C. trunca*; *f*, *C. bulbosa*; *g*, *C. disjuncta*; *h*, *C. femorata*; *i*, *C. fuscipennis*; *j*, *C. gubernans*. Scale bars = 1 mm.

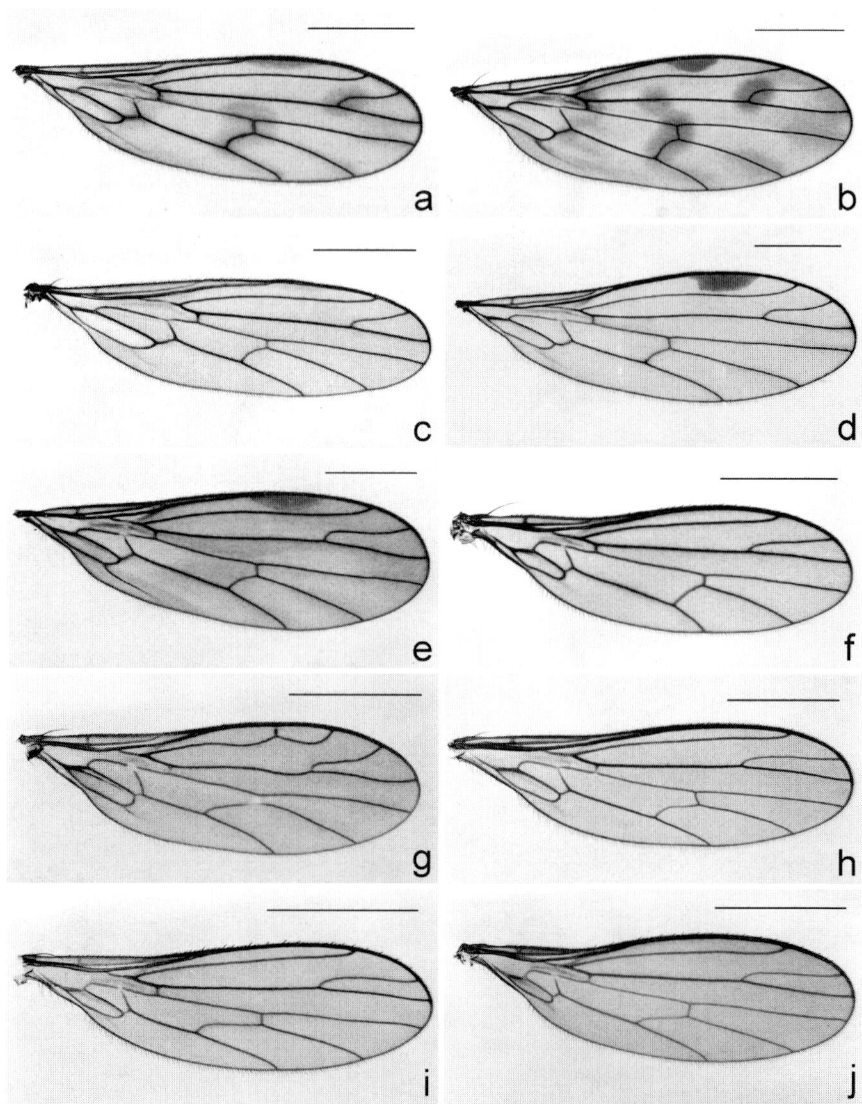

Fig. 31. Wings of *Clinocera*: *a*, *C. guttipennis*; *b*, *C. lecta*; *c*, *C. schlinkerti*; *d*, *C. stigma*; *e*, *C. wallowa*; *f*, *C. brevicauda*; *g*, *C. lineata*; *h*, *C. macdonaldi*; *i*, *C. madera*; *j*, *C. nigriscutata*. Scale bars = 1 mm.

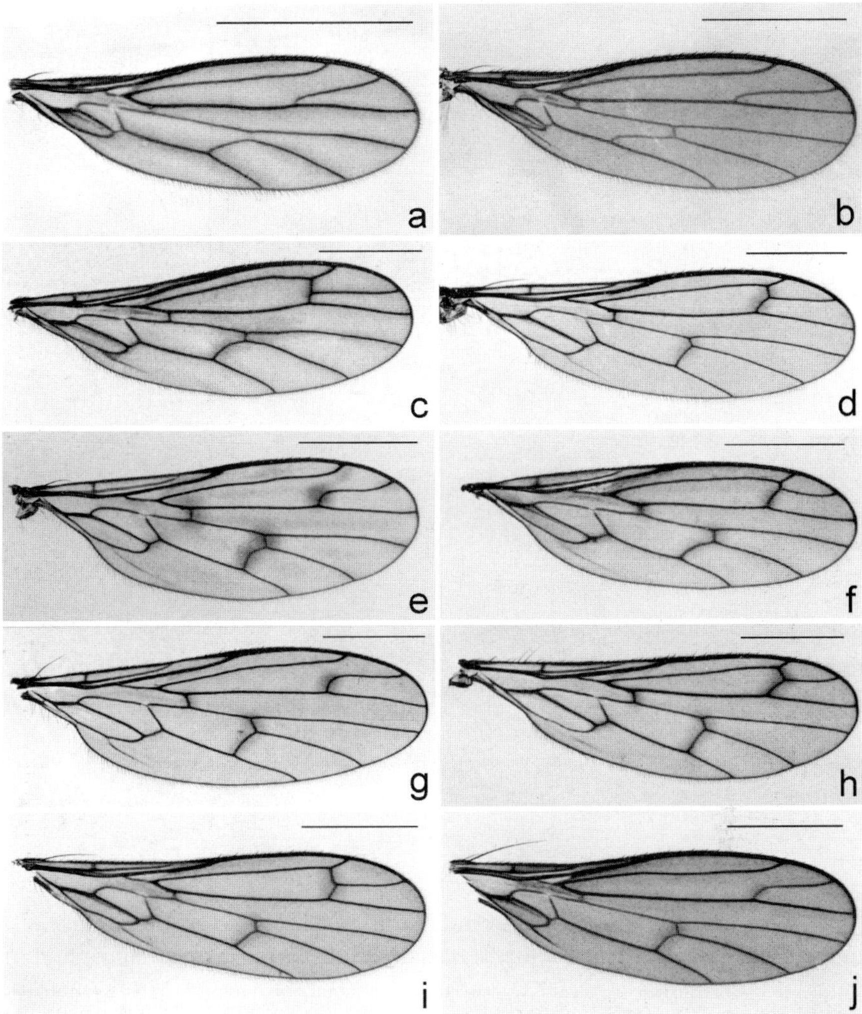

Fig. 32. Wings of *Clinocera*: *a*, *C. sublineata*; *b*, *C. woodi*; *c*, *C. aucta*; *d*, *C. binotata*; *e*, *C. maculata*; *f*, *C. nivalis*; *g*, *C. stagnalis*; *h*, *C. substagnalis*; *i*, *C. isabellina*; *j*, *C. marshalli*. Scale bars = 1 mm.

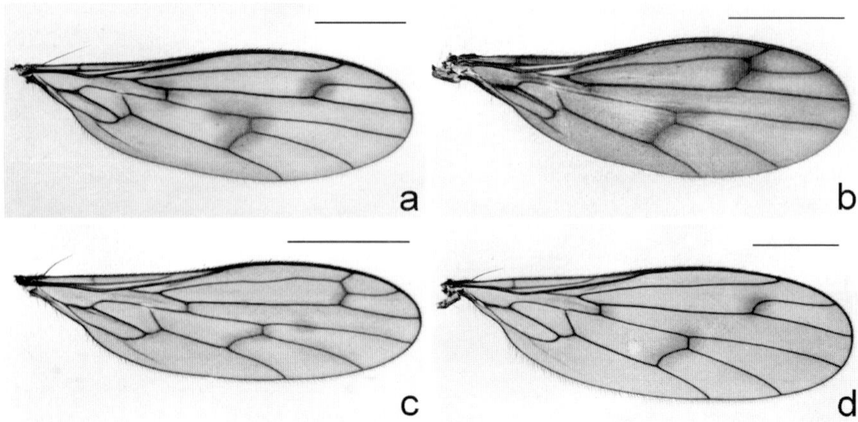

Fig. 33. Wings of *Clinocera*: *a*, *C. meridana*; *b*, *C. miradorana*; *c*, *C. naomina*; *d*, *C. oriunda*. Scale bars = 1 mm.

7. Southwestern–Mexican elements

There are six species in the *C. lineata* group (Maps 21, 23, 24), which range from Arizona and southern Utah south to Mexico north of the Isthmus of Tehuantepec. These species are confined to humid upland canyons and forested mountain ranges. The very rare *Clinocera* collection records in Mexico inhibit assignment of the above six species to the specific subregions defined by Morrone (2006).

8. Glacial relicts

Some species with otherwise northern distributions also occur as isolated southern populations referred to as glacial disjuncts, relicts, or remnants. For example, *C. stagnalis* is widespread across the Boreal zone of North America, with populations distributed southward down the Rocky Mountains. In the east, a glacial relict population has been identified on Isle Royale in Lake Superior (Map 28). This island and the north shore of Lake Superior are well known for the presence of arctic–alpine plants that have survived in isolated niches since the retreat of the continental glacier (Soper and Maycock 1963). They have persisted in favourable microclimatic niches that provided refuge areas when conditions changed, blocking migration routes northward. Isle Royale also harbours a relict population of the clinocerine *Wiedemannia simplex* (Loew) (Sinclair 1998).

The Mount Washington region of New Hampshire is also well known for a large number of isolated populations of otherwise northern or western plants and animals. Alexander (1940) summarized and reviewed the insect collection history of this region. Populations of two species of *Clinocera* (Maps 15, 25, *C. dis-*

juncta, *C. aucta*) found in the upper regions of Mount Washington can be classified as glacial relicts. It is believed that these cold-adapted species became isolated and separated by the northward retreat of the lowland population following the melting continental ice sheets (Danks 1981). Among aquatic insects, isolated remnants of northern species of Plecoptera occur in the White Mountains (Hitchcock 1969). An isolated population of *C. aucta* also occurs in Highlands National Park on Cape Breton Island (Nova Scotia).

Summary

In the Nearctic Region there are 3 species that are restricted to the east, 32 species that are restricted to the west, and 3 species that are Holarctic in distribution. Only *C. nivalis* can be viewed as confined to the arctic, where it is widespread beyond treeline, though more widespread in the Palearctic Region (see Collin 1961). *Clinocera aucta* should also be classified as an arctic-confined species in North America, but its distribution in eastern Canada requires further investigation, given the record from Nova Scotia. *Clinocera stagnalis* is a more southerly species and probably not a true arctic form — it usually reaches the tundra only at sites close to treeline or in alpine habitats, especially in the west (for additional insect examples see Danks 1981). No Beringian elements are known among species of *Clinocera*.

The greatest Nearctic species-level diversity of *Clinocera* (Table 9) occurs in the northwest (British Columbia, Washington, Oregon), with British Columbia (27 species) possessing the highest number of species found in North America. In addition to species-level diversity, the generic-level diversity of Empidoidea (exclusive of Dolichopodidae *s.str.*) in the northwest is high, with nine endemic genera.

In the east the greatest number of species occur in Quebec (10 species), likely because of the more northerly latitude of this province, where many Holarctic species occur. New Hampshire has the highest number of species (7) in the eastern United States because of a number of glacial relicts on Mount Washington. Ontario is probably the best surveyed province in Canada and the low diversity (5 species) is likely an accurate estimate.

Kavanaugh (1980) summarized the diversity of a montane genus of Carabidae and showed that the highest diversity was also centred in western Washington, northwestern Oregon, and southwestern British Columbia. This great diversity of species in the Western Cordillera compared with the Appalachian region is likely the result of the higher number of mountain ranges separated from each other by arid or warm lowland areas, coupled with greater dispersal or interchange between eastern Asia and western montane regions of North America (Ross 1956). Irwin and Kavanaugh (1981) suggested that the species in a group of cool- or cold-adapted beetles evolved following the cooling in the Late Eocene and diversified initially through vicariance created by

Table 9. Number of species of *Clinocera* per province (Canada) or state (USA).

State or province	No. of species
Eastern	
Canada	
New Brunswick	1
Newfoundland and Labrador	3
Northwest Territories	4
Nova Scotia	4
Ontario	5
Quebec	10
USA	
Alabama	1
Arkansas	1
Washington, D.C.	1
Georgia	5
Indiana	3
Kentucky	5
Maine	4
Maryland	1
Massachusetts	3
Michigan	4
Minnesota	2
New Hampshire	7
New York	6
North Carolina	6
Ohio	1
Pennsylvania	6
South Carolina	2
Tennessee	6
Vermont	1
Virginia	4
West Virginia	6
Wisconsin	3
Western	
Canada	
Alberta	17
British Columbia	27
Manitoba	2
Northwest Territories	4
Nunavut	2
Yukon	6
USA	
Alaska	14
Arizona	8
California	15
Colorado	15
Idaho	12
Kansas	1

Table 9 (*concluded*)

Montana	5
Nebraska	1
Nevada	8
New Mexico	6
Oregon	18
Utah	14
Washington	21
Wyoming	10

Miocene orogenic events. These authors considered that further speciation occurred during early glacial episodes. The great diversity of western *Clinocera* is likely due to similar climatic and topographic changes.

Geographical patterns of Neotropical species of Clinocera

Only two of six Nearctic species-groups of *Clinocera* range south into the Neotropical Region: the *C. lineata* and *C. stagnalis* groups. Only occasional species from the other species-groups are known south of the United States border. For example, there are only a few scattered records of *C. conjuncta* and *C. fuscipennis* (Maps 8, 16) from the Nearctic Region of northern Mexico.

1. Mesoamerican Highlands elements

Four species are known from the Mesoamerican Highlands (Map 33), all of which also occur in Costa Rica, the most extensively surveyed region in Central America. The species were collected from 1500–2800 m elevation. Three of these species are assigned to the *C. lineata* group, which is not known south of the Panamanian Isthmus (Maps 21, 24). The few known locality records for these species are too inadequate to resolve their distributions in this region, but all appear to be restricted to the highland regions of Mesoamerica.

2. South American elements

Only the *C. stagnalis* group is present in South America. The patterns of species distributions are very similar to the general patterns in Simuliidae (Coscarón and Coscarón-Arias 1995). This is not surprising because the two groups inhabit similar cool cascading streams, although simuliids also occur in warmer and slow-water habitats and consequently are much more extensively distributed across the continent. Within South America the following general elements or patterns for species of *Clinocera* are currently observed.

2.1. South American Transition zone (sensu *Morrone 2006*)

This zone extends through the highlands of the Andes between Venezuela, northern Chile, and west-central Argentina (Morrone 2006). There are currently four species of *Clinocera* confined to the northern region of this zone and three species located in the southern subregions (Maps 30, 31, 34).

2.2. Andean region

The Andean region encompasses the southern tip of South America, extending to central Chile and Argentina (Map 34), and is included in the Austral kingdom (Morrone 2006). Many species of insects are common to this region and the South American Transition zone (Morrone 2006), including *C. oriunda*, which is widespread from Ecuador to Tierra del Feugo. Many of these species are thought to have evolved in Patagonia and gradually spread northward during the Tertiary and Pleistocene (Morrone 2006). The Andean region also includes many groups with Austral or Gondwanan elements, where there are possible relationships with Australia, New Zealand, and southern Africa (for a review of Empidoidea see Sinclair 2003*a*). See below for possible Gondwanan patterns in *Clinocera*. The Andean region comprises several subregions, including the Subantarctic subregion.

2.2.1. Subantarctic subregion

This region comprises the austral Andes, including the islands of southern Chile and Argentina (Morrone 2006). There are three species restricted to the austral Andes and one widespread species (Maps 30–32).

Within this subregion, six provinces are recognized, including the Juan Fernandez Islands Province, where four species of *Clinocera* are currently considered endemic, possibly representing two colonization events. In other words, they represent two separate dispersals of ancestral species from, presumably, mainland Chile that reached this island group and radiated. The three principal Juan Fernandez Islands are of volcanic origin, ranging in age from 1 to 5 million years, and lie about 660 km west of continental Chile (Stuessy et al. 1984). These islands exhibit high endemism: 67% of the species of vascular plants (Stuessy et al. 1990) and 53% of the 193 species of flies are restricted to the islands (Kuschel 1963). The estimated diversity of the Diptera of the Juan Fernandez Islands requires reassessment. For example, at the time of Kuschel's (1963) compilation, all five known species of Sphaeroceridae were considered endemic (Marshall 1997). Following two surveys of Robinson Crusoe Island, Marshall (1997) estimated that 53% of the 19 species of Sphaeroceridae were endemic. The closest relative of the seven endemic species of the *Phthitia venosa* group occurs on Saint Helena rather than in Chile (Marshall and Smith 1995). The four species of *Clinocera* are known only from one of the three main Juan Fernandez Islands (Robinson Crusoe Island), though additional species may be present on at least one of the other two. The sister-group relationship of these island species to the remaining Neotropical species of *Clinocera* remains unknown.

2.3. Paraná subregion (Morrone 2006)

The Paraná subregion is spread over northeastern Argentina, eastern Paraguay, and southern and eastern Brazil (Map 34). This subregion includes the Brazilian Atlantic forest, the Paraná Forest, and the *Araucaria angustifolia* forest (Morrone 2006). There are five species of *Clinocera* confined to the Paraná subregion of Brazil. This monophyletic subgroup of species (*C. attenuata, C. cubatao, C. cucama, C. cummingi*, and *C. parana*) possess a highly modified distiphallus and some have reduced thoracic chaetotaxy.

Today this region is rather isolated from the other temperate regions of South America. It is most probable that the presence of *Clinocera* in the Paraná subregion was due to isolation from western populations following the Oligocene and Miocene cooling and aridification and climatic changes resulting from the uplift of the Andes (Morrone 2006).

Summary

The genus *Clinocera* is included among a large number of genera of Clinocerinae primarily confined to the Northern Hemisphere, with only four or five clinocerine genera inhabiting mostly former Gondwanan continents (Sinclair 1995). This northern clade dates at least from the time of the formation of Baltic amber (Eocene, 40 mya), based on the discovery of an undescribed inclusion of *Proclinopyga* from the Baltic amber collections of Dr. Hans Ulrich (Museum Koenig, Bonn, Germany). Based on this minimum age and the present distribution of this northern clade, it is hypothesized that this group of genera is of Laurasian origin, and the presence of any of these genera south of the equator today is due to dispersal (e.g., *Wiedemannia* and *Dolichocephala*). Specifically for *Clinocera*, I consider all species in former Gondwanan land masses to be the result of dispersal. Consequently the presence of two *Clinocera* species-groups in Mesoamerica and South America is viewed as the result of dispersal southward from North America. The studies of the interchange of biota between the Americas have focused mainly on vertebrates and flowering plants (Stehli and Webb 1985) and only a limited number have explored insect groups (e.g., McCafferty 1998). The establishment of the Panamanian land bridge during the Pliocene is the most often cited interchange period, known as the Great American Interchange, but additional linkages also occurred at much earlier dates. The dispersal southward of the two species-groups of *Clinocera* likely occurred at two separate times. The following scenario is highly speculative, owing to the lack of phylogenetic information and fossil evidence, but the mounting distributional data provide some support for this analysis:

(1) The limited southern extent of the *C. lineata* group as far south as the Panamanian Isthmus suggests that this is a more recent dispersal event, likely occurring since the formation of the link between North America and South America in the Late Pliocene (3.1–3.7 Ma; Hoernle et al. 2002). The limited

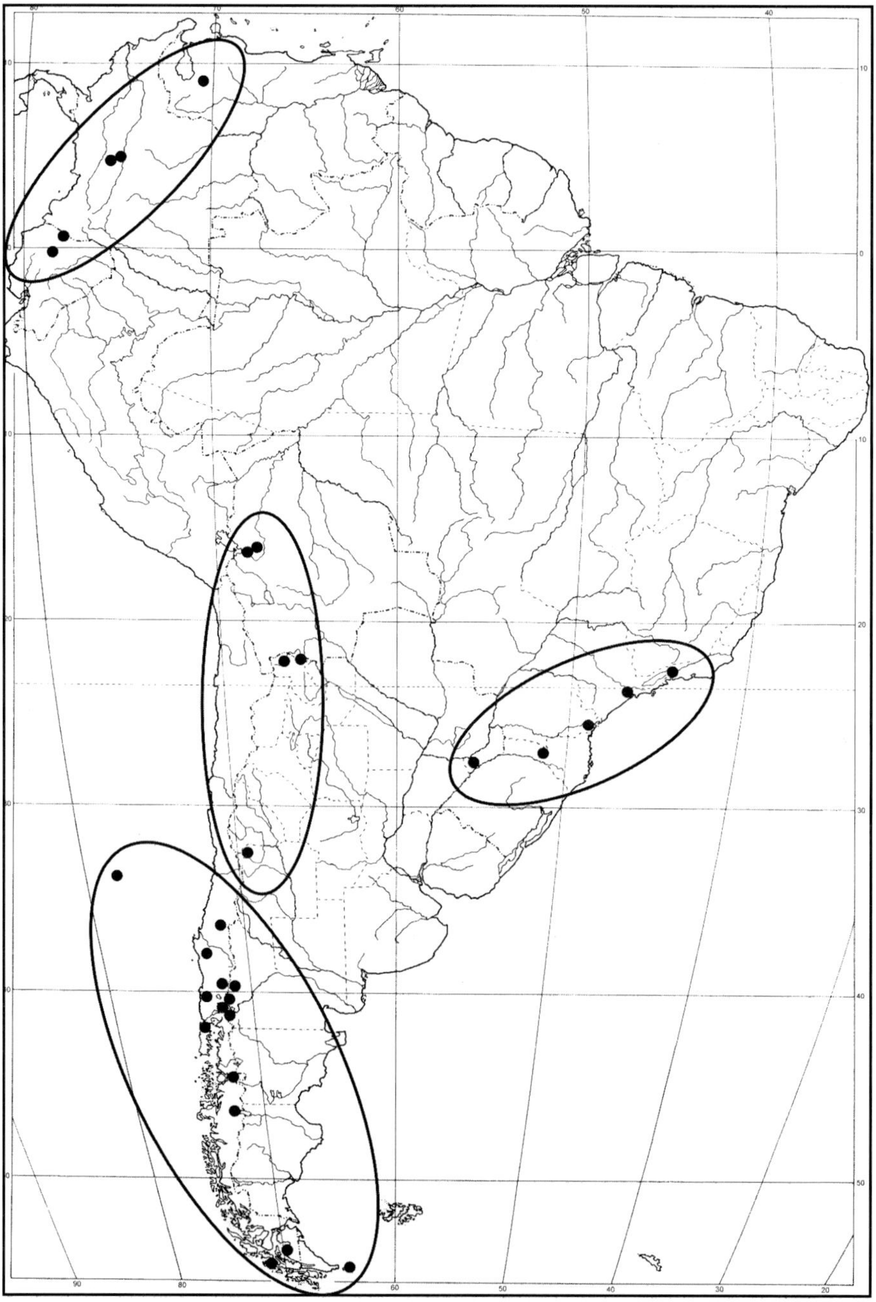

Map 34. Known distribution of *Clinocera*: composite map of all South American species. See the text for an explanation of encircled localities.

southward dispersal of the group and its absence in the South American continent suggests post-linkage dispersal.

(2) The presence of the *C. stagnalis* group throughout South America (and several records from Costa Rica; Map 33), especially in the Paraná region, and the isolated species from Campbell Island strongly suggest that this group successfully dispersed into South America at a much earlier date than the *C. lineata* group.

Ancestral populations of *Clinocera* must have existed in the Paraná region prior to the cooling, aridification, and other climatic changes resulting from the uplift of the Andes, which occurred during the Oligocene and Miocene. These changes would have effectively isolated populations of *Clinocera* in the Paraná region from related groups in the western regions of South America.

Additional distributional evidence supporting this earlier invasion is the interesting record of *Clinocera* from Campbell Island in the subantarctic region of New Zealand. *Clinocera gressitti* Smith is known only from this island and is assigned to the *C. stagnalis* group (Sinclair 2000*b*; Smith 1964). By the Late Cretaceous, this island was possibly submerged; the exposed portions are presently 6–11 million years old, and during glacial episodes only coastal zones were ice-free (Sinclair 2000*b*). No species of this group are known from Australia (all Australian *Clinocera* are in the *C. lineata* group), and the closest relatives are likely in South America (Sinclair 2000*b*). Is this a Gondwanan relict or a postdrift colonist that arrived via oversea dispersal? Dispersal across oceanic gaps is certainly possible, given the presence of *Clinocera* on the Juan Fernandez Islands and the presence of *C. stagnalis*, a Holarctic element, on several mid-Atlantic islands (i.e., Iceland and the Faeroes) (see Sinclair 2000*b*, table 1), but the distance is considerably greater between South America and Campbell Island.

If we assume that oversea dispersal is unlikely, the presence of this species in the Subantarctic region south of New Zealand suggests that the *C. stagnalis* group dispersed from the Nearctic Region via an intercontinental land bridge considered to have formed during a late Campanian / early Maastrichtian uplift (78 Ma), and which possibly existed into the early Paleocene (Hoernle et al. 2002; Iturralde-Vinent and MacPhee 1999). This scenario has been suggested for several groups of Diptera of Laurasian origin that were able to disperse into Gondwanan continents (Amorim et al. 2006). A Late Cretaceous interchange would have facilitated dispersal through South America to Australia (+ New Zealand) via Antarctica. This scenario explains how the ancestral population of the *C. stagnalis* group reached the Australasian Region, but today survives only on Campbell Island.

References

Alexander, C.P. 1940. The Presidential Range of New Hampshire as a biological environment, with particular reference to the insects. The American Midland Naturalist, **24**: 104–132.

Amorim, D.S., Santos, C.M.D., and Oliveira, S.S. 2006. Circumantarctic relationships in Neotropical Diptera: old and new. Pp. 9–10 *in* The 6th International Congress of Dipterology, Fukuoka, Japan. Abstract volume. 354 pp.

Ball, G.E., and Currie, D.C. 1997. Ground beetles (Coleoptera: Tachypachidae and Carabidae) of the Yukon: geographical distribution, ecological aspects, and origin of the extant fauna. Pp. 445–489 *in* H.V. Danks and J.A. Downs (eds.). Terrestrial arthropods. Insects of the Yukon. Biological Survey of Canada Monograph No. 2.

Bezzi, M. 1899. Contribuzioni alla fauna ditterologica italiana. II. Ditteri delle marche e degli abruzzi. Bullettino della Società Entomologica Italiana, **30**: 121–164.

Bezzi, M. 1909. Beiträge zur Kenntnis der südamerikanischen Dipterenfauna auf Grund de Sammelergebnisse einer Reise in Chile, Peru und Bolivia, ausgeführt in den Jahren 1902–1904 von W. Schnuse. Fam. Empididae. Nova Acta Leopoldino-Carolinae, **91**: 299–406.

Brindle, A. 1964. Taxonomic notes on the larvae of British Diptera. No. 18. The Hemerodromiinae (Empididae). The Entomologist, **91**: 162–165.

Brindle, A. 1973. Taxonomic notes on the larvae of British Diptera. No. 28. The larva and pupa of *Hydrodromia stagnalis* (Haliday). The Entomologist, **106**: 249–252.

Campbell, J.M., and Peck, S.B. 1990. *Omalonomus relictus*, an unusual new genus and new species (Coleoptera: Staphylinidae, Omaliinae) of blind rove beetle: a preglacial (Tertiary?) relict in the Cypress Hills, Alberta–Saskatchewan, Canada. The Canadian Entomologist, **122**: 949–961.

Caspers, N., and Wagner, R. 1982. Emergenz-Untersuchungen an einem Mittelgebirgsbach bei Bonn. VII. Empididen- und Dolichopodiden-Emergenz 1976 (Insecta, Diptera, Brachycera). Archiv für Hydrobiologie, **93**: 209–237.

Chvála, M. 1983. The Empidoidea (Diptera) of Fennoscandia and Denmark. II. General part. The families Hybotidae, Atelestidae and Microphoridae. Fauna Entomologica Scandinavica, **12**: 1–279.

Chvála, M., and Wagner, R. 1989. Empididae. Pp. 228–336 *in* Á.Soós and L. Papp (eds.). Catalogue of Palaearctic Diptera. Volume 6. Therevidae – Empididae. Elsevier Science Publishing, Amsterdam, the Netherlands. 435 pp.

Collin, J.E. 1933. Empididae. Pp. 1–334 *in* Diptera of Patagonia and south Chile. Part 4. British Museum (Natural History), London, UK.

Collin, J.E. 1935. Results of the Oxford University Hudson Strait Expedition, 1931. XXXIII. Diptera (Cyclorrhapha) from Akpatok Island, Ungava Bay, Canada. Annals and Magazine of Natural History, London, (10), **15**: 369–381.

Collin, J.E. 1961. Empididae. *In* British flies. Volume 6. University Press, Cambridge, UK. 782 pp.

Coquillett, D.W. 1903. The genera of the dipterous family Empididae, with notes and new species. Proceedings of the Entomological Society of Washington, **5**: 245–272.

Coquillett, D.W. 1910. New genera and species of North American Diptera. Proceedings of the Entomological Society of Washington, **12**: 124–131.

Coscarón, S., and Coscarón-Arias, C.L. 1995. Distribution of Neotropical Simuliidae (Insecta, Diptera) and its areas of endemism. Revista de la Academia Colombiana de Ciencias Exactas, Físicas y Naturales, **19**: 717–732.

Courtney, G.W. 1990. Revision of Nearctic mountain midges (Diptera: Deuterophlebiidae). Journal of Natural History, **24**: 81–118.

Courtney, G.W. 1994. Revision of Palaearctic mountain midges (Diptera: Deuterophlebiidae), with phylogenetic and biogeographic analyses of world species. Systematic Entomology, **19**: 1–24.

Crosskey, R.W., and White, G.B. 1977. The Afrotropical Region: a recommended term in zoogeography. Journal of Natural History, **11**: 541–544.

Cumming, J.M., Sinclair, B.J., and Wood, D.M. 1995. Homology and phylogenetic implications of male genitalia in Diptera – Eremoneura. Entomologica Scandinavica, **26**: 121–151.

Currie, D.C. 1997. Black flies (Diptera: Simuliidae) of the Yukon, with reference to the black-fly fauna of northwestern North America. Pp. 563–614 *in* H.V. Danks and J.A. Downs (eds.). Terrestrial arthropods. Insects of the Yukon. Biological Survey of Canada Monograph No. 2.

Danks, H.V. 1981. Arctic arthropods: a review of systematics and ecology with particular reference to the North American fauna. Entomological Society of Canada, Ottawa, Ont. 608 pp.

Danks, H.V., and Downs, J.A. (eds.) 1997. Terrestrial arthropods. Insects of the Yukon. Biological Survey of Canada Monograph No. 2. 1034 pp.

Delfinado, M.D., and Hardy, D.E. (eds.). 1973. A catalog of the Diptera of the Oriental Region. Volume 1. The University Press of Hawai´i, Honolulu, Hawai´i. 618 pp.

Downes, J.A., and Smith, S.M. 1969. New or little known feeding habits in Empididae (Diptera). The Canadian Entomologist, **101**: 404–408.

Engel, E.O. 1931. XV. Empididae (Dipt.) aus Java. Archiv für Hydrobiologie (Supplement), **8**: 486–489.

Evenhuis, N.L. 1989. Introduction. Pp. 17–18 *in* N.L. Evenhuis (ed.). Catalog of the Diptera of the Australasian and Oceanic Regions. Special Publication 86, The Bernice P. Bishop Museum, Honolulu, Hawai´i. 1155 pp.

Farr, T.H. 1953. Studies on North American Empididae. Ph.D. thesis, University of Massachusetts, Amherst, Mass. 123 pp.

Frey, R. 1913. Zur Kenntnis der Dipterenfauna Finlands. II. Empididae. Acta Societatis Pro Fauna et Flora Fennica, **37**(10): 1–89.

Griffiths, G.C.D. 1980. Preface. Pp. v–vii *in* G.C.D. Griffiths (ed.). Flies of the Nearctic Region. Volume 1. Handbook. Part 1. History of Nearctic dipterology. E. Schweizerbart'sche Verlagsbuchhandlung, Stuttgart, Germany. 62 pp.

Haliday, A.H. 1833. Catalogue of Diptera occurring about Holywood in Downshire. Entomological Magazine, London, **1**: 147–180.

Harper, P.P. 1980. Phenology and distribution of aquatic dance flies (Diptera: Empididae) in a Laurentian watershed. American Midland Naturalist, **104**: 110–117.

Hitchcock, S.W. 1969. Plecoptera from high altitudes and a new species of *Leuctra* (Leuctridae). Entomological News, **80**: 311–316.

Hoernle, K., van den Bogaard, P., Werner, R., Lissinna, B., Hauff, F., Alvarado, G., and Garbe-Schönberg, D. 2002. Missing history (16–71 Ma) of the Galápagos hotspot: implications for the tectonic and biological evolution of the Americas. Geology, **30**: 795–798.

Huber, B.A., Sinclair, B.J., and Schmitt, M. 2007. The evolution of asymmetric genitalia in spiders and insects. Biological Reviews of the Cambridge Philosophical Society, **82**: 647–698.

International Commission on Zoological Nomenclature. 1963. Opinion 678. The suppression under the plenary powers of the pamphlet published by Meigen, 1800. The Bulletin of Zoological Nomenclature, **20**: 339–342.

Irwin, T.L., and Kavanaugh, D.H. 1981. Systematics and zoogeography of *Bembidion* Latreille: I. The *carlhi* and *erasum* groups of western North America (Coleoptera: Carabidae, Bembidiini). Entomologica Scandinavica Supplement, **15**: 22–72.

Iturralde-Vinent, M.A., and MacPhee, R.D.E. 1999. Paleogeography of the Caribbean region: implications for Cenozoic biogeography. Bulletin of the American Museum of Natural History, **238**. 95 pp.

Jonassen, T. 1987. New Norwegian Empididae (s.str.) (Dipt.). Fauna Norvegica Series B, **34**: 37–40.

Jonassen, T. 1998. Recent additions to the Norwegian fauna of Empidoidea (Diptera, Brachycera). Fauna Norvegica Series B, **45**: 63–68.

Joost, W. 1981. Beitrag zur Kenntnis der Hemerodromiinae des Kaukasus (1) (Diptera, Empididae). Reichenbachia, **19** (31): 183–191.

Joost, W. 1982. Beitrag zur Kenntnis der Hemerodromiinae (Diptera, Empididae) Thüringens. Entomologische Nachrichten und Berichte, **26**: 81–86.

Kavanaugh, D.H. 1980. Insects of western Canada, with special reference to certain Carabidae (Coleoptera): present distribution patterns and their origins. The Canadian Entomologist, **112**: 1129–1144.

Kavanaugh, D.H. 1988. The insect fauna of the Pacific Northwest coast of North America: present patterns and affinities and their origins. Memoirs of the Entomological Society of Canada, **144**: 125–149.

Kuschel, G. 1963. Composition and relationship of the terrestrial faunas of Easter, Juan Fernandez, Desventuradas, and Galápagos Islands. Occasional Papers of the California Academy of Sciences, **44**: 79–95.

Landry, B., and Harper, P.P. 1985. The aquatic dance fly fauna of a subarctic river system in Québec, with the description of a new species of *Hemerodromia* (Diptera: Empididae). The Canadian Entomologist, **117**: 1379–1386.

Laurence, B.R. 1953. On the feeding habits of *Clinocera (Wiedemannia) bistigma* Curtis (Diptera: Empididae). Proceedings of the Royal Entomological Society, London, Series A, General Entomology, **28**: 139–144.

Lindroth, C. 1931. Die Insektenfauna Islands und ihrer Probleme. Zoologiska Bidrag fran Uppsala, **13**: 105–599.

Loew, H. 1860. Diptera americana ab Osten-Sackenio collecta. Wiener Entomologische Monatsschrift, **4**: 79–84.

Loew, H. 1862. Diptera Americae septentrionalis indigena. Berliner Entomologische Zeitschrift, **6**: 185–232.

Loew, H. 1864. *Clinocera bivittata* nov. sp. Wiener Entomologische Monatsschrift, **8**: 258–260.

Loew, H. 1876. Beschreibungen neuer amerikanischer Dipteren. Zeitschrift für die Gesamte Naturwissenschaft, **48**: 317–340.

Lundbeck, W. 1898. Diptera groenlandica. Videnskabelige Meddelelser fra den Naturhistorisk Forening i Kjøbenhavn, 1898: 236–314, 2 pls.

Lyneborg, L. 1968. XLII a. Diptera Brachycera et Cyclorrhapha. Zoology of the Faroes, **2**(2): 1–79.

Macquart, J. 1835. Histoire naturelle des insectes. Diptères. Tome 2. Ouvrage accompagné de planches. Roret, Paris. 703 or 710 pp.

Maddison, W.P., and Maddison, D.R. 2003. MacClade 4: analysis of phylogeny and character evolution. Version 4.06. Sinauer Associates, Sunderland, Mass.

Marshall, S.A. 1997. Sphaeroceridae of the Juan Fernandez Islands, Chile. Studia dipterologica, **4**: 165–171.

Marshall, S.A., and Smith, I.P. 1995. A review of the *Phthitia* of the Juan Fernandez Islands, with descriptions of four new flightless species from Robinson Crusoe Island (Diptera: Sphaeroceridae). Acta Entomologica Chilena, **9**: 31–50.

Masteller, E.C., and Flint, O.S. 1984. Trichoptera emergence patterns from a small stream in northwestern Pennsylvania influences by sewage effluent. Pp. 225–233 *in* J.C. Morse (ed.). Proceedings of the Fourth International Symposium on Trichoptera. Series Entomologica 30. Dr. W. Junk Publishers, The Hague, the Netherlands.

McAlpine, J.F. 1981. Morphology and terminology — adults. [Chapter] 2. Pp. 9–63 *in* J.F. McAlpine, B.V. Peterson, G.E. Shewell, H.J. Teskey, J.R. Vockeroth, and D.M. Wood (coords.). Manual of Nearctic Diptera. Volume 1. Agriculture Canada Monograph 27. vi + 1–674 pp.

McCafferty, W.P. 1998. Ephemeroptera and the great American interchange. Journal of the North American Benthological Society, **17**: 1–20.

McCreadie, J.W., Adler, P.H., and Masteller, E.C. 1994. Long-term emergence patterns of black flies (Diptera: Simuliidae) in northwestern Pennsylvania. Hydrobiologia, **288**: 39–46.

Meigen, J.W. 1800. Nouvelle classification des mouches à deux ailes (Diptera L.) d'après un plan tout nouveau. Perronneau, Paris. 40 pp.

Meigen, J.W. 1803. Versuch einer neuen Gattungs Eintheilung der europäischen zweiflügeligen Insekten. Magazin für Insektenkunde, **2**: 259–281.

Meigen, J.W. 1804. Klassifikazion und Beschreibung der europäischen zweiflügeligen Insekten (Diptera Linn.). Band. Abt. I. xxviii + 152 pp. Abt. II. vi + pp. 153–314. Reichard, Braunschweig.

Melander, A.L. 1902. Monograph of the American Empididae. Transactions of the American Entomological Society, **28**: 195–368.

Melander, A.L. 1928 (1927). Diptera, Fam. Empididae. Pp. 1–434 *in* P. Wytsman (ed.). Genera Insectorum. Volume 185.

Melander, A.L. 1965. Family Empididae (Empidae, Hybotidae). Pp. 446–481 *in* A. Stone, C.W. Sabrosky, W.W. Wirth, R.H. Foote, and J.R. Coulson (eds.). A catalog of the Diptera of America north of Mexico. United States Department of Agriculture, Agriculture Handbook 276. 1696 pp.

Mik, J. 1880. Beschreibung neuer Dipteren. I. Eilf neue europäische *Clinocera*-Arten. Verhandlungen der kaiserlich–königlichen zoologisch–botanischen Gesellschaft, **30**: 347–358.

Morrone, J.J. 2006. Biogeographic areas and transition zones of Latin America and the Caribbean Islands based on panbiogeographic and cladistic analyses of the entomofauna. Annual Review of Entomology, **51**: 467–494.

Newsome, R.O., and Dix, R.L. 1968. The forests of the Cypress Hills, Alberta and Saskatchewan, Canada. The American Midland Naturalist, **80**: 118–185.

Niesiołowski, S. 1992. Empididae Aquatica Wodne wujkowate (Insecta: Diptera). Fauna Polski, **14**: 1–128.

Peterson, B.V. 1960. Notes on some natural enemies of Utah black flies (Diptera: Simuliidae). The Canadian Entomologist, **92**: 260–274.

Peterson, B.V., and Davies, D.M. 1960. Oberservations on some insect predators of black flies (Diptera: Simuliidae) of Algonquin Park, Ontario. Canadian Journal of Zoology, **38**: 9–18.

Plant, A.R. 2003. Phenology of Empididae and Hybotidae (Diptera) in Great Britain. Dipterists' Digest, **10**: 13–30.

Pont, A.C. 1995. The type-material of Diptera (Insecta) described by G.H. Verrall and J.E. Collin. Oxford University Museum Publication 3, Clarendon Press, Oxford, UK. 223 pp.

Rafael, J.A. 1996. Revisão des espécies neotropicais de Empidae (Diptera) descritas por Mario Bezzi. V. Clinocerinae. Revista Brasileira de Entomologia, **40**(3/4): 437–440.

Ricker, W.E. 1964. Distribution of Canadian stoneflies. Gewässer und Abwässer, **34/35**: 50–71.

Rogers, E. 1981. Mating in *Wiedemannia* (Diptera: Empididae). Journal of Natural History, **15**: 1013–1018.

Ross, H.H. 1956. Evolution and classification of the mountain caddisflies. The University of Illinois Press, Urbana, Ill. 213 pp.

Ross, H.H. 1965. Pleistocene events and insects. Pp. 583–596 *in* H.E. Wright, Jr. and D.G. Frey (eds.). The Quaternary of the United States. Princeton University Press, Princeton, N.J. 922 pp.

Séguy, E. 1941. Récoltes de R. Paulian et A. Villiers dans le haut Atlas marocain, 1938 (XVIIe note). Diptères. Revue de Françaises d'Entomologie, **8**: 25–33 and 82–88.

Sinclair, B.J. 1994. Revision of the Nearctic species of *Trichoclinocera Collin* (Diptera: Empididae; Clinocerinae). The Canadian Entomologist, **126**: 1007–1059.

Sinclair, B.J. 1995. Generic revision of the Clinocerinae (Empididae), and description and phylogenetic relationships of the Trichopezinae, new status (Diptera: Empidoidea). The Canadian Entomologist, **127**: 665–752.

Sinclair, B.J. 1998 (1997). A review of the Nearctic species of *Wiedemannia* Zetterstedt (Diptera: Empididae: Clinocerinae). Studia dipterologica, **4**: 337–352.

Sinclair, B.J. 1999*a*. Review of the Holarctic *Clinocera appendiculata* complex (Dipt., Empididae, Clinocerinae). Entomologist's Monthly Magazine, **135**: 223–232.

Sinclair, B.J. 1999*b*. Review of the Clinocerinae of southern Africa (Diptera: Empididae). Annals of the Natal Museum, **40**: 103–125.

Sinclair, B.J. 2000*a*. Morphology and terminology of Diptera male genitalia. Pp. 53–74 *in* L. Papp and B. Darvas (eds.). Contributions to a manual of Palaearctic Diptera. Volume 1: General and applied dipterology. Science Herald, Budapest, Hungary.

Sinclair, B.J. 2000*b*. Revision of the genus *Clinocera* Meigen from Australia and New Zealand (Diptera: Empididae: Clinocerinae). Invertebrate Taxonomy, **14**: 347–361.

Sinclair, B.J. 2003*a*. Southern African Empidoidea (Diptera) — phylogenetic patterns and biogeographic implications. Cimbebasia, **19**: 205–213.

Sinclair, B.J. 2003*b*. Taxonomy, phylogeny and zoogeography of the subfamily Ceratomerinae of Australia (Diptera: Empidoidea). Records of the Australian Museum, **55**: 1–44.

Sinclair, B.J. 2007. The status of *Clinocera rufipes* Bezzi, a new junior synonym of *Clinocera nigra* Meigen (Diptera: Empididae: Clinocerinae). Zootaxa, **1554**: 67–68.

Sinclair, B.J., and Harkrider, J.R. 2004. The immature stages and rearing observations of the aquatic dance fly, *Roederiodes wirthi* Chillcott (Diptera: Empididae: Clinocerinae), with taxonomic notes on the genus. Studia dipterologica, **11**: 51–61.

Sinclair, B.J., and Marshall, S.A. 1987 (1986). The madicolous fauna in southern Ontario. Proceedings of the Entomological Society of Ontario, **117**: 9–14.

Sinclair, B.J., and Saigusa, T. 2005 (2004). Revision of the *Trichoclinocera dasyscutellum* group from East Asia (Diptera: Empididae: Clinocerinae). Bonner zoologische Beiträge, **53**(1–2): 193–209.

Smith, I.M. 1991. Water mites (Acari: Parasitengona: Hydrachnida) of spring habitats in Canada. Memoirs of the Entomological Society of Canada, **155**: 141–167.

Smith, I.M., and Oliver, D.R. 1986. Review of parasitic associations of larval water mites (Acari: Parasitengona: Hydrachnida) with insect hosts. The Canadian Entomologist, **118**: 407–472.

Smith, K.G.V. 1962. Studies on the Brazilian Empididae (Diptera). Transactions of the Royal Entomological Society of London, **114**(7): 195–266.

Smith, K.G.V. 1964. Insects of Campbell Island. Diptera: Empididae. Pacific Insects Monograph, **7**: 325–328.

Smith, K.G.V. 1965. Diptera from Nepal. Empididae. Bulletin of the British Museum (Natural History), Entomology, **17**: 61–112.

Smith, K.G.V. 1967. Family Empididae (Empidae, Hybotidae). Pp. 39.1–39.67 *in* A catalogue of the Diptera of the Americas south of the United States. Departamento de Zoologia, Secretaria da Agricultura, São Paulo, Brazil.

Smith, K.G.V. 1969. The Empididae of southern Africa (Diptera). Annals of the Natal Museum, **19**: 1–342.

Soós, Á., and Papp, L. (eds.) 1984. Catalogue of Palearctic Diptera. Volume 9. Micropezidae–Agromyzidae. Elsevier Science Publishers, Amsterdam, the Netherlands. 460 pp.

Soper, J.H., and Maycock, P.F. 1963. A community of arctic–alpine plants on the east shore of Lake Superior. Canadian Journal of Botany, **41**: 183–198.

Stehli, F.G., and Webb, S.D. (eds). 1985. The great American biotic interchange. Plenum Press, New York. 532 pp.

Steyskal, G.C., and Knutson, L.V. 1981. Empididae. [Chapter] 47. Pp. 607–624 *in* J.F. McAlpine, B.V. Peterson, G.E. Shewell, H.J. Teskey, J.R. Vockeroth, and D.M. Wood (coords.). Manual of Nearctic Diptera. Volume 1. Agriculture Canada Monograph 27. vi + 674 pp.

Stuckenberg, B.R. 1999. Antennal evolution in the Brachycera (Diptera), with a reassessment of terminology relating to the flagellum. Studia dipterologica, **6**: 33–48.

Stuessy, T.F., Foland, K.A., Sutter, J.F., Sanders, R.W., and Silva O. M. 1984. Botanical and geological significance of potassium–argon dates from the Juan Fernandez Islands. Science (Washington, D.C.), **225**: 49–51.

Stuessy, T.F., Crawford, D.J., and Marticorena, C. 1990. Patterns of phylogeny in the endemic vascular flora of the Juan Fernandez Islands, Chile. Systematic Botany, **15**: 338–346.

Swofford, D.L. 2002. PAUP*. Phylogenetic analysis using parsimony (*and other methods). Version 4.0b10. Sinauer Associates, Sunderland, Mass.

Tangelder, I.R.M. 1988. The biogeography of the Holarctic *Nephrotoma dorsalis* species-group (Diptera, Tipulidae). Beaufortia, **38**: 1–35.

Tucker, E.S. 1907. Some results of desultory collecting of insects in Kansas and Colorado. The Kansas University Science Bulletin, **4**: 51–112.

Ulrich, H. 1972. Zur Anatomie des Empididen-Hypopygiums (Diptera). Veröffentlichungen der zoologischen Staatssammlung München, **16**, 1–28.

Vaillant, F. 1952a (1951). Un empidide destructeur de simulies. Bulletin de la Société zoologique de France, **76**: 371–379.

Vaillant, F. 1952b. Les larves hygropétriques et les nymphes de quelques Empididae Atalantinae (Diptères). Travaux du Laboratoire d'Hydrobiologie et de Pisciculture de l'Université de Grenoble, **43–44**: 39–46.

Vaillant, F. 1956. Recherches sur la faune madicole (hygropétrique s.l.) de France, de Corse et d'Afrique du Nord. Mémoires du Museum National d'Histoire Naturelle Série A, Zoologie, **11**: 1–258.

Vaillant, F. 1957 (1956). Contribution à l'étude des Diptères Empididae du Grand-Atlas marocain. I. Hemerodromiinae et Atalantinae. Bulletin de la Société des Sciences naturelles et physiques du Maroc, **36**: 61–71.

Vaillant, F. 1960. Quelques Empididae Atalantinae d'Asie russe [Dipt.]. Bulletin de la Société entomologique de France, **65**: 170–186.

Vaillant, F. 1965 (1964). Revision des Empididae Hemerodromiinae de France, d'Espagne et d'Afrique du Nord. [Dipt.]. Annales de la Société entomologique de France, **133**: 143–173.

Vockeroth, J.R. 1966. A method of mounting insects from alcohol. The Canadian Entomologist, **98**: 69–70.

Wagner, R. 1980. Die Dipterenemergenz am Breitenbach (1969–1973). Spixiana, **3**: 167–177.

Wagner, R. 1982. Dipteren-Emergenz zweier Lunzer Bäche 1972–1974 nebst Beschreibung einer neuen Empidide (Diptera). Archiv für Hydrobiologie, **95**: 491–506.

Wagner, R. 1997. Diptera Empididae, dance flies. Pp. 333–344 *in* A. Nilsson (ed.). Aquatic insects of North Europe. Volume 2. Apollo Books, Stenstrup, Denmark. 440 pp.

Wagner, R., and Gathmann, O. 1996. Long-term studies on aquatic dance flies (Diptera, Empididae) 1983–1993: distribution and size patterns along the stream, abundance changes between years and the influence of environmental factors on the community. Archiv für Hydrobiologie, **137**: 385–410.

Wagner, R., Leese, F., and Panesar, A.R. 2004. Aquatic dance flies from a small Himalayan Mountain stream (Diptera: Empididae: Hemerodromiinae, Trichopezinae and Clinocerinae). Bonner zoologische Beiträge, **52**: 3–32.

Werner, D., and Pont, A.C. 2003. Dipteran predators of simuliid blackflies: a worldwide review. Medical and Veterinary Entomology, **17**: 115–132.

Wirth, W.W., and Stone, A. 1956. Aquatic Diptera. Pp. 372–482 *in* R.L. Usinger (ed.). Aquatic insects of California. University of California Press, Berkeley, Calif. 508 pp.

Wootton, R.J., and Ennos, A.R. 1989. The implications of function on the origin and homologies of the dipterous wing. Systematic Entomology, **14**: 507–520.

Yang D., and Yang J. 1995. Diptera: Empididae. Pp. 235–240 *in* T. Zhu (ed.). Insects and macrofungi of Gutianshan, Zhejiang. Zhejiang Science and Technology Press, Hangzhou. 327 pp. [In Chinese with English summary.]

Yang D., Grootaert, P., and Horvat, B. 2005. A new species of *Clinocera* Meigen (Diptera: Empididae) from China. Zootaxa, **908**: 1–4.

Yang D., Zhu F., and An S. 2006. Diptera: Empididae. Pp. 304–308 *in* D.-c. Jin and Z.-z. Li (eds.). Insects from Chishui spinulose tree fern landscape. [Also Chishui Suoluo Jingguan Kunchong]. Insect Fauna from National Nature Reserve of Guizhou Province, III. Guizhou Science and Technology Publishing House, Guiyang, People's Republic of China. 409 pp. [In Chinese with English summary.]

Zetterstedt, J.W. 1838. Diptera. Sectio tertia. Dipterologis scandinaviae: amicis et popularibus carissimis. Pp. 477–868 *in* Insecta Lapponica. L. Voss, Lipsiae [= Leipzig]. vi + 1140 pp.

Zetterstedt, J.W. 1849. Diptera Scandinaviae deposita et descripta. Volume 8 pp. 2935–3366. Lundae (Lund), Sweden.

Appendix 1

Checklist of the species of Clinocera

There are currently 103 described species of *Clinocera* worldwide.

adesa Sinclair, 1999*a*: 112.
amarilla **sp. nov.**
appendiculata (Zetterstedt, 1838: 559).
appalachia **sp. nov.**
arizonica **sp. nov.**
arnaudi **sp. nov.**
attenuata **sp. nov.**
aucta (Zetterstedt, 1849: 3019).
australiana Sinclair, 2000*b*: 349.
bickeli Sinclair, 2000*b*: 350.
binotata Loew, 1876: 325.
borkenti **sp. nov.**
brevicauda **sp. nov.**
brunnea **sp. nov.**
bulbosa **sp. nov.**
caerulea **sp. nov.**
chilamche Smith, 1965: 107.
chilensis **sp. nov.**
collini **sp. nov.**
colombiana **sp. nov.**
coloradensis **sp. nov.**
conjuncta Loew, 1860: 80.
cordillerana **sp. nov.**
cubatao Smith, 1962: 264.
cucama Smith, 1962: 265.
cummingi **sp. nov.**
cuspidata Wagner & Leese, 2004: 16.
daniellae **sp. nov.**
disjuncta **sp. nov.**
ditaeniata Bezzi, 1909: 387.
empodiata Collin, 1933: 322.
femorata **sp. nov.**
fontinalis (Haliday, 1833: 160).
fuscipennis Loew, 1876: 324.
gressitti Smith, 1964: 327.
guangdongensis Yang, Grootaert & Horvat, 2005: 2.
gubernans (Melander, 1928: 234).
guizhouensis Yang, Zhu & An, 2006: 306.
guttipennis **sp. nov.**
insularis **sp. nov.**
irrorata Sinclair, 2000*b*: 353.
isabellina **sp. nov.**
lapazensis **sp. nov.**
lecta Melander, 1902: 243.
lineata Loew, 1862: 207.
longicauda **sp. nov.**
longicercus Wagner & Leese, 2004: 16.
lunata Wagner & Leese, 2004: 18.
lunatoides Wagner & Leese, 2004: 18.
macdonaldi **sp. nov.**
maculata Loew, 1860: 79.
madera **sp. nov.**
marginesetosa Wagner & Leese, 2004: 16.
maroccana (Séguy, 1941: 29).
marshalli **sp. nov.**
megalatlantica (Vaillant, 1957: 65).
melanderi **sp. nov.**
meridana **sp. nov.**
mexicana **sp. nov.**
minutissima (Vaillant, 1960: 172).
miradorana **sp. nov.**
monticola Sinclair, 2000*b*: 355.
nadi Smith, 1965: 108.
naomina **sp. nov.**
nigra Meigen, 1804: 292.
nigra orientalis (Vaillant, 1965: 148).
nigriscutata **sp. nov.**
nivalis (Zetterstedt, 1838: 545).
notialis **sp. nov.**
obunca **sp. nov.**
olivacea Melander, 1928: 233.
oriunda Collin, 1933: 323.
pacifica **sp. nov.**
pani Smith, 1965: 108.
parana **sp. nov.**
penai **sp. nov.**
prasinata Melander, 1928: 232.
queenslandica Sinclair, 2000*b*: 355.
rainiericola **sp. nov.**
robinsoni **sp. nov.**
rubriventris Sinclair, 2000*b*: 356.
sabroskyi **sp. nov.**
schlinkerti **sp. nov.**
schumanni Joost, 1981: 184.
setitibia **sp. nov.**
setosa Wagner & Leese, 2004: 18.
sinclairi Wagner & Leese, 2004: 19.
sinensis Yang & Yang, 1995: 235.
sinica (Yang & Yang, 1995: 237).
smithi Sinclair, 1999*a*: 113.
spinosa Sinclair, 2000*b*: 358.
stackelbergi (Vaillant, 1960: 174).
stagnalis (Haliday, 1833: 159).
stigma **sp. nov.**
storchi Mik, 1880: 348.
sublineata **sp. nov.**
substagnalis **sp. nov.**
subtrunca **sp. nov.**
tripunctata Smith, 1969: 291.
trunca Melander, 1928: 230.
velutina (Engel, 1931: 488).
wallowa **sp. nov.**
wesmaeli (Macquart, 1835: 656).
woodi **sp. nov.**
wui Yang & Yang, 1995: 236.

Appendix 2

Key to Nearctic species of *Clinocera* north of Mexico

The following key is included to aid in the identification of all Nearctic species and separate keys to species-groups can be found under each corresponding section. This key is written for dried specimens (see Methods section for drying options). Specimens stored in alcohol, especially females, are very difficult to identify; however, the male terminalia of wet specimens can be easily exposed, allowing a better view of the clasping cercus and apex of the phallus.

1 Stigma absent (Figs. 31*g*, 32*e*); face flat or convex, never with tubercle. 2

— Stigma in middle of cell r_1, elongate and usually distinct (Figs. 30*i*, *h*); if absent, face with distinct tubercle 14

2 Fifth tarsomere with dorsoapical extension; wings with spots at base of R_4, apex of cell dm, r-m crossvein and apex of cell bm (Figs. 32*d–h*); phallus split at base of distiphallus (Figs. 21*d*, 22*a–d*) (*C. stagnalis* group) 3

— Fifth tarsomere lacking extension; wing spots present or absent; phallus slender, lacking split at base of distiphallus (females are difficult to determine without associated male) 7

3 R_{2+3} and R_4 not connected by an auxiliary crossvein (Figs. 32*e*, *g*), although occasionally present on one wing in some specimens of *C. stagnalis* 4

— R_{2+3} and R_4 connected by an auxiliary crossvein (Figs. 32*d*, *f*, *h*) 5

4 Legs dark except for extreme tip of femora; face entirely clothed in blue pruinescence; forefemur lacking basal setae (northern, Rocky Mountains) . . . . ***C. stagnalis*** (Haliday)

— Femur and tibia bright orange–brown, tarsomeres dark; lower half of face clothed in blue pruinescence; forefemur with stout basal setae (Figs. 9*c*, *d*) (eastern). . . . . ***C. maculata*** Loew

5 Extreme apex of femora dark, not pale; anterior spiracle encircled by black ring; pleura dark, pruinescence faint (northern). . . . . ***C. nivalis*** (Zetterstedt)

— Extreme apex of femora reddish brown; anterior spiracle not encircled by black ring; pleura with distinct blue pruinescence 6

6 Forefemur with stout basal setae; clasping cercus expanded apically (Fig. 21*d*) (widespread). . . . . ***C. binotata*** Loew

— Forefemur without stout basal setae; clasping cercus parallel-sided, not expanded apically (Fig. 22*d*) (western) . . . . ***C. substagnalis*** sp. nov.

7 Scutum with pair of black stripes or lacking distinct vittae. Clasping cercus strongly tapered; apex of surstylus round (Figs. 21*b*, *c*) (*C. nigra* group) 8
— Scutum with dark median stripe and lateral pale brownish stripes, lacking pair of black stripes. Clasping cercus not strongly tapered; apex of surstylus narrowed to bek-like projection (Figs. 19*a–c*) (*C. lineata* group) 9

8 R_{2+3} and R_4 connected by an auxiliary crossvein; faint clouding about veins and crossveins (Fig. 32*c*); scutum with pair of distinct dark stripes (northern) . . . . ***C. aucta*** (Zetterstedt)
— R_{2+3} and R_4 not connected by crossvein; wings infuscate, clouding absent; scutum lacking distinct vittae (eastern) . . . . ***C. appalachia*** sp. nov.

9 Face entirely clothed in blue pruinescence; scutum with narrow dark median stripe, flanked by wide shiny stripes 10
— Face with blue pruinescence on lower three-quarters, brown below antenna, or pruinescence reduced to narrow lateral stripes on lower half; scutum with wide dull, dark, median stripe, flanked by narrower pale brownish stripes (females generally difficult to distinguish) 11

10 Veins wavy (Fig. 31*g*); R_{2+3} often with auxiliary crossvein to costa; lateral stripes of scutum blue (widespread) . . . . ***C. lineata*** Loew
— Veins straight (Fig. 32*a*); R_{2+3} lacking auxiliary crossvein; lateral stripes of scutum grey (eastern) . . . . ***C. sublineata*** sp. nov.

11 Face with pale lateral stripes on lower half (sometimes partially fused), upper half with median stripe; clasping cercus bearing setulae, with slightly hooked apex (Fig. 19*c*); if face mostly blue, clasping cercus narrow to somewhat expanded at midlength (southwestern) . . . . . ***C. madera*** sp. nov.
— Face with blue pruinescence; clasping cercus broad, bearing stout setae, apex straight to tapered (Figs. 18*c*, *d*, 19*b*) 12

12 Thoracic chaetotaxy not reduced; ocellar bristle long and slender 13
— Thoracic chaetotaxy greatly reduced; ocellar bristle very short and slender, shorter than postocular bristles (southwestern) . . . . ***C. macdonaldi*** sp. nov.

13 Clasping cercus extending beyond surstylus, expanded apically (Fig. 18*c*) (southwestern) . . . . ***C. arizonica*** sp. nov.
— Clasping cercus short, extending slightly beyond surstylus, apex tapered (Fig. 18*d*) (southwestern) . . . . ***C. brevicauda*** sp. nov.

14 Two or more (rarely 1, sometimes very slender and inconspicuous) postsutural supra-alar bristles; facial tubercle on lower face sometimes present; distiphallus arising from narrow shaft; apex of clasping cercus tapered to ventroapical nipple-like tip (Figs. 16*a–f*) (*C. fuscipennis* group) 15
— One postsutural supra-alar bristle (Fig. 5*b*); facial tubercle lacking; distiphallus arising from expanded apex of shaft; apex of clasping cercus truncate (Figs. 13*a–d*) or tapered to slender round tip (Figs. 11*a–d*) 24

15 Fifth tarsomere lacking dorsoapical extension; facial swelling variable 16

— Fifth tarsomere with dorsoapical extension (Fig. 9*b*) (very short in *C. wallowa*); facial swelling distinct. 20

16 Wing with distinct circular spots (Fig. 31*b*) (western) . . . . **C. lecta** Melander

— Wing hyaline, spots absent or with cloudy regions around veins 17

17 Forefemur swollen, especially at base, with long basal setae, some longer than width of femur; forefemur with anteroventral row of closely spaced short stout setae (western) . . . . **C. femorata** sp. nov.

— Forefemur slender, not especially swollen at base, lacking basal setae; anteroventral row of widely spaced short setae present or absent 18

18 Stigma broad, conspicuous, often dark with proximal margin truncate (Fig. 31*d*); if faint and narrow proximately, then face with small distinct tubercle above palpus and biserial row of stout ventral setae of male forefemur absent; clasping cercus pointed apically (Fig. 18*a*). . . . . **C. stigma** sp. nov.

— Stigma normally faint, elliptical, with proximal margin pointed; face lacking distinct tubercle above palpus; male forefemur with biserial row of ventral setae; clasping cercus rounded apically (Figs. 16*a*, *c*). 19

19 Wing with cloudy regions around veins, especially crossveins (Fig. 30*g*); R_{2+3} often distinctly dipped beneath stigma; scutellum with blue pruinescence; legs dark; scutum bivittate (northern). . . . . **C. disjuncta** sp. nov.

— Wing hyaline to infuscate cloudy regions absent around veins (Fig. 30*i*); R_{2+3} gradually arched, not dipped beneath stigma; scutellum usually dark, rarely with blue pruinescence; legs dark to reddish; scutum bivittate in some western populations, especially in pale-legged forms (widespread) . . . . **C. fuscipennis** Loew

20 Scutum with 5 dorsocentral bristles, lacking interspersed shorter setae. . . . 21

— Scutum with 5 dorsocentral bristles with interspersed long and short setae 22

21 Wing spots usually distinct and circular (Fig. 31*a*); clasping cercus truncate (Fig. 17*b*) (western) . . . . **C. guttipennis** sp. nov.

— Wing lacking distinct spots (Fig. 31*e*); clasping cercus tapered apically (Fig. 18*b*) (western). . . . . **C. wallowa** sp. nov.

22 Lateral setae on proepisternum stout and dark; wing lacking distinct spots (Fig. 31*c*); clasping cercus widely expanded apically (Fig. 17*d*) (western) . . . . **C. schlinkerti** sp. nov.

— Lateral setae on proepisternum slender and pale; wing with or without spots (Figs. 30*f*, *j*); clasping cercus expanded apically or slender (Figs. 15*d*, 17*a*) 23

23 Wing spots indistinct, reduced to clouding along crossveins or absent (Fig. 30*f*); distiphallus relatively short, not extending beyond base of surstylus (Fig. 15*d*); clasping cercus narrow subapically (western) . . . . ***C. bulbosa*** sp. nov.

— Wing with distinct circular spots (Fig. 30*j*); distiphallus long, extending well beyond base of surstylus (Fig. 17*a*); clasping cercus expanded subapically (western) . . . . ***C. gubernans*** (Melander)

24 Scutum lacking pair of black stripes; clasping cercus tapering to round tip, inner stout setae distributed on apical half; surstylus truncate apically, lacking notch (Fig. 11); specimens usually reddish brown (*C. caerulea* group) 25

— Scutum with pair of black stripes (bivittate); apical margin of the clasping cercus truncate, bearing stout setae on apical third of inner face; surstylus often with subapical notch (Figs. 13*a*, *d*); specimens generally greyish, often with extensive blue pruinescence (*C. conjuncta* group) 28

25 Face brown, lacking blue pruinescence 26

— Face with blue pruinescence. 27

26 Head darker than thorax, especially pleura; hind first tarsomere with 2 or 3 stout erect setae beneath; distiphallus long and coiled (Fig. 11*d*) (western) . . . . ***C. longicauda*** sp. nov.

— Head concolorous with thorax; hind first tarsomere lacking erect setae beneath; distiphallus shorter than phallic shaft (Fig. 11*a*) (widespread) . . . . ***C. brunnea*** sp. nov.

27 Face with central blue pruinescent spot; lower half of pleura with bluish pruinescence; distiphallus arising from C-shaped arch; apex of clasping cercus straight to slightly curved (Fig. 11*b*); R_{2+3} mostly straight beneath stigma (western) . . . . ***C. caerulea*** sp. nov.

— Face mostly with blue pruinescence, except immediately below antenna; pleura with faint pruinescence; distiphallus arising straight from phallic shaft; apex of clasping cercus distinctly curved (Fig. 11*c*); R_{2+3} arched beneath stigma (western) . . . . ***C. daniellae*** sp. nov.

28 Legs and halter bright to pale yellowish brown 29

— Legs and halter dark reddish brown to black 30

29 Legs and forecoxa bright orange; mid and hind coxae clothed in pale-blue pruinescence; lacking apical distiphallus (Fig. 12*a*) (western) . . . . ***C. amarilla*** sp. nov.

— Legs, including coxae, brownish orange; all coxae clothed in pale-blue pruinescence; short apical distiphallus present (Fig. 14*a*) (western). . . . . ***C. pacifica*** sp. nov.

30 Costa with 2 or more basal costal bristles, often with a series of long costal setae distal to basal costal bristles; males lacking distiphallus 31
— Costa with 1 basal costal bristle, lacking additional long costal setae; males rarely lacking distiphallus. 37

31 Seven or more dorsocentral bristles (or numerous setae interspersed); apex of phallus with small membranous flap and pair of small subapical plates (Fig. 12*d*); female face brown (western) . . . . ***C. coloradensis*** sp. nov.
— Five dorsocentral bristles, rarely more; apex of phallus lacking small membranous flap and subapical plates; female face brown or blue 32

32 Male foretibia with long erect setae beneath, longer than half width of tibia; lower half of female face with blue pruinescence (western) . . . . ***C. setitibia*** sp. nov.
— Male foretibia with short erect setae beneath, shorter than half width of tibia; female face brown or entire face with blue pruinescence. 33

33 Male 34
— Female. 36

34 Clasping cercus short and truncate; apex of phallus with pair of narrow converging plates (Fig. 12*c*) (western) . . . . ***C. borkenti*** sp. nov.
— Clasping cercus long and arched (Figs. 12*b*, 13*b*); apex of phallus with single plate. 35

35 Apex of phallus with outwardly arching, spear-shaped pointed tip; clasping cercus arched medially to opposing cercus, apex rounded bearing stout pointed setae (Fig. 12*b*) (western) . . . . ***C. arnaudi*** sp. nov.
— Apex of phallus with inwardly bent, narrow tip; clasping cercus arched dorsally, lacking pointed apical setae (Fig. 13*b*) (western) . . . . ***C. cordillerana*** sp. nov.

36 Female face with blue pruinescence (western). . . . ***C. cordillerana*** sp. nov.
— Female face brown (western) . . . . ***C. borkenti*** sp. nov. and ***C. arnaudi*** sp. nov.

37 Auxiliary crossvein between R_{2+3} and R_4 present on at least one wing (Figs. 29*h*, 30*e*) 38
— Auxiliary crossvein between R_{2+3} and R_4 absent (Fig. 30*b*) 41

38 Lower half of face with blue pruinescence 39
— Lower half of face brown 40

39 Wing with numerous spots on crossveins (Figs. 29*g*, *h*); R_{4+5} with stump vein near midlength; distiphallus with basal notch; apex of phallic shaft lacking ventral process (Fig. 13*a*) (widespread) . . . . . . ***C. conjuncta*** Loew

— Wing lacking spots; R_{4+5} lacking stump vein (Fig. 30*e*); distiphallus lacking basal notch; apex of phallic shaft with triangular ventral process (Fig. 15*c*) (western) . . . . ***C. trunca*** Melander

40 Body with golden iridescent sheen; distiphallus present; surstylus with subapical notch (Fig. 15*b*) (western) . . . . ***C. subtrunca*** sp. nov.

— Body brown, lower half of pleura with faint blue pruinescence; distiphallus absent; surstylus with pointed apex, lacking subapical notch (Fig. 14*c*) (western). . . . . ***C. rainiericola*** sp. nov.

41 Male face with blue pruinescence, except brown just below antenna; notopleuron with blue pruinescence; female face brown; female forefemur lacking rows of ventral setae. 42

— Male and female face brown or with blue pruinescence on lower half only; notopleuron brown; female forefemur with biserial row of short setae beneath 43

42 Wing length 4–4.5 mm; clasping cercus truncate; distiphallus bent back at right angles at base; lacking expansion at base of male clasping cercus (Fig. 14*b*) (western). . . . . ***C. prasinata*** Melander

— Wing length 3–3.5 mm; clasping cercus rounded apically; distiphallus straight; basal expansion of male clasping cercus covered with stout setae (Fig. 13*c*) (western) . . . . ***C. melanderi*** sp. nov.

43 Face brown; distiphallus arising subapically from phallus; clasping cercus with notch on posterior margin (Fig. 13*d*) (western) . . . . ***C. olivacea*** Melander

[Note: some specimens of *C. subtrunca* with auxiliary crossvein lacking will key out here, but are distinguished by a golden iridescent sheen and small size (wing length 4–4.5 mm)].

— Lower half of face with blue pruinescence; distiphallus arising dorsally from phallus; clasping cercus with straight posterior margin (Fig. 14*d*) (known only from Colorado). ***C. sabroskyi*** sp. nov.

Index

Page numbers of principal entries are in boldface type. Synonyms are in italic type, family-group names are in capital letters, and original combinations are in square brackets.

NRC·CNRC

From Discovery to Innovation...
De la découverte à l'innovation...

La Science à l'œuvre pour le
at work for Canada

Coming soon / À paraître bientôt
December 2008 / décembre 2008

Manual of Central American Diptera

Volume 1

Edited by / Publié sous la direction de

B.V. Brown, A. Borkent, J.M. Cumming, D.M. Wood, N.E. Woodley, M. Zumbado

ISBN-13: 978-0-660-19833-0
Hardcover / Édition reliée
NRC / CNRC : 49728
$94.95 CAN; Other countries $94.95 US
94,95 $CAN; À l'étranger : 94,95 $US

This two-volume book, a collaboration of over 70 experts on Diptera, is the first-ever resource for the identification and understanding of Central American flies.

Volume 1 covers 42 of the 106 families of flies–sometimes termed the Lower Diptera–and includes introductory chapters dealing with morphology and terminology, natural history, economic importance and phylogeny of Diptera as well as keys to all the Central American fly families for both the adult and larval stages. The text is accompanied by over 1500 line drawings and 144 color photographs of flies.

Cet ouvrage, présenté en deux volumes, est le fruit d'un effort concerté de plus de 70 spécialistes du domaine. C'est le premier livre sur l'identification et l'étude des diptères de l'Amérique centrale.

Le volume 1 traite de 42 des 106 familles de diptères, quelques fois appelées diptères primitifs, et renferme des chapitres d'introduction sur la morphologie et la terminologie, l'histoire naturelle, le rôle dans l'économie, la phylogénie des diptères, ainsi que des clés d'identification de toutes les familles de diptères (stades larvaire et adulte) de l'Amérique centrale. Le texte est accompagné de plus de 1 500 dessins au trait et de 144 photographies couleur.

To order books / Pour commander

Order Office / Bureau des commandes
NRC Research Press / Presses scientifiques du CNRC, National Research Council / Conseil national de recherches
Ottawa, ON K1A 0R6, Canada. Telephone / Téléphone : 613-990-2254; Fax / Télécopieur : 613-952-7656
E-mail / Courriel : pubs@nrc-cnrc.gc.ca; Order online / Commandes en ligne : http://pubs.nrc-cnrc.gc.ca

Shipping and handling extra.
Canadian customers add GST. Prepayment required unless order is accompanied by a purchase order number. Cheques or money orders should be made payable to: **"Receiver General for Canada, Credit NRC"**. For all orders outside Canada, rates are in U.S. dollars. All sales are final.

Frais d'expédition et de manutention en sus.
Clients du Canada, ajoutez la TPS. Les commandes doivent être payées d'avance à moins qu'elles ne soient accompagnées d'un numéro de bon de commande. Les chèques ou mandats doivent être libellés à l'ordre du « **Receveur général du Canada, crédit du CNRC** ». Pour les commandes hors-Canada, les tarifs sont en dollars américains. Vente ferme.

01/09/2008

National Research Council Canada | Conseil national de recherches Canada

Canada